onal Imaging and Vision

nprehensive book series embraces state-of-the-art expository works and advanced monographs on any aspect of this interdisciplinary field.

covered by the series fall in the following four main categories:

- Imaging Systems and Image Processing
- Computer Vision and Image Understanding
- Visualization
- Applications of Imaging Technologies

y monographs or multi-authored books that have a distinct subject area, that is where each pter has been invited in order to fulfill this purpose, will be considered for the series.

Volume 41

For further volumes:
www.springer.com/series/5754

Mathematical Methods
Analysis and Represent

Computati

Managing Edit

MAX VIERG
Utrecht Unive

Series Edito

GUNILLA
DANIEL C
RACHID D
KATSUSH
REINHAR
ALES LE
STAN Z.
DIMITRI
HEINZ-
JOHN K

This co
research

Topics

Onl
cha

Luc Florack · Remco Duits · Geurt Jongbloed ·
Marie-Colette van Lieshout · Laurie Davies
Editors

Mathematical Methods for Signal and Image Analysis and Representation

Springer

Editors
Prof. Luc Florack
Dept. Mathematics & Computer Science
Eindhoven University of Technology
Eindhoven
The Netherlands

Dr. Remco Duits
Dept. Mathematics & Computer Science
Eindhoven University of Technology
Eindhoven
The Netherlands

Prof. Geurt Jongbloed
Dept. Applied Mathematics
Delft University of Technology
Delft
The Netherlands

Dr. Marie-Colette van Lieshout
Probability & Stochastic Networks (PNA)
Centrum Wiskunde & Informatica
Amsterdam
The Netherlands

Prof. Laurie Davies
Fakultät für Mathematik
Universität Duisburg-Essen
Essen
Germany

ISSN 1381-6446 Computational Imaging and Vision
ISBN 978-1-4471-2352-1 e-ISBN 978-1-4471-2353-8
DOI 10.1007/978-1-4471-2353-8
Springer London Dordrecht Heidelberg New York

British Library Cataloguing in Publication Data
A catalogue record for this book is available from the British Library

Library of Congress Control Number: 2012930491

Mathematics Subject Classification: 62H35, 68U10

Printed on acid-free paper

Springer is part of Springer Science+Business Media (www.springer.com)

Preface

The research institute EURANDOM (European Institute for Statistics, Probability, Stochastic Operations Research and its Applications) was established in 1997 on the campus of Eindhoven University of Technology, The Netherlands. Its mission is to foster research in the area of stochastics and its applications. It achieves this mission by recruiting and training talented young researchers and helping them to find their way to tenured positions in academia and industry, and by carrying out and facilitating research through postdoctoral and graduate appointments, visitor exchange and workshops. Its chief mission statement has been given nationwide support in The Netherlands by a recently installed national cluster called STAR (Stochastics—Theoretical and Applied Research), for which EURANDOM acts as coordinating and facilitating node.

As part of its workshop programme, EURANDOM organized a series of international workshops on image processing and analysis. The third one in this series was the workshop on *Locally Adaptive Filters in Signal and Image Processing*, November 24–26, 2008, focusing specifically on locally adaptive methods. The ability of a system to adapt to the local state is important in many problems in image analysis. Many renowned young experts were invited to give overview talks on this theme covering state-of-the-art and novel research.

Despite the high quality of contributions, no proceedings of this workshop have been issued. Instead, the workshop initiated a collaborative effort, focusing more generally on mathematical methods for signal and image analysis and representation. The results of this effort are described in this book.

Contributions have been carefully selected to be representative for a variety of generic approaches as well as to illustrate formal connections among these. Roughly speaking deterministic methods are central to the first half of the book, whereas the second half considers mainly statistical methods. However, some chapters in the middle of the book clearly encompass both approaches, and more than a hundred cross-references throughout the book emphasize the many formal connections and analogies that exist between seemingly different paradigms.

This book differs from most existing books on medical signal and image analysis or computer vision to the extent that it does not focus on specific applications (although some are detailed for the sake of illustration), but on *methodological frameworks* on which such applications may be built. This book should therefore be of interest to all those in search of a suitable methodological basis for specific applications, as well as to those who are interested in fundamental methodologies per se.

Eindhoven, Netherlands Luc Florack

Acknowledgements

The institute EURANDOM of Eindhoven University of Technology has funded and facilitated the visitor exchange programme and workshop for the international collaboration that has resulted in this edited book. Further funding was obtained from the Netherlands Organisation for Scientific Research (NWO) through the programme for Incidental Financial Support, through a personal grant in the Innovational Research Incentives Scheme to Luc Florack, and through a grant from the Royal Netherlands Academy of Arts and Sciences (KNAW).

Special thanks go to Lucienne Coolen-van Will, Marèse Wolfs-van de Hurk and Enna van Dijk for administrative support, the Department of Mathematics & Computer Science, the Department of Biomedical Engineering, and the Executive Board of Eindhoven University of Technology for supporting the cross-divisional Imaging Science & Technology Eindhoven High Potential Research Program (IST/e).

Contents

Contributors

Maxime Boucher School of Computer Science, McGill University, Montreal, Canada

Bernhard Burgeth Mathematical Image Analysis Group, Faculty of Mathematics and Computer Science, Saarland University, Saarbruecken, Germany

Laurent Demaret HelmholtzZentrum München, Institut für Biomathematik und Biometrie (IBB), Neuherberg, Germany

Xavier Descombes Laboratoire d'Informatique, Signaux et Systèmes de Sophia-Antipolis I3S, UMR6070, UNS CNRS 2000, Sophia Antipolis Cedex, France

Stephan Didas Abteilung Bildverarbeitung, Fraunhofer-Institut für Techno- und Wirtschaftsmathematik, Kaiserslautern, Germany

Remco Duits Department of Mathematics and Computer Science & Department of Biomedical Engineering, Eindhoven University of Technology, Eindhoven, The Netherlands

Alan Evans McConnell Brain Imaging Center, Montreal Neurological Institute, McGill University, Montreal, Canada

Michael Felsberg Computer Vision Laboratory, Department of Electrical Engineering, Linköping University, Linköping, Sweden

Luc Florack Department of Mathematics and Computer Science & Department of Biomedical Engineering, Eindhoven University of Technology, Eindhoven, The Netherlands

Hartmut Führ Lehrstuhl A für Mathematik, RWTH Aachen, Aachen, Germany

Yaniv Gur SCI Institute, University of Utah, Salt Lake City, UT, USA

Michal Haindl Institute of Information Theory and Automation of the ASCR, Prague, Czech Republic

Armin Iske Department of Mathematics, University of Hamburg, Hamburg, Germany

Bart Janssen Department of Biomedical Engineering, Eindhoven University of Technology, Eindhoven, The Netherlands

Wahid Khachabi Department of Mathematics, University of Hamburg, Hamburg, Germany

Kai Krajsek Institute for Chemistry and Dynamics of the Geosphere, ICG-3, Forschungszentrum Jülich GmbH, Jülich, Germany

Jean-Michel Loubes Institut de Mathématiques de Toulouse, UMR 5219, Université Toulouse 3, Toulouse cedex 9, France

Michal Matuszak Faculty of Mathematics & Computer Science, Nicolaus Copernicus University, Toruń, Poland

Ofer Pasternak Department of Psychiatry, Brigham and Women's Hospital Harvard Medical School, Boston, MA, USA

Luis Pizarro Department of Computing, Imperial College London, London, UK

Jörg Polzehl Weierstrass Institute for Applied Analysis and Stochastics, Berlin, Germany

Markus Reiß Institute of Mathematics, Humboldt University Berlin, Berlin, Germany

Paul Rochet Institut de Mathématiques de Toulouse, UMR 5219, Université Toulouse 3, Toulouse cedex 9, France

Yves Rozenholc MAP5, UMR CNRS 8145, University Paris Descartes, Paris, France

Hanno Scharr Institute for Chemistry and Dynamics of the Geosphere, ICG-3, Forschungszentrum Jülich GmbH, Jülich, Germany

Tomasz Schreiber Faculty of Mathematics & Computer Science, Nicolaus Copernicus University, Toruń, Poland

Nir Sochen Department of Applied Mathematics, Tel Aviv University, Tel Aviv, Israel

Karsten Tabelow Weierstrass Institute for Applied Analysis and Stochastics, Berlin, Germany

Markus van Almsick Department of Biomedical Engineering, Eindhoven University of Technology, Eindhoven, The Netherlands

Joachim Weickert Mathematical Image Analysis Group, Faculty of Mathematics and Computer Science, Saarland University, Saarbruecken, Germany

Chapter 1
A Short Introduction to Diffusion-Like Methods

Hanno Scharr and Kai Krajsek

Abstract This contribution aims to give a basic introduction to diffusion-like methods. There are many different methods commonly used for regularization tasks. Some of them will be briefly introduced and their connection to diffusion shown. In addition to this we will go into some detail for diffusion-like methods in a narrower sense, i.e. methods based on PDEs similar to diffusion PDEs known from physics. Main issues highlighted here are which PDE to use, how diffusivities in such a PDE are constructed, and which discretization is suitable for a given task.

1.1 Introduction

There are quite a few methods for regularization tasks like noise reduction, inpainting, super-resolution, or interpolation described in literature. Many of them, if not all, can somehow be brought into connection with diffusion. Obviously we cannot visit all of them in this paper making this introduction incomplete. Our focus here will be on nonlinear averaging, mainly used for noise reduction, even though currently best performing denoising algorithms on natural grey value images are not diffusions in a narrower sense (see e.g. [346]). Nevertheless one goal beneath basic introduction is to mention at least some of the major contributions to this field.

We will nearly completely ignore the fact that diffusion can be used to build a scale-space. First discovered in Japan [231, 442] scale-space filtering is a topic of its own (see e.g. [290, 450]). A linear scale space is built by applying linear diffusion (see Sect. 1.2.1) to a signal in short time steps and recording the more and more smoothed signal. Other scale spaces can be derived by applying other diffusion-like schemes, cf. Chaps. 7 and 9.

Regularization schemes are represented in literature from different view-points. Diffusion schemes typically start with the formulation of a continuous partial dif-

H. Scharr (✉) · K. Krajsek
Institute for Chemistry and Dynamics of the Geosphere, ICG-3, Forschungszentrum Jülich GmbH, 52425 Jülich, Germany
e-mail: h.scharr@fz-juelich.de

K. Krajsek
e-mail: k.krajsek@fz-juelich.de

L. Florack et al. (eds.), *Mathematical Methods for Signal and Image Analysis and Representation*, Computational Imaging and Vision 41,
DOI 10.1007/978-1-4471-2353-8_1,

ferential equation (PDE) describing a process which changes data over time [334, 433]. This PDE is then discretized yielding an iterative update scheme. The classical diffusion defined by the heat equation known from physics involves linear filtering of the input data by derivatives

$$\partial_t s(\mathbf{x}, t) = \operatorname{div}(\mathbf{D}\nabla s(\mathbf{x}, t)), \tag{1.1}$$

where $s(\mathbf{x}, t)$ is the evolving signal or image with $s(\mathbf{x}, 0) = r(\mathbf{x})$ and the initially observed data $r(\mathbf{x})$; $\nabla = (\partial_{x_1}, \dots, \partial_{x_N})$ is the vector of spatial derivatives. Diffusion tensor $\mathbf{D}$ is a symmetric, positive definite tensor which may vary with space and evolution time and may depend on local data. Adaptivity of a diffusion scheme is achieved via adaptation of the diffusion tensor. This makes the scheme nonlinear. In computational physics diffusion is typically simulated using e.g. finite differences on a sampling grid. This grid is refined when the result is not accurate enough, making discretization simple. In image processing no such refinement is typically applied, giving more influence to discretization details. How to discretize an anisotropic nonlinear diffusion process will be subject of Sect. 1.5.

There are different naming conventions in the literature for diffusion schemes. Especially the term *anisotropic diffusion* is inconsistently used. Following [433] we use the term isotropic diffusion, when $\mathbf{D}$ is proportional to the identity matrix $\mathbf{D} = c\mathbb{1}$, i.e. when Eq. (1.1) collapses to

$$\partial_t s(\mathbf{x}, t) = \operatorname{div}(c\nabla s(\mathbf{x}, t)). \tag{1.2}$$

If *diffusivity* or *edge stopping function* c depends on the image $s(\mathbf{x}, t)$, we call a diffusion scheme *isotropic nonlinear diffusion*. We call diffusions with general $\mathbf{D}$, not proportional to $\mathbb{1}$, *anisotropic diffusion*, in contrast to several publications using this term for isotropic nonlinear diffusion (e.g. [41, 334]). This naming inconsistency originates from the fact that the overall effect of isotropic nonlinear diffusion on the evolved data is anisotropic. We call a diffusion scheme *linear* if $\mathbf{D}$ does not depend on the evolving image. The simplest case is *linear homogenous diffusion* with a constant edge stopping function c, where Eq. (1.2) simplifies to

$$\partial_t s(\mathbf{x}, t) = c\Delta s(\mathbf{x}, t), \tag{1.3}$$

where $\Delta = \sum_i^N \partial_{x_i}^2$ is the spatial Laplacian. A time step applying linear diffusion to s on the unbounded domain is solved by convolution with a Gaussian kernel (see Sect. 1.2.1).

Diffusion equations with many different edge stopping functions c and diffusion tensors D have been proposed in literature (see e.g. [41, 364, 369, 433]). They depend on gray value or color gradients, curvatures, or other image features (see [376, 440] for possible dependencies when regularizing optical flow). There is rich ongoing work on diffusions for vector, matrix or tensor-valued data (see e.g. [114, 266, 337, 381, 414], and elsewhere in this volume, cf. Chaps. 3, 4, and 5, where the main problem is to select a useful metric and discretize operators respecting it. We will show an example in Sect. 1.5.2. Diffusion methods explicitly focusing on metrics induced by known group structures are shown e.g. in this volume, Chaps. 5, 7, 8, 9, and 10, and also elsewhere [114].

A framework for diffusions in generalized image spaces has been defined using the Beltrami operator, a natural generalization of the Laplacian on non-flat manifolds [250]. E.g. an RGB-color image is a 2D manifold the 5D space (x, y, I^r, I^g, I^b) spanned by 2 spatial dimensions x and y and 3 color intensities I^r, I^g, and I^b. The Beltrami framework will be introduced in Sect. 1.2.3.

Equation (1.1) can be seen as a first order Taylor approximation describing a dynamic process as often used in physics. This is sufficient in continuous time and space as used in physics, but in discrete time and/or space richer representations of a process are sometimes advantageous. First order derivatives ∇ may therefore be exchanged by other and/or more operators [361, 366, 465]. We introduce the isotropic nonlinear case in Sect. 1.3.2 and give an example for the anisotropic nonlinear case in Sect. 1.6.2.

Some, but not all of these PDEs may be derived from suitable energy functions via calculus of variations. The PDE then changes the data such that the energy is minimized. Consequently a diffusion scheme optimizes a property our image data is assumed to fulfill. Energy functions may be designed assuming data models motivated from physics underlying the imaging process, e.g. modeling step edges explicitely via a line process [42, 316]. Such energy functions also occur in robust statistics and can be expressed in terms of probabilities via Gibbs distributions or Markov Random Fields (MRFs), cf. [183] and Chaps. 13, 14 and 15. The edge stopping function then corresponds to a robust error function and diffusion is then related to M-estimation [41]. Probability distributions forming potential functions in the energy can be learned from training data using this relation to image statistics [371, 465] (cf. also Chap. 11). Robust error functions are derived from histograms of filtered images. In the classical diffusion case filter kernels are spatial derivatives, but other kernels may be used as well. The kernels may even be learned from training data [361]. The histograms are treated as observed statistics or empirical marginal distributions defining a probability to observe a certain image. Maximizing this probability means minimizing an energy. We will go into more detail in Sect. 1.3.1.

There are algorithmic approaches presenting and evaluating *discrete schemes* used for filtering, e.g. nonlinear Gaussian Filtering [17, 187, 445], Mean-Shift Filtering [81] or Bilateral Filtering [409]. Typically they may also be formulated in terms of minimization of a *cost functional* corresponding to the energy functions formulated in continuous time and space. We show some prominent examples and their relation to diffusion in Sect. 1.4. Here it is important to note that diffusion in a strict sense is only defined in continuous space and time. Therefore a discrete scheme is called a *consistent* diffusion scheme, if it becomes the diffusion equation in the limit $h \to 0$ and $\tau \to 0$ for spatial sampling step h and temporal sampling step τ. This is the definition for consistency known from numerics. Associating some discrete scheme with a discrete scheme for diffusion does *not* show that the first scheme is a diffusion. We will elaborate this for the case of bilateral filtering in Sect. 1.4.1.

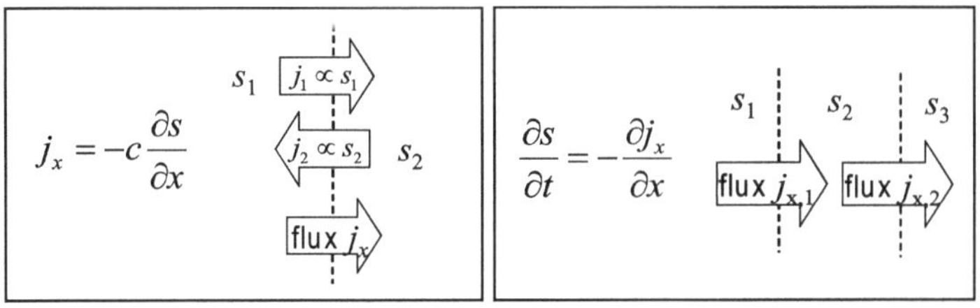

Fig. 1.1 Fick's Law. *Left*: Number of particles crossing a given border is proportional to the number of particles s per unit volume, i.e. proportional to the density. Overall flux j depends on the density difference (or continuously: gradient). *Right*: Change of particle number in a given volume equals in-flux minus out-flux (continuously: divergence). Combining the two laws yields the heat equation (by means of the Gauss divergence theorem) with diffusion coefficient or diffusivity c

1.2 Diffusion in a Narrow Sense

In this section, we introduce different types of diffusion. We start with the simplest case well-known as Fick's law from physics since 1855 [142], continue with formulation of isotropic nonlinear diffusion (sometimes called Perona-Malik Diffusion [334]) and anisotropic diffusion (sometimes called Coherence [433] or Edge-Enhancing Diffusion). Finally we show the currently most general formulation, the Beltrami framework [250].

1.2.1 Diffusion in Physics, Basic Solution and Numerics

Diffusion occurs in statistical physics and thermodynamics where random 'Brownian' motion of particles leads e.g. to heat transport or mixing of liquids or gases. Densities or temperature then evolve with time as described by the heat equation (1.3): $\partial_t s(\mathbf{x}, t) = c\Delta s(\mathbf{x}, t)$ (cf. Fig. 1.1), in which $s(\mathbf{x}, t)$ is the evolving density and c a diffusion constant or *diffusivity*. The evolution of s by linear isotropic diffusion, i.e. diffusion with $c =$ const. is given by convolution of s with a Gaussian kernel with variance $\sigma^2 = 2ct$.

The nonlinear heat equation (1.2) may be solved by finite differences. In the simplest case we exchange the time derivative on the left hand side by a forward difference (Euler forward), and derivatives by neighbor differences

$$\frac{s^{t+\tau} - s^t}{\tau} = \begin{pmatrix} [1, -1] \\ \begin{bmatrix} 1 \\ -1 \end{bmatrix} \end{pmatrix}^T * \left(c \begin{pmatrix} [1, -1] \\ \begin{bmatrix} 1 \\ -1 \end{bmatrix} \end{pmatrix} * s^t \right) \tag{1.4}$$

and get an update scheme

$$s^{t+\tau} = \begin{bmatrix} 0 & \tau c_{x,y+\frac{h}{2}} & 0 \\ \tau c_{x+\frac{h}{2},y} & 1 - \tau(c_{x+\frac{h}{2},y} + c_{x-\frac{h}{2},y} + c_{x,y+\frac{h}{2}} + c_{x,y-\frac{h}{2}}) & \tau c_{x-\frac{h}{2},y} \\ 0 & \tau c_{x,y-\frac{h}{2}} & 0 \end{bmatrix} * s^t \tag{1.5}$$

or $s^{t+\tau} = (1 + \tau A_{x,y}) * s^t$. This is called an explicit scheme. It boils down to convolution of the signal with a spatially (and temporally) varying kernel $(1 + \tau A_{x,y})$. In the case of spatially constant c this simplifies to

$$s^{t+\tau} = \begin{pmatrix} 0 & \tau c & 0 \\ \tau c & 1 - 4\tau c & \tau c \\ 0 & \tau c & 0 \end{pmatrix} * s^t, \tag{1.6}$$

where A is the so-called 5-point-star times c. This scheme has positive entries only, i.e. is a convex regularizer and features absolute stability, if and only if $\tau c < 0.25$. It becomes instable if $\tau c > 0.5$, as then frequencies at the Nyquist border are amplified by a factor < -1. For small τc the convolution kernel applied to s^t is a reasonable discretization of a Gaussian. The same bounds can be derived by application of the Gershgorin circle theorem on the spectrum which is supposed to be contained within $(-1, 1)$.

Discretizing the left hand side by a backward difference quotient (Euler backward) we get an implicit scheme $(s^t - s^{t-\tau})/\tau = A_{x,y} * s^t$ or equivalently $s^{t+\tau} = (1 - \tau A_{x,y})^{-1} * s^t$ boiling down to a recursive filter applied to s^t.

1.2.2 *Anisotropic Diffusion*

Anisotropic diffusion typically acts along measured local image orientations (cf. Fig. 1.2). They are described by the structure tensor $\mathbf{J}_\rho$ [36]

$$\mathbf{J}_\rho = \int w_\rho(\mathbf{x}) \nabla s(\mathbf{x}) \nabla^T s(\mathbf{x})\, d\mathbf{x}, \tag{1.7}$$

where $w_\rho(x)$ are Gaussian weights with standard deviation ρ. Being symmetric $\mathbf{J}_\rho$ can be diagonalized, i.e. $\mathbf{M}$ is a diagonal matrix with eigenvalues $\mathbf{M}_{ii} = \mu_i$ and

$$\mathbf{J}_\rho = (e_1, \dots, e_N)\mathbf{M}(e_1, \dots, e_N)^T. \tag{1.8}$$

Anisotropic diffusion filtering evolves the initial noisy image $s(x, 0)$ via

$$\partial_t s = \nabla \cdot (\mathbf{D} \nabla s) \tag{1.9}$$

(cf. Eq. (1.1)). $\mathbf{D}$ is the diffusion tensor, a positive definite symmetric matrix, and $s(\mathbf{x}, t)$ is the evolving spatio-temporal image. Diffusion time t is sometimes used as the scale parameter in a scale-space. It should not be confused with the time coordinate x_3 of a 2D image sequence. The diffusion tensor $\mathbf{D}$ usually applied in anisotropic diffusion uses the same eigenvectors e_i as the structure tensor $\mathbf{J}_\rho$ (see Eq. (1.7)). Thus smoothing is applied according to the spatio-temporal image structure. Smoothing strengths along these structures are given by eigenvalues λ_i of $\mathbf{D}$. Given a diagonal matrix $\mathbf{L}$ with $\mathbf{L}_{ii} = \lambda_i$, the diffusion tensor $\mathbf{D}$ is transformed into the image coordinate system given by the eigenvectors e_i:

$$\mathbf{D} = (e_1, \dots, e_N)\mathbf{L}(e_1, \dots, e_N)^T. \tag{1.10}$$

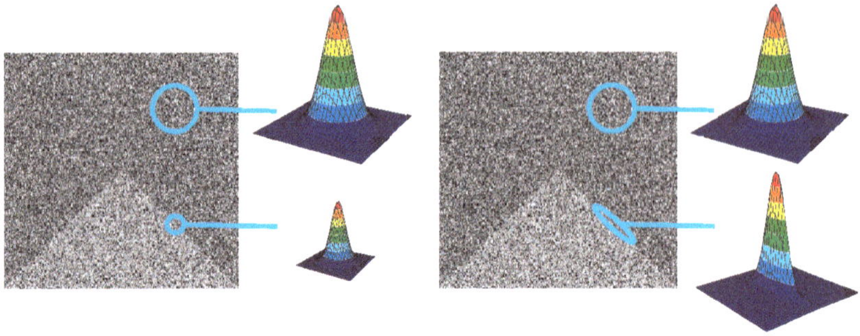

Fig. 1.2 Isotropic nonlinear versus anisotropic diffusion. *Left*: Isotropic nonlinear diffusion reduces diffusivity in all directions when an image structure is present. *Right*: Anisotropic diffusion reduces diffusivities across edges only

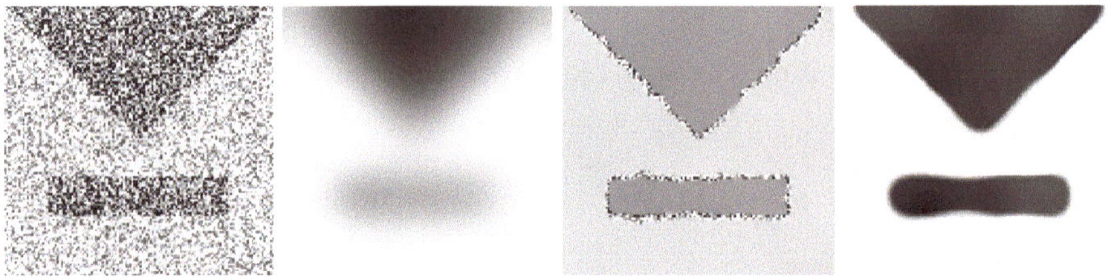

Fig. 1.3 Effect of diffusion filtering. Illustrative examples exaggerating dominating smoothing effects. *From left to right*: Noisy input image, smoothing result with isotropic linear, isotropic nonlinear and anisotropic diffusion

The directional diffusivities $\lambda_i, i \in \{1, \dots, N\}$ determine the behavior of the diffusion. For image enhancing they shall be high for low values of μ_i and vice versa.

Different possible choices for **L** include isotropic linear, isotropic nonlinear and anisotropic processes. There are anisotropic choices with fixed smallest or fixed largest directional diffusivities λ_i and choices where all diffusivities vary. Results demonstrating the different smoothing effects are shown in Fig. 1.3.

Common choices for directional diffusivities are

Isotropic-Linear The standard linear diffusion using a Gaussian kernel corresponds to $D = \alpha\mathbb{1}$, with $\alpha \geq 0$.

Isotropic Non-linear Perona-Malik [334] type diffusion seeks to adapt the smoothing strength to the absolute value of the gray value gradient. Tensor D is given by $D = f(\nabla g)\mathbb{1}$. Among the choices for the diffusivity f the following is given:

$$f(\nabla g) = \exp(-\|\nabla g\|/K). \tag{1.11}$$

The considered diffusivities have in common that they decrease with increasing gradient magnitude. Thus smoothing across edges is prevented.

Edge Enhancing This is basically an anisotropic version of the previous. Following [164] we extend the original 2D formulation [433, 434] to nD as follows:

$$\begin{aligned} \lambda_i = \lambda_2 &:= f(\nabla g) \quad \text{for } i \neq N, \\ \lambda_N &:= 1. \end{aligned} \tag{1.12}$$

The largest diffusivity fixed to 1 enforces strong smoothing in the direction of the corresponding eigenvector even when there is no clear linear structure present.

Coherence Enhancing In this type of diffusion the eigenvalues of the diffusion tensor are chosen as $\lambda_i = \alpha$ for $i \neq N$ and [435]:

$$\lambda_N := \begin{cases} \alpha & \text{if } \kappa = 0, \\ \alpha + (1-\alpha)\exp(\frac{-C}{\kappa}) & \text{else.} \end{cases} \tag{1.13}$$

With a small positive parameter α and the coherence κ measured by:

$$\kappa = \sum_{i=1}^{N-1} \sum_{j=i+1}^{N} (\mu_i - \mu_j)^2. \tag{1.14}$$

This process is designed to smooth only when there is a large spread in the eigenvalues, enhancing line-like, coherent structures.

Orientation-Enhancing In order to fully exploit the information provided by the structure tensor and to facilitate orientation estimation (= optical flow in image sequences) the eigenvalues of the diffusion tensor can be chosen [369, 431]:

$$\lambda_i := \begin{cases} 1 & \text{if } \mu_i \leq \sigma^2, \\ 1 - \exp(-\frac{c}{(\mu_i - \sigma^2)^2}) & \text{else,} \end{cases} \tag{1.15}$$

where $c > 0$ regulates the transition and σ is related to the noise variance of the image derivatives. This exponential function has been used in [334, 433].

The functions f used to calculate each λ_i are often chosen ad hoc, e.g. selected from robust error functions [41]. As we will see in the next section (Sect. 1.3.1) they may in principle be learned from image statistics. However no energy functional exists for anisotropic nonlinear diffusion, where **D** depends on signal s. Therefore in this case generative learning is not possible in a strict sense (cf. [371]). An energy function for anisotropic nonlinear diffusion can be given, if **D** in principle should not depend on s, even though s is used to calculate an approximation of the true **D** [366]. We derive and use this energy function in Sect. 1.6.1.

1.2.3 Beltrami Framework

The basic concept behind Beltrami flow is to consider an image as a (curved) surface embedded into a higher dimensional space [250]. This concept is frequently used in

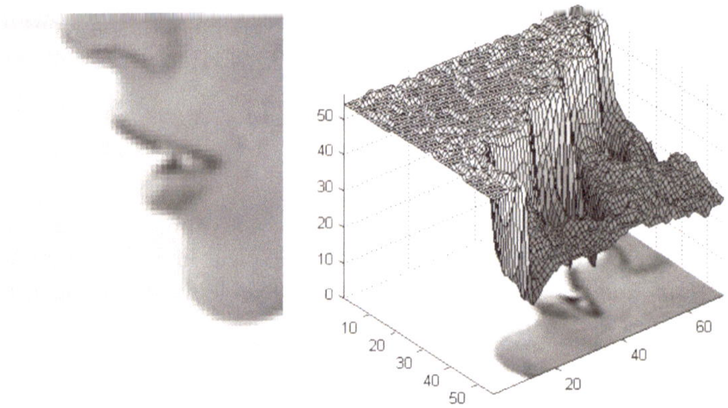

Fig. 1.4 Embedding of a gray value image in a feature space

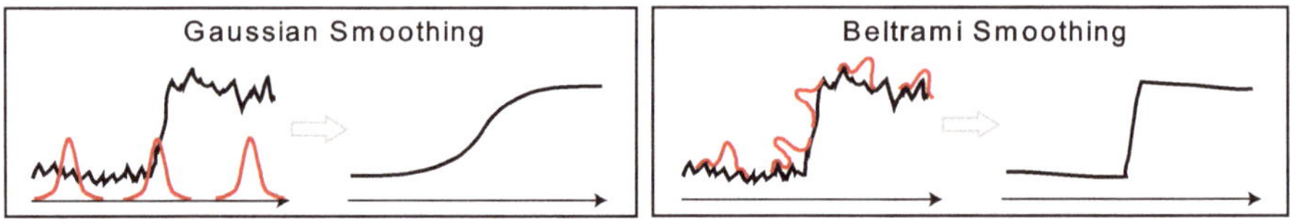

Fig. 1.5 Linear diffusion (i.e. Gaussian smoothing) compared to Beltrami smoothing. Linear diffusion averages a signal respecting spatial distances only. Beltrami smoothing averages a signal respecting distances along the curved manifold

literature, especially when working on spaces with underlying non-flat group structures, cf. Chaps. 5, 8 and 9, and elsewhere [114, 157]. Results from differential geometry are then used to process this surface. We will illustrate this concept by means of denoising gray valued images, however this concept can be generalized to other image types (e.g. color images or tensor valued images) in a straightforward manner. Instead of considering a gray valued image as a function $s(\mathbf{x})$ from the image domain $\Omega \subset \mathbb{R}^2$ into a one dimensional feature space $I \subset \mathbb{R}$, an image is considered as a surface M embedded in the product space (trivial fiber bundle) $E = \Omega \times I$ (cf. Figs. 1.4 and 1.5). The embedding is described by the map $X(\mathbf{x}) = (\mathbf{x}, s(\mathbf{x}))$. In order to be able to define energies on the image, M is considered as a Riemannian manifold: at each tangent space $T_{\mathbf{x}}(M)$ on a manifold M a (positive definite) inner product $g(\mathbf{x}) : TM \times TM \to \mathbb{R}$ is given by $g(\mathbf{x}) = g_{ij}(\mathbf{x})\, dx^i \otimes dx^j$, in which $\otimes$ denotes tensor outer product. The metric defines the length of an "infinitesimal line element" via $ds^2 = g_{ij}(\mathbf{x})\, dx^i\, dx^j$ as well as the "infinitesimal volume element" $\sqrt{|g|}\, dx\, dy$, where $|g(\mathbf{x})| = \det[g_{ij}(\mathbf{x})]$. The geometrical framework allows to relate the metric of the embedding space E with the metric of the image surface M via the so-called *pullback metric* $g_{ij} = h_{uv} \partial_i X^u \partial_j X^v$ assuring that infinitesimal distances defined by h_{uv} in E equal infinitesimal distances in M. If we consider the usual Euclidean metric in Ω and I, the metric in the embedding space reads $d\ell^2 = dx^2 + dy^2 + ds^2(x, y)$ leading to following pullback metric on M: $g_{11} = 1 + (\partial_x s)^2$, $g_{12} = g_{21} = \partial_x s \partial_y s$ and $g_{22} = 1 + (\partial_y s)^2$. Based on this mathematical structure an image can now be characterized by an energy based on distances on the image surface M. For instance, a denoised image may have minimal surface, i.e. the denoised image minimizes the energy

$$S(M) = \int \sqrt{|g|}\, dx\, dy. \tag{1.16}$$

The corresponding diffusion equation can then be obtained with calculus of variation, i.e. setting the negative functional derivative of the energy functional equal to the temporal derivative of the signal (cf. Eq. (1.22))

$$\partial_t s = \operatorname{div}\left(\frac{1}{\sqrt{|g|}} \nabla s \right) \tag{1.17}$$

which is isotropic nonlinear diffusion (or Perona Malik diffusion [334]) with the edge stopping function $\phi(x) = 1/\sqrt{|g|}$. The geometrical framework thus allows to relate an ad hoc chosen edge stopping function with a metric of the image surface M. However the choice of a suitable edge stopping function has only been shifted to the choice of a suitable metric (in the embedding space). A more general energy functional has been proposed [250]:

$$S(X^u, g_{ij}, h_{uv}) = \int \sqrt{|g|} g^{ij} h_{uv} \partial_i X^u \partial_j X^v\, d^n x, \tag{1.18}$$

where n denotes the dimension of the image domain, h_{uv} the embedding space metric and g_{ij} the metric of the image manifold. Depending on the interpretation of the different entities, different well known diffusion schemes can be reproduced by this energy functional. These include the reparameterization invariant linear scale-space by Florack et al. [155], Perona Malik anisotropic diffusion [334], Mumford-Shah segmentation models [316], Rudin-Osher-Fatemi total variation TV method for image enhancement based on the L1 norm [364], and the different Blake-Zisserman membrane models [42]. Also diffusion schemes for vector valued images or images whose feature space constitute itself a nonlinear manifold arise in a natural way, cf. Sect. 1.5.2 as well as Chaps. 3, 5, 8, 9, as well as [114].

1.3 Diffusion and Image Statistics

Diffusion and diffusion-like methods can be derived from image statistics, probability distribution functions (PDFs), or energy functionals. This section introduces the main statistical concepts needed, pathways to diffusion and extensions to diffusion.

A considerable advantage of a statistical point of view is an airtight justification of noise reduction. In scientific applications changing measured data e.g. by denoising it or by rejecting outliers is only allowed if such a change improves the data. Obviously, improving means to optimize some criterion which needs to be given explicitly. What is more, the criterion must be the right one for the data-set at hand.

However, what is the right criterion and in what sense is it right? From a statistical point of view the denoised or otherwise reconstructed data should be the most likely one given the data and prior knowledge! Consequently we formulate suitable probabilities and show in which way optimization schemes correspond to diffusion.

1.3.1 From Probability Distributions to Diffusion

An isotropic nonlinear diffusion process can be derived from an energy function, that itself may be derived from a probability distribution. The smooth image s is the maximizer of the posterior probability distribution $p(s|r)$, i.e. the probability that $s := s(\cdot, t) : \mathbb{R}^N \to \mathbb{R}$ for some fixed $t > 0$ is the desired smooth image when $r : \mathbb{R}^N \to \mathbb{R}$ has been observed

$$\hat{s} = \arg\max_{s} p(s|r) \quad \text{with } p(s|r) \propto \prod_{i} (p(r_i|s_i) p(\|\nabla s_i\|)). \tag{1.19}$$

The sampling distribution (likelihood function for fixed r_i) $p(r_i|s_i)$ at every pixel i, with $s_i = s(\mathbf{x}_i)$, may be defined by a measured image statistics, i.e. a normalized histogram of observed noise. In image processing it typically is modeled to only depend on intensity differences $\varepsilon_i = r_i - s_i$, i.e. on measurement noise.[1] The spatial term $p(\|\nabla s_i\|)$ formulates prior knowledge about the solution s. It exploits a Markov Random Field (MRF) assumption [183], which defines the prior in terms of local neighbor properties. For a 1D signal the assumption used here is that if we know a signal value s_i at a position i, then we can give a probability to observe a certain s_{i+1} at neighbor position $i+1$, and vice versa.

Please note that the likelihood term depends on measured data r and is therefore often called *data term*. The prior term only depends on the sought for smooth solution s and is therefore often called *smoothness term*.

We may interpret $p(s|r)$ to be a Gibbs distribution

$$p(s) = \frac{1}{Z} e^{-E(s)},$$

where E denotes the energy corresponding to p and Z is the partition function normalizing the integral of p over all s. Maximizing $p(s|r)$ is equivalent to minimizing its energy, i.e. its negative logarithm

$$\hat{s} = \arg\min_{s} E(s) \quad \text{with } E(s) = -\sum_{i} (\rho_0(s_i - r_i) + \lambda \rho_1(\|\nabla s_i\|)), \tag{1.20}$$

where we used the notation $\rho(x) = -\log p(x)$, added indices to stress that sampling and prior are different distributions, and introduced weight λ which accounts for the

[1] Considering e.g. Poisson or shot noise and low intensities, this is not a good approximation.

confidence one has in the different model terms. The smoothness term in (1.20) can be interpreted as nonlinear isotropic diffusion [41, 375]. As diffusion is defined in continuous domain, we rewrite the smoothness term as energy functional

$$E(s) = \int \rho(\|\nabla s\|)\, dx. \tag{1.21}$$

We denote with δE the functional derivative of E if the differential

$$\langle \delta E(u), w \rangle := \lim_{\varepsilon \to 0} \frac{E(u + \varepsilon w) - E(u)}{\varepsilon}, \tag{1.22}$$

exists for all test functions $w : \mathbb{R}^N \to \mathbb{R}$. The functional derivative can be seen as a generalization of the gradient of a multivariate function in vector calculus to a functional defined on a function space. Consequently, $\delta E = 0$ is a necessary condition for a minimizer of E, known as the Euler-Lagrange equation. If we embed the signal into a 1-parameter family $s : \mathbb{R}^N \times \mathbb{R}^+ \to \mathbb{R}$, then the stationary point can be interpreted as the steady state solution, if it exists, of the following evolution equation:

$$\langle \partial_t s(\cdot, t), w \rangle = -\langle \delta E(s(\cdot, t)), w \rangle, \tag{1.23}$$

which, in a weak sense, amounts to the following gradient flow PDE:

$$\partial_t s = \mathrm{div}(\psi(\|\nabla s\|)\nabla s) \quad \text{with } \psi(\alpha) = \rho'(\alpha)/\alpha. \tag{1.24}$$

This means that nonlinear isotropic diffusion can be interpreted in terms of classical or Bayesian statistics. The term ρ is denoted as an *error norm* in the context of classical robust statistics and *potential function* in a Bayesian interpretation. In a robust statistical approach, the main data is assumed to follow a certain distribution, e.g. a Gaussian distribution and there are a few outliers whose distribution is not explicitly known. The challenge is to choose an error norm such that outliers do not influence the estimate. An energy function in a robust statistical approach does not necessarily belong to a valid probability distribution. In contrast to this in the Bayesian approach the complete probability distribution of main data and outliers is modeled by a probabilistic distribution, i.e. the robust error norm directly follows from the statistical properties of the complete data. A collection of well-known error norms and the corresponding diffusivities[2] are depicted in Fig. 1.6.

1.3.2 Gibbs Reaction-Diffusion

The prior term above is based on the absolute value of the image gradient $|\nabla s|$. This choice is ad hoc and the partial derivatives in $|\nabla s|$ may be exchanged by a set of

[2]Diffusivities calculated by e.g. Tuckey or Cup functions may become 0 and thus a diffusion tensor based on them is not guaranteed to be positive definite, but positive semi-definite.

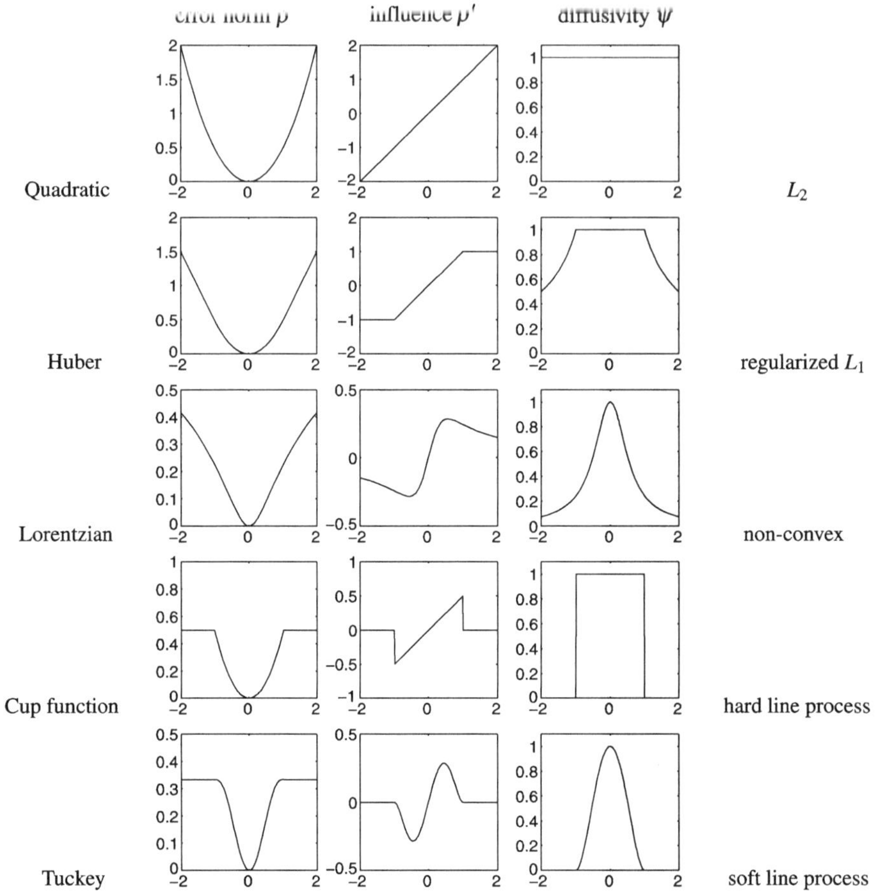

Fig. 1.6 Different (robust) error norms

linear filters. Zhu and Mumford [465] proposed to use Gibbs distributions of the form $p(s,R,F)=\frac{1}{Z}e^{-E(s,R,F)}$, where $R=\rho_1,\dots,\rho_J$ and F is the set of filters applied, $F=\{F_1,\dots,F_J\}$, Z a normalization factor making p integrate to 1 and

$$E(s,R,F)=\sum_i\sum_{j=1}^{J}\rho_j(F_j*s_i), \tag{1.25}$$

where $*$ denotes convolution. Filters F_j on different scales are used and the respective ρ_j are learned from training data. Zhu and Mumford observe that ρ-functions are well modeled by $\rho(\xi)=a(1-(1+(|\xi-\xi_0|/b)^{-\gamma})^{-1})$ for different parameters a, b, ξ_0, and γ. When $a>0$ we get a typical potential function with minimum at ξ_0. In most cases one gets $a>0$ for filters like ∇ or Δ which capture the general smoothness of an image (cf. Fig. 1.7, left). Interestingly for filters characterizing prominent features, e.g. Gabor filters at various orientations and scales one gets

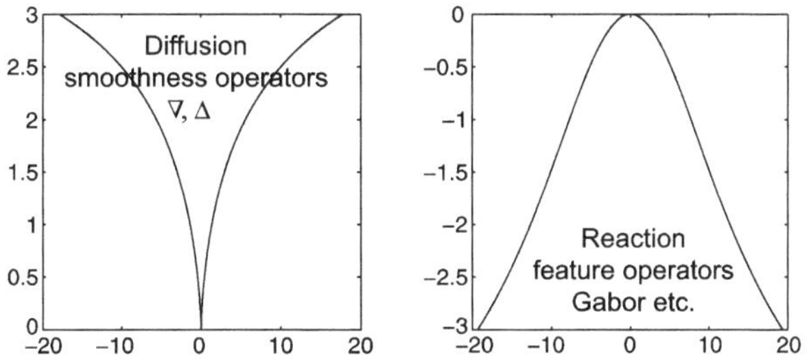

Fig. 1.7 Typical ρ-functions as derived by Zhu and Mumford for diffusion and reaction terms

$a < 0$, i.e. destabilizing behavior, resulting e.g. in edge-enhancement (cf. Fig. 1.7, right). Zhu and Mumford call these terms *reaction*, while the smoothing terms are called *diffusion*. All these terms are generalized isotropic nonlinear diffusions in our nomenclature. Roth and Black [361] propose a framework for also learning the filters F.

1.3.3 Steerable Random Fields and Anisotropic Diffusion

Following [371] anisotropic diffusion with a diffusion tensor can be derived using Zhu and Mumford's [465] approach (see Sect. 1.3.2). With the special filter choice $F = \mathbf{n}_1\nabla, \ldots, \mathbf{n}_J\nabla$, i.e. directional derivatives along the normalized vectors $\mathbf{n}_j$ we get the posterior probability (cf. Eqs. (1.19) and (1.25))

$$\hat{s} = \arg\max_s p(s|r) \quad \text{with } p(s|r) \propto \prod_i \left(p(r_i|s_i) \prod_{j=1}^{J} p_j(\mathbf{n}_j \nabla s_i) \right). \tag{1.26}$$

Maximizing $p(s|r)$ is equivalent to minimizing its negative logarithm, i.e. the energy

$$\hat{s} = \arg\min_s E(s) \quad \text{with } E(s) = -\sum_i \left(\rho_0(s_i - r_i) + \lambda \sum_{j=1}^{J} \rho_j(\mathbf{n}_j \nabla s_i) \right), \tag{1.27}$$

where we used $\rho_0(s_i - r_i) = -\log p(r_i|s_i)$ and $\rho_j(\mathbf{n}_j\nabla s_i) = -\log p_j(\mathbf{n}_j\nabla s_i)$, added indices to stress that likelihood and prior are different distributions, and introduced weight λ which accounts for the confidence one has in the smoothness terms (as before in Sect. 1.3.1). We set up a gradient descent minimization scheme using the functional derivative of E (cf. Eq. (1.22))

$$\partial_t s = -\rho_0'(s_i - r_i) + \lambda \nabla^T \sum_i \sum_{j=1}^{J} \psi_j(\mathbf{n}_j \nabla s_i)\mathbf{n}_j \mathbf{n}_j^T \nabla s_i \quad \text{for all } i \tag{1.28}$$

with $\psi_j(\alpha) = \rho_j'(\alpha)/\alpha$. Comparing the second term in Eq. (1.28) with anisotropic diffusion (Eq. (1.1)) reveals the relation between the diffusion tensor

$$\mathbf{D} = \sum_{i,j} \psi_j(\mathbf{n}_j \nabla s_i)\mathbf{n}_j \mathbf{n}_j^T \tag{1.29}$$

and the derivatives of the potential functions $\psi_j(\mathbf{n}_j \nabla s_i)$. Consequently, the diffusion tensor can be learned from training data.

Unfortunately the diffusion tensor from Eq. (1.28) is not constructed from a structure tensor as commonly done (cf. Sect. 1.2.2). Defining the prior as $\prod_j p(\mu_j)$, where μ_j are the eigenvalues of the structure tensor sorted by size yields a scheme similar to structure-tensor-based anisotropic diffusion [371]. However, as eigenvalues of the structure tensor are *smoothed* squared directional derivatives, the gradient in the spatial term of the diffusion operator (rightmost ∇ in Eqs. (1.1) and (1.28)) is also smoothed. If this smoothing is taken out of the structure tensor, the resulting scheme reduces to isotropic nonlinear diffusion [371]. This problem can be circumvented, when only the orientation of the diffusion is derived via the structure tensor, but directional diffusivities depend on unsmoothed directional derivatives [362]. Again this is not structure-tensor-based anisotropic diffusion, but called steerable random fields. An energy functional yielding structure-tensor-based anisotropic diffusion can be derived in a strict sense [366], however the structure tensor then is only used as a proxy for an orientation tensor coming from a linear model. This tensor does not depend on the signal s. We show this approach in Sect. 1.6.1.

1.3.4 Robust Statistics and Kernel Estimation

In the last section, we discussed the relation between probabilistic and diffusion based denoising methods. In particular we showed that the prior term in a Bayesian approach is directly linked to the energy of a robust estimator which then leads to classical diffusion schemes. Examining the likelihood term in (1.19) in the same manner reveals similar relations between different denoising methods as shown next. Let us consider several observations r_j in a local neighborhood. Assuming that all these observations belong to the same signal value[3] corrupted with identical independently distributed Gaussian noise, the corresponding likelihood function reads

$$p(\{r_j\}|s) \propto \exp\left\{-\frac{1}{2\sigma^2}\sum_j (s - r_j)^2\right\} \tag{1.30}$$

with the corresponding energy

$$E(s) = \sum_j (s - r_j)^2. \tag{1.31}$$

[3]In the case of an image, where r_j are spatially distributed on the pixel grid this is of course a smoothness assumption on the underlying signal s.

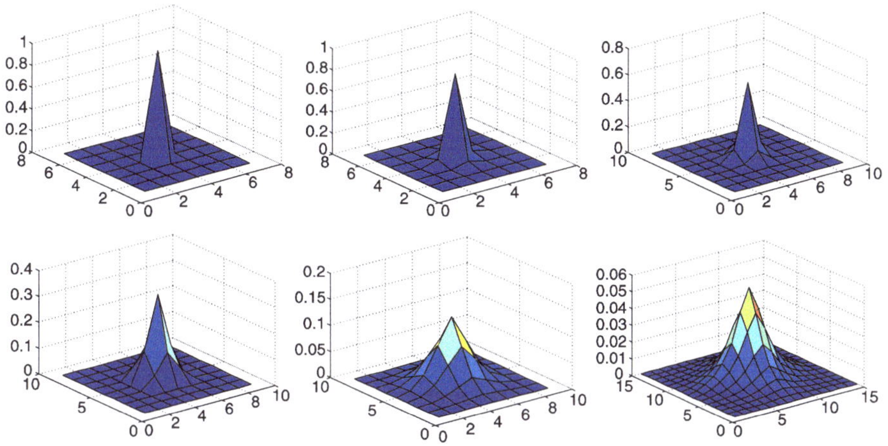

Fig. 1.8 Optimal spatial filter masks in a data term depending on an assumed noise level. *From left to right and top to bottom*: Noise level 1, 5, 10, 20, 50, 100

The maximum likelihood estimator is thus given by the *linear least squares estimator*, i.e. the signal is estimated by the mean of all observations. The assumption of a constant signal model might be too restrictive and one is attempted to give pixel values closer to a position x_k of interest more weight leading to the *weighted least squares estimator*

$$E(s_k) = \sum_j w_x(x_k - x_j)(s_k - r_j)^2 \quad \Rightarrow \quad s_k = \frac{\sum_j w_x(x_k - x_j) r_j}{\sum_j w_x(x_k - x_j)}, \tag{1.32}$$

where x_j and x_k denote the (pixel) position of r_j and s_k, respectively, and w_x is a weight function. The weighted least squares scheme here is usually implemented by a convolution with a (normalized) smoothing kernel w_x, equivalent to linear diffusion if a Gaussian kernel is used (cf. Sect. 1.2.1). Optimal weights depending on pixel position can be chosen from statistical characteristics of the model error (a constant signal here) and are typically not of Gaussian shape (cf. [267] and Fig. 1.8).

If the observed signal contains outliers, i.e. either the signal model or the noise model is severely violated, the estimator can be made robust. Such a *robust estimator* is obtained by exchanging the quadratic energy function by a robust error metric, i.e. the corresponding estimator is the minimum of the energy

$$E(s_k) = \sum_j \rho(s_k - r_j). \tag{1.33}$$

It may be minimized by iterating

$$s_k^{t+1} = \frac{\sum_j w_s(s_k^t - r_j) r_j}{\sum_j w_s(s_k^t - r_j)}, \tag{1.34}$$

Likelihood (data term)	Prior (smoothness term)
• May be learned from data.	• May be learned from data.
• Smoothness assumption on signal only if r spatially distributed.	• Smoothness assumption on signal.
• Term contains measured data r and solution s.	• Term contains only solution s.
• Initialized with mean or r (typically).	• Initialized with r, if only one term.
• Solution non-constant.	• Solution (piece-wise) constant.
• Scheme iterated till convergence.	• Only few iterations, then stopped.
• Variance of spatial kernel constant wrt. $h \Rightarrow 0$, $\tau \Rightarrow 0$.	• Variance of spatial Gaussian decreases with $h \Rightarrow 0$, $\tau \Rightarrow 0$ ($\sigma^2 = 2c\tau$).
• Consistency check: No diffusion.	• Consistent diffusion.

Fig. 1.9 Summary: Likelihood versus prior. First iteration may be identical in both cases

where $w_s(u) = \rho'(u)/u$, and the upper index t is the iteration number. Please note that if ρ is a negative Gaussian $\rho(u) = -\exp(-u^2)/2$, then $\rho'(u) = u\exp(-u^2)$ and $w_s(u) = \exp(-u^2)$ is a Gaussian as well.

In a Bayesian framework, the error metric is interpreted as a potential function that directly encodes the statistical properties of signal model and noise. As for the weighted least squares estimator, we may introduce a further weight w_x reducing the influence of estimates being further away from the central position x_k. The minimizer of the corresponding energy

$$E(s_k) = \sum_j w_x(\mathbf{x}_k - \mathbf{x}_j)\rho(s_k - r_j) \tag{1.35}$$

is denoted as the robust M-smoother in literature (cf. e.g. [75, 449]). This energy may be minimized by iterating

$$s_k^{t+1} = \frac{\sum_j w_x(\mathbf{x}_k - \mathbf{x}_j) w_s(s_k^t - r_j) r_j}{\sum_j w_x(\mathbf{x}_k - \mathbf{x}_j) w_s(s_k^t - r_j)}. \tag{1.36}$$

The equivalence to nonlinear diffusion becomes apparent, when we select a spatial neighborhood consisting of nearest neighbors only for w_x and start with $s_k^0 = r_k$. The first iteration step then is equivalent to a diffusion step, where w_s models diffusivities. This equivalence can also be shown for larger neighborhoods [20, 21].

Please note When we start with a smoothness or constancy assumption formulated on the measured data r, we construct a *data term*. In the respective estimation schemes r is *never updated*. When we start with a similar assumption on the underlying signal s as in Sect. 1.3.1, we get diffusion schemes, where r_j stands for the initial value s_j^0 and is updated. Here, we introduced spatial weights ad hoc, in Sect. 1.3.1 they come from an MRF assumption.

In Fig. 1.9 we summarize properties of likelihood-based data terms and prior-based smoothness terms.

1.4 Some Diffusion-Like Nonlinear Regularization Schemes

There are several regularization techniques using averaging kernels, usually Gaussians, together with a nonlinearity (for overviews see e.g. [55, 75, 449] as well Chap. 4 in this volume). Local M-smoothing [75, 449] and related robust statistics-based methods have already been shown in Sect. 1.3.4. We only show two prominent examples here: bilateral filtering [409] also called cascaded nonlinear Gaussian filtering [17, 187, 445], and channel smoothing [139], cf. Chap. 2. They are closely connected to nonlinear isotropic diffusion. A false friend in the list of diffusion-like methods is mean shift filtering [81], a mode-seeking method. Iterations occurring in this approach are similar to diffusion, but only the first iteration really is equivalent to a diffusion step.

1.4.1 Bilateral Filtering and Nonlinear Gaussian Filtering

Bilateral Filtering [409] as well as nonlinear Gaussian filtering [17, 187, 445] operate on the input data given by $r(\mathbf{x})$ and filter it via

$$\hat{s}(\mathbf{x}) = k^{-1}(\mathbf{x}) \sum_{\boldsymbol{\xi}} w_1(\mathbf{x} - \boldsymbol{\xi}) w_2(r(\mathbf{x}) - r(\boldsymbol{\xi})) r(\mathbf{x}), \tag{1.37}$$

$\mathbf{x} \in \mathbb{R}^N$, where $k(\mathbf{x}) = \sum_{\boldsymbol{\xi}} w_1(\mathbf{x} - \boldsymbol{\xi}) w_2(r(\mathbf{x}) - r(\boldsymbol{\xi}))$ is a normalization, and $\hat{s}(\mathbf{x}) : \mathbb{R}^N \to \mathbb{R}$ is the filtered image. The filter weights w_1 and w_2 may be Gaussians as suggested in [17, 187, 409, 445] but other filter weights may also be applied. Positions $\mathbf{x}_i$ may be restricted to a spatial local neighborhood with size depending on the standard deviation σ_1 of w_1, typically $3\sigma_1$.

Equation (1.37) is simple linear Gaussian smoothing if the second kernel $w_2 \equiv 1$ and thus a direct solution of the heat equation (1.3) for one given time step, cf. Sect. 1.2.1. The second kernel $w_2(r(\mathbf{x}) - r(\xi))$ down-weights the contribution of a value $r(\boldsymbol{\xi})$ if it differs from the value $r(\mathbf{x})$ at the current position $\mathbf{x}$. This is equivalent to reducing the diffusivity between the points $\mathbf{x}$ and $\boldsymbol{\xi}$. Iterating Bilateral Filtering by applying it to the filtered data is therefore similar to isotropic nonlinear diffusion in relatively coarse time steps. An investigation based on a detailed analysis of discrete schemes also shows this connection between diffusion and bilateral filtering [20, 21].

So-called cascaded nonlinear Gaussian filtering changes standard deviations σ_1 and σ_2 of w_1 and w_2, respectively, in every iteration step. Typically one starts with small σ_1 (space) and large σ_2 (range) and doubles σ_1 while halving σ_2 in every iteration step.

We will now check for numerical consistency of bilateral filtering with isotropic non-linear diffusion. It is not sufficient to compare discrete schemes in order to decide whether or not bilateral filtering is a consistent numerical scheme for isotropic nonlinear diffusion. We need to known how it behaves in the limit of continuous signals s and r. It is clear that if we only go to continuous domain, without associating r with the initial (i.e. $t = 0$) signal $s(x,t)|_{t=0}$ this scheme cannot become diffusion—the diffusion equation contains no r. In this case the energy associated with the respective likelihood term reads

$$E(s_k) = \sum_j w(\mathbf{x}_k - \mathbf{x}_j)\rho(r_k - r_j)(s_k - r_j)^2 \tag{1.38}$$

with $r_j = r(\mathbf{x}_j)$, which can be interpreted as the energy function of a Gaussian distribution with the precision matrix $\Lambda_{kj} = w(\mathbf{x}_k - \mathbf{x}_j)\rho(r_k - r_j)$. Rewriting Eq. (1.37) as

$$\hat{s}(\mathbf{x}, t+\tau) = k^{-1}(\mathbf{x}) \sum_{\boldsymbol{\xi}} w_1(\mathbf{x} - \boldsymbol{\xi}) w_2(s(\mathbf{x}, t) - r(\boldsymbol{\xi})) s(\mathbf{x}, t) \tag{1.39}$$

we do not end up with diffusion either, but perform robust averaging of multiple measurements r. This update scheme corresponds to a robust M-smoother.

Only if we exchange also $r(\xi)$ by $s(\xi, 0)$, we may end up with diffusion if we do the limiting process right

$$\hat{s}(\mathbf{x}, t+\tau) = k^{-1}(\mathbf{x}) \sum_{\boldsymbol{\xi}} w_1(\mathbf{x} - \boldsymbol{\xi}) w_2(s(\mathbf{x}, t) - s(\boldsymbol{\xi})) s(\mathbf{x}, t). \tag{1.40}$$

This update scheme is proposed in [409]. If the variance of kernel w_1 decreases with selected time step τ and τ is small enough, only nearest neighbors need to be addressed. Comparing Eq. (1.40) with the update scheme in Eq. (1.5) reveals that w_2 may be interpreted as diffusivity c in that scheme. This interpretation of bilateral filtering is consistent with isotropic nonlinear diffusion.

1.4.2 Mean Shift Filtering

Noise reduction may be regarded as estimation of the most likely measurement value (or other feature) at a given position. Density of features in feature space may be regarded as empirical probability function (PDF) of the represented parameter. The modes of a feature space, i.e. maximal dense regions, may therefore be identified as local maxima of the unknown PDF. Mean shift filtering locates maxima, i.e. stationary density points by gradient ascent without estimating the density. How does this work?

A kernel density estimator in a flat space is given by

$$\hat{f}(\mathbf{x}) = \frac{1}{n} \sum_{i=1}^{n} K(\mathbf{x} - \mathbf{x}_i), \tag{1.41}$$

where $\hat{f}$ is the estimated density, n data points $\mathbf{x}_i \in \mathbb{R}^N$, $i \in \{1, \ldots, n\}$, are given and K is some normalized smoothing kernel typically radially symmetric $K(\mathbf{x}) =$

$k(\|\mathbf{x}\|^2)$. Standard selections for K are a sphere around the origin or a multivariate Gaussian. The gradient of the density is then

$$\nabla \hat{f}(\mathbf{x}) = \frac{1}{n}\sum_{i=1}^{n}(\mathbf{x}_i - \mathbf{x})g(\|\mathbf{x} - \mathbf{x}_i\|^2)$$
$$= \frac{1}{n}\left[\sum_{i=1}^{n} g(\|\mathbf{x} - \mathbf{x}_i\|^2)\right]\left[\frac{\sum_{i=1}^{n}\mathbf{x}_i g(\|\mathbf{x} - \mathbf{x}_i\|^2)}{\sum_{i=1}^{n} g(\|\mathbf{x} - \mathbf{x}_i\|^2)} - \mathbf{x}\right], \quad (1.42)$$

where $g(x) := -k'(x)$, i.e. when K is a Gaussian, the kernel $G(\mathbf{x}) = g(\|\mathbf{x}\|^2)$ is also a Gaussian. While the first term in Eq. (1.42) is a density estimate calculated with kernel G, the second term is the *mean shift* $\mathbf{m}$

$$\mathbf{m} := \frac{\sum_{i=1}^{n}\mathbf{x}_i g(\|\mathbf{x} - \mathbf{x}_i\|^2)}{\sum_{i=1}^{n} g(\|\mathbf{x} - \mathbf{x}_i\|^2)} - \mathbf{x}. \quad (1.43)$$

This is the vector pointing from the filter center $\mathbf{x}$ to the position of the weighted mean. An iteration scheme may be defined by

$$\mathbf{y}_{j+1} = \frac{\sum_{i=1}^{n}\mathbf{x}_i g(\|\mathbf{y}_j - \mathbf{x}_i\|^2)}{\sum_{i=1}^{n} g(\|\mathbf{y}_j - \mathbf{x}_i\|^2)}, \quad (1.44)$$

where $\mathbf{y}_j$ is the jth location where the kernel G is applied at. Comparing Eq. (1.44) to Eq. (1.34) we see that mean shift filtering is equivalent to location M-estimation, where $g(u) = \rho'(u)/u$, $u \in \mathbb{R}^+$.

Mean shift filtering for image denoising applies Eq. (1.44) to an image, where $\mathbf{x}$, with $t > 0$ fixed, is an element of the product space $E = \Omega \times I$ of image domain Ω and intensity (or feature) domain I (the same product space is used in the Beltrami framework, cf. Sect. 1.2.3). When a stationary point ($\in E$) is reached, the feature ($\in I$) of this point is assigned to the starting pixel position ($\in \Omega$).

Please note that mean shift filtering always operates on the unchanged input data and therefore really reaches a mode of the PDF. Diffusions do not seek for modes, as they end in flat images. However their first step is towards the nearest mode.

1.4.3 Channel Smoothing

Channel smoothing explicitly represents the product space $E = \Omega \times I$ (cf. Sect. 1.2.3 as well as Chap. 2 in this volume) by so-called channels. Usual intensity coding represents scalar values s, i.e. intensities by a single value. In contrast to this, channels use a concept similar to fuzzy encoding, where a set of N channels c_n with weights w_n, $n \in 1, \dots, N$ encode a value. Each weight w_n represents how much the encoded value belongs to the channel value c_n the weight stands for. E.g. 10 channels may represent the numbers 1 to 10, i.e. $c_n = n$, $N = 10$.

A simple encoding scheme then would represent the intensity, e.g. 3.5 by setting channels weights $w_3 = w_4 = 0.5$ and all other weights to zero. A suitable decoding step calculates intensities from channels, e.g. by weighted averaging of channels $s = \sum_n n w_n$. If more than one value is encoded in the channels, summation is performed locally, e.g. summing over a channel and its direct neighbors $s_n = (n-1)w_{n-1} + n w_n + (n+1)w_{n+1}$, and selection of the value represented by the highest sum of weights $w_{n-1} + w_n + w_{n+1}$. This selection typically is the only nonlinear step in the scheme.

Channel smoothing consists of 3 steps: (1) encoding, (2) spatial smoothing of channels, (3) decoding. Encoding and decoding is done as written above. For smoothing each channel is convolved individually by a Gaussian kernel, i.e. linear diffusion is applied to channel weights. This means input intensities contribute to a local averaging only if their value contributes to a channel weight which is later on included in the decoding step.

It has been shown that this is equivalent to local M-smoothing, where also input values are excluded (or down-weighted) if they are too far away from the *solution* [139]. This is different from bilateral filtering or nonlinear Gaussian filtering, where values are not excluded based on the solution, but if they are too far away from the initial value of a reference (central) pixel.

1.5 Discretization

Discretization schemes differ in the way spatial and temporal terms in Eq. (1.1) are treated. The most simple and common choice is two-level explicit finite-difference schemes. But besides this, there are e.g. three-level methods [173], semi-implicit approaches [68, 441], multigrid methods [2], adaptive finite element techniques [19], schemes with wavelets as trial functions [173], spectral methods [173], and stochastic simulations [353]. Methods for anisotropic diffusion filters with a diffusion tensor include finite elements with grid adaptation [350] or multigrid acceleration [108], lattice Boltzmann techniques [237] and explicit finite difference schemes [82, 83, 434]. An excellent semi-analytic 3×3-pixel scheme for 2D input images has been presented in [443]. The core idea is to discretize derivatives in 2×2-cells and calculate D for the center of this cell. For each 2×2-cell the dynamical system defined by Eq. (1.1) is solved analytically for fixed diffusion tensor D. Resulting pixel solutions are averaged yielding a 3×3-pixel scheme. This locally semi-analytic scheme features good rotation invariance and absolute stability, i.e. minima do not drop and maxima do not raise. For higher dimensions this scheme is not yet available.

The anisotropic diffusion process in Eq. (1.1) can be solved using finite differences when data lives in a flat space. An Euler forward approximation for $\frac{\partial s}{\partial t}$ yields

$$\frac{s_i^{l+1} - s_i^l}{\tau} = \sum_{j \in \Omega_i} A_{j,i}^l s_j^l \quad \Leftrightarrow \quad s_i^{l+1} = s_i^l + \tau \sum_{j \in \Omega_i} A_{j,i}^l s_j^l, \tag{1.45}$$

where Ω_i is a neighborhood of voxel i, τ is a time step size and s_i^l denotes the approximation of $s(x,t)$ in pixel i at (diffusion) time $l\tau$. The expression $\sum_{j\in\Omega_i} A_{j,i}^l s_j^l$ is a discretization of $\nabla\cdot(D\nabla s)$. It comes down to the convolution of the image with a spatially and temporally varying mask A_i^l. Hence, we may calculate s at time level $l+1$ directly from s at level l. We present and compare different implementations of A below. All these schemes have consistency order 2 and are therefore numerically equivalent when refining spatial discretization. As one can show [433] finite difference schemes feature absolute stability only for diffusion tensors with small enough condition, i.e. ratio between largest and smallest eigenvalues.

1.5.1 Rotation Invariance

Rotation invariance for a diffusion scheme means that smoothing behavior is the same, independent of the orientation of the structure it smoothes along. For 2D images several schemes exist providing rotation invariant behavior for tensor-based anisotropic diffusion [369, 439, 443]. One of the most prominent applications of orientation estimation is optical flow estimation in image sequences. We will now focus on this application.

The *standard discretization* is a 3-tab finite difference scheme introduced in [434] in 2D. A 3D version is shown in [369]. It is derived by rewriting $\nabla\cdot(D\nabla s)$ as a sum of terms, and replacing derivatives in terms with derivatives along the same direction by a filter $[1,-1]$. In mixed derivative terms one uses $[0.5, 0, -0.5]$. Parameters $D_{i,j}$ are interpolated linearly by $[0.5, 0.5]$ when needed at inter pixel positions. This results in a spatial discretization with a support of 3 pixels in every dimension.

Rotation invariance of a scheme can be increased using separable first order derivative filters optimized for the calculation of gradient *direction* (see e.g. [368]). To use these optimized filters we rewrite the spatial differential operator in Eq. (1.1) as

$$\nabla\cdot(D\nabla s) = (\partial_{x_1},\ldots,\partial_{x_N})D(\partial_{x_1},\ldots,\partial_{x_N})^T s. \tag{1.46}$$

This expression is evaluated in an explicit way, i.e. using only known values from the old time level l. The key point is to use the optimized filters as derivative operators. They are composed of a common one dimensional derivative filter (e.g. $[1,0,-1]/2$, denoted $\mathscr{D}$) and one dimensional smoothing kernels (e.g. $[3,10,3]/16$, denoted $\mathscr{B}$) in all other directions. The derivatives are then $\partial_{x_i} = \mathscr{D}_{x_i} * \mathscr{B}_{x_j} * \mathscr{B}_{x_k}$ where $\{i,j,k\}$ is a permutation of $\{1,2,3\}$, lower indices give the direction of the kernel and $*$ is a convolution. Optimized choices for $\mathscr{D}$ and $\mathscr{B}$ are given in Table 1.1. These filters approximate rotation invariance significantly better than related filters like the Sobel filters or scaled Gaussian derivatives of the same size.

The following algorithm is repeated for every time step of size τ

1. Calculate the structure tensor J from signal s (Eq. (1.7)).
2. Get the diffusion tensor D by J (Eqs. (1.10) and (1.15)).

Table 1.1 Derivative filters, optimized for gradient direction

Name	$\mathscr{D}$	$\mathscr{B}$
2-tab	$[1, -1]$	$[0.5, 0.5]$
3-tab	$[0.5, 0, -0.5]$	$[3, 10, 3]/16$
4-tab	$[-1, 27, -27, 1]/24$	$[-0.0876, 0.5876, 0.5876, -0.0876]$
5-tab	$[-1, 8, 0, -8, 1]/12$	$[-0.0439, 0.1646, 0.7586, 0.1646, -0.0439]$

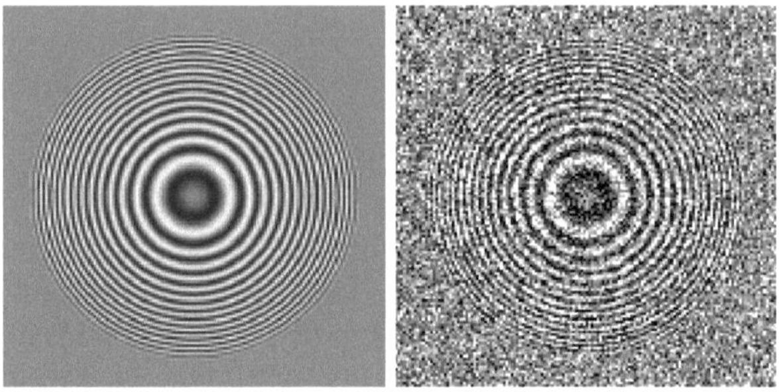

Fig. 1.10 Synthetic sinusoidal sphere with $143 \times 143 \times 143$ voxel. *Left*: x–y-slice at $z = 71$. *Right*: With added Gaussian noise of standard deviation $\sigma = 64$

3. Calculate the flux $j_i := \sum_{m=1}^{3} D_{i,m} \partial_{x_m} s, \ \forall i \in \{1, 2, 3\}$.
4. Calculate $A * s$ (Eq. (1.45)) by $A * s = \sum_{m=1}^{3} \partial_{x_m} j_m$.
5. Update s in an explicit way (Eq. (1.45)).

In the performance tests depicted in Fig. 1.11 the filters named 3-tab of Table 1.1 are always used for step 1. This is done in order to always construct the diffusivity tensor D identically. Only at steps 3 and 4 other filters are used. Schemes are named according to these filters. We observe that severe blurring of fine structures occurs when using standard discretization (STD), 2tab and 4tab-schemes. Significantly less over-smoothing is obtained when using 3-tab and 5-tab schemes. Quantitative and qualitative tests with similar 2D schemes in [370, 439] validate this result.

We introduce normal distributed noise of standard deviation $\sigma_n = 20$ to the rubic cube sequence, Fig. 1.12 left in order to illustrate the improvement of an optical flow field under heavy noise on a real sequence. With this amount of noise the optical flow field (see also Eq. (1.59)) estimated with the structure tensor[4] $\mathbf{J}_\rho$ has not much resemblance to the flow field estimated on the original sequence, Fig. 1.12 middle left and middle right. With a stopping criterion based on confidence of optical flow estimation presented in [369] diffusion terminates at diffusion time $t = 14$. The displacement field computed at this time step appears even better than the one from the original data, see Fig. 1.12 right. Note that by adjusting the scale ρ in the original structure tensor $\mathbf{J}_\rho$ one can also cope with high noise but only at the cost of locality.

[4]Optical flow is derived as ratio of the first and third, and second and third components of the eigenvector e_3 to the smallest eigenvalue μ_3 of a structure tensor calculated from a spatio-temporal gradient.

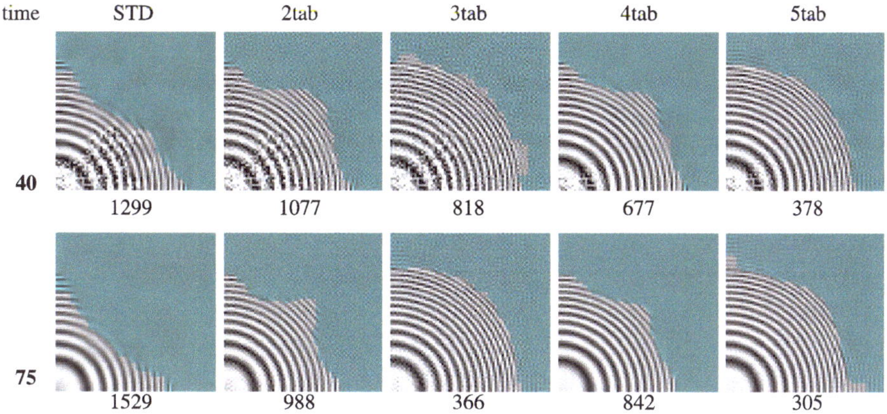

Fig. 1.11 Restoration of the noisy image from Fig. 1.10. *The overlayed color* indicates the low signal area for ease of viewing blurring artifacts. *The numbers below the images* are their mean square differences to the original noise free image Fig. 1.10

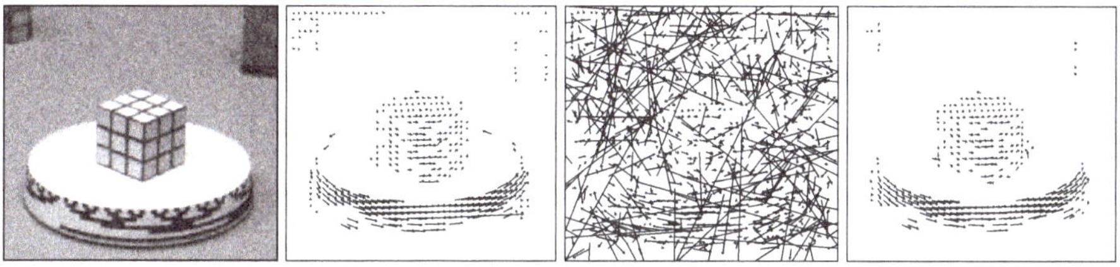

Fig. 1.12 Rubic cube sequences. *From left to right*: Noisy image, flow on noiseless sequence, flow on noisy sequence and flow after diffusion

1.5.2 Anisotropic Diffusion for Tensor Valued Images

Due to the rapid technological development of diffusion tensor imaging (DTI), the processing of tensor valued images have become more and more important in recent years (see e.g. [252–254] for a comprehensive introduction to DTI).

The differential geometric framework introduced in [250] (Sect. 1.2.3) allows a direct generalization to tensor valued images, i.e. instead of a gray value $s(\mathbf{x})$, a tensor $\Sigma(\mathbf{x})$ (positive definite matrix) is assigned to each pixel. Tensor valued images are considered as sections in the product space $\mathbb{R}^n \times P(n)$, of the image domain and the space of positive definite tensors. In principle one could work with local coordinates and finite difference approximations of the derivatives in this space. However, finite difference approximations do not (always) feature absolute stability (see above for the case of scalar valued images). Thus convex filtering is not guaranteed using such discretizations. For tensor valued images this means that positive definiteness of the tensors might not be preserved. In order to prevent such shortcomings, a global perspective could be used as introduced in [145] for nonlinear isotropic diffusion. Here we present the generalization to nonlinear anisotropic diffusion (see also [266]).

The main idea is to formulate the diffusion equation on the tangent space of the image section and obtain one step of the flow by means of the exponential map. Except for the nonlinear anisotropic diffusion scheme, the diffusion equation can be derived from a corresponding energy functional $E(u)$ via calculus of variation, i.e. the gradient descent scheme of these energy functionals can be identified with a diffusion equation (cf. Sects. 1.3.1 and 1.6.1).

The functional derivative of an energy functional of a tensor field can be defined in the same way as in Sect. 1.3.1, i.e. by the differential

$$\langle \delta E, W \rangle := \int \langle \delta E(\mathbf{x}), W(\mathbf{x}) \rangle_{\Sigma(\mathbf{x})} \, d\mathbf{x}, \tag{1.47}$$

where $W : \mathbb{R}^n \to P(n)$ denotes a tensor valued test function and

$$\langle \cdot, \cdot \rangle_{\Sigma(\mathbf{x})} : TP(n) \times TP(n) \to \mathbb{R} \tag{1.48}$$

denotes the Riemannian scalar product at $\Sigma(\mathbf{x})$. The corresponding diffusion equation (1.23) generalizes accordingly.

Let us now consider the linear anisotropic diffusion equation, i.e. Eq. (1.1) with $\boldsymbol{D}$ not depending on the evolving signal. The corresponding energy function is

$$E(u) = \frac{1}{2} \int \nabla u^T D \nabla u \, d\mathbf{x}. \tag{1.49}$$

Let $\partial_i \boldsymbol{\Sigma}(\mathbf{x})$, $i = 1, \ldots, N$ denote the partial derivative of the tensor field in direction i, elements of the tangent space at $\boldsymbol{\Sigma}$. We define the energy functional

$$E(\boldsymbol{\Sigma}) = \frac{1}{2} \int \sum_{i,j} d_{ij} \langle \partial_i \boldsymbol{\Sigma}, \partial_j \boldsymbol{\Sigma} \rangle_{\boldsymbol{\Sigma}} \, d\mathbf{x} \tag{1.50}$$

with

$$\langle \partial_i \boldsymbol{\Sigma}, \partial_j \boldsymbol{\Sigma} \rangle_{\boldsymbol{\Sigma}} = \mathrm{Tr}((\partial_i \boldsymbol{\Sigma}) \boldsymbol{\Sigma}^{-1} (\partial_j \boldsymbol{\Sigma}) \boldsymbol{\Sigma}^{-1}). \tag{1.51}$$

The components of the diffusion tensor d_{ij} (please do not confuse d_{ij} with the elements of the tensor field) locally controls the direction of smoothing and for the moment being does not depend on the evolving tensor field. Inserting Eqs. (1.50–1.51) in the right hand side of Eq. (1.47) yields, after some algebraic manipulations, the functional derivative

$$\delta E = -\sum_{i,j} d_{ij} (\partial_i \partial_j \boldsymbol{\Sigma} - (\partial_i \boldsymbol{\Sigma}) \boldsymbol{\Sigma}^{-1} (\partial_j \boldsymbol{\Sigma})) - \sum_{i,j} (\partial_i d_{ij})(\partial_j \boldsymbol{\Sigma}). \tag{1.52}$$

The terms on the right side, with $i = j$, i.e.

$$\Delta_i \boldsymbol{\Sigma} = \partial_i^2 \boldsymbol{\Sigma} - (\partial_i \boldsymbol{\Sigma}) \boldsymbol{\Sigma}^{-1} (\partial_i \boldsymbol{\Sigma}) \tag{1.53}$$

are components of the Laplace Beltrami operator $\Delta = \sum_i \Delta_i$ derived in [332]. In addition to the work in [145, 332], we also state mixed components

$$\Delta_{ij}\boldsymbol{\Sigma} = \partial_i \partial_j \boldsymbol{\Sigma} - (\partial_i \boldsymbol{\Sigma})\boldsymbol{\Sigma}^{-1}(\partial_j \boldsymbol{\Sigma}), \quad i \neq j \tag{1.54}$$

needed for the linear anisotropic diffusion equation.

The nonlinear anisotropic diffusion equation is defined exchanging the diffusion tensor components in Eq. (1.53) with components depending on the evolved tensor field. So we have all components to define an anisotropic diffusion equation on the space of positive definite matrices in an intrinsic way. To this end, only the second order derivatives ∂_i^2 and $\partial_i \partial_j$ occurring in (1.1) need to be exchanged by their counterparts Δ_i and Δ_{ij}.

So far we have assumed the tensor to be defined on a continuous domain. In the experiential setting tensor fields are defined on discrete grids. The application of Riemanian anisotropic diffusion requires a discrete approximation for the derivatives derived above. In principle, we could use matrix differences to approximate the derivatives but this would contradict our effort to derive an intrinsic expression of the anisotropic diffusion equation. The finite differences are extrinsic since they are based on Euclidean differences between tensors, i.e. they use the difference in the space of symmetric matrices and not the Riemannian metric of the space $P(n)$.

We need discrete approximations of derivatives of first and second order to approximate the functional derivative δE in Eq. (1.52) on a discrete grid. Intrinsic approximations to first order derivatives have already proposed in [145] and are reviewed here with the following propositions. To this end we define

$$\overrightarrow{AB} := A^{\frac{1}{2}} \log(A^{-\frac{1}{2}} B A^{-\frac{1}{2}}) A^{\frac{1}{2}}, \tag{1.55}$$

cf. [145] for details.

Proposition I *The first order discrete approximation of the first order derivative of* $\boldsymbol{\Sigma}$ *in direction* j *reads*

$$\partial_j \boldsymbol{\Sigma}(x) = \frac{1}{2\varepsilon}(\overrightarrow{\boldsymbol{\Sigma}(x)\boldsymbol{\Sigma}(x+\varepsilon e_j)} - \overrightarrow{\boldsymbol{\Sigma}(x)\boldsymbol{\Sigma}(x-\varepsilon e_j)}) + O(\varepsilon). \tag{1.56}$$

A second order discrete approximation scheme to the second order derivative in direction e_j *has been derived in* [332]. *We state it here as a second preposition.*

Proposition II *The second order discrete approximation of the second order derivative in direction* e_j *is*

$$\Delta_j \boldsymbol{\Sigma}(x) = \frac{1}{\varepsilon^2}(\overrightarrow{\boldsymbol{\Sigma}(x)\boldsymbol{\Sigma}(x+\varepsilon e_j)} + \overrightarrow{\boldsymbol{\Sigma}(x)\boldsymbol{\Sigma}(x-\varepsilon e_j)}) + O(\varepsilon^2). \tag{1.57}$$

Fig. 1.13 Denoising experiment. *Upper row (from left to right)*: Original tensor field, noise corrupted tensor field. *Lower row (from left to right)*: Diffusion with flat metric; diffusion with Riemannian metric

For the anisotropic diffusion equation we also need mixed derivatives $\Delta_{ij}\boldsymbol{\Sigma}$ *that can be approximated according to Proposition* III *derived in* [266].

Proposition III *The second order discrete approximation of the second order mixed derivative in direction* i *and* j *is given by*

$$\begin{aligned}\Delta_{ij}\boldsymbol{\Sigma}(x)+\Delta_{ji}\boldsymbol{\Sigma}(x) &= \frac{1}{\varepsilon^2}(\overrightarrow{\boldsymbol{\Sigma}(x)\boldsymbol{\Sigma}(x+\varepsilon e_n)}+\overrightarrow{\boldsymbol{\Sigma}(x)\boldsymbol{\Sigma}(x-\varepsilon e_n)}\\ &\quad -\overrightarrow{\boldsymbol{\Sigma}(x)\boldsymbol{\Sigma}(x+\varepsilon e_p)}-\overrightarrow{\boldsymbol{\Sigma}(x)\boldsymbol{\Sigma}(x-\varepsilon e_p)})\\ &\quad +O(\varepsilon^2), \end{aligned} \tag{1.58}$$

with the abbreviation $e_n = \frac{1}{\sqrt{2}}(e_i + e_j)$, $e_p = \frac{1}{\sqrt{2}}(e_i - e_j)$.

In Fig. 1.13 we see denoising results on a synthetic tensor field with noise added. We observe that a standard finite difference discretization based on the Euclidean metric leads to swelled tensors, known as the *eigenvalue swelling effect*. The discretization scheme based on Riemannian geometry does not show this effect.

1.6 Model-based Diffusion

1.6.1 An Energy Functional for Anisotropic Diffusion

As we have seen above, diffusion can be derived from smoothness assumptions on the 'true' signal s. For some diffusions an energy functional like Eq. (1.20) can be given, but for structure-tensor-based anisotropic diffusion as described by Weickert [433] we have not shown such an energy so far. The problem was that no such energy can be defined, when structure estimates depend on the signal s.

However, the well known brightness constancy constraint equation used for optical flow estimation (BCCE, defined below) leads to tensor-based anisotropic diffusion, when used as a smoothness constraint in a cost function. Although our experiments below are done using 2D images, we use a 3D formulation here, because in this form the BCCE is most well known.

Let us assume that s is a densely sampled image sequence, thus $x_1 = x$, $x_2 = y$, and $x_3 = t$. In contrast to Eq. (1.20) let us further assume that s should fulfill the BCCE

$$\partial_x s\, dx + \partial_y s\, dy + \partial_t s\, dt = 0 \quad \Leftrightarrow \quad (\nabla^T s)\mathbf{u} = 0, \tag{1.59}$$

where $\mathbf{u} = (dx, dy, dt)^T$ is a parameter vector. This is a typical linear model, where linear means linear in $\mathbf{u}$. In order to avoid the trivial solution $\mathbf{u} = 0$ one usually either defines $|\mathbf{u}| = 1$ or $\mathrm{d}t = 1$. Here we prefer the first assumption, because it is more clearly related to orientation estimation. From this vector optical flow, i.e. x- and y-displacements u_x and u_y, respectively, can be calculated via $u_x = dx/dt$ and $u_y = dy/dt$. There are many ways to estimate such a parameter vector (see e.g. [23]). For now let us assume we know the probability distribution $p(\mathbf{u})$ at each pixel and thus could give its expectation value $\langle \mathbf{u} \rangle$. We can then formulate the cost function

$$E(s(\cdot,t)) = \int_\Omega (s(\cdot,t) - r)^2\, d\mathbf{x} + \alpha \int_\Omega \langle (\nabla^T s(\cdot,t)\mathbf{u})^2 \rangle\, d\mathbf{x}. \tag{1.60}$$

As before in Sects. 1.3.1 and 1.3.3 we set up a gradient descent minimization scheme using the functional derivative of E (cf. Eq. (1.22)). We get

$$\partial_t s = (r - s) - \alpha \bar{\nabla}^T \langle \mathbf{u}\mathbf{u}^T \rangle \nabla s, \tag{1.61}$$

where s is treated as a volume and ∂_t is the derivative with respect to the 'iteration time' of the scheme (not to be confused with the time in an image sequence), and $\bar{\nabla} = -\nabla$ is a mirrored version of ∇ (cf. also [465]). The term $\langle \mathbf{u}\mathbf{u}^T \rangle$ is a square, symmetric, positive definite matrix. As shown in [366] this diffusion tensor is well approximated by a regularized inverse of the structure tensor. Equation (1.61) describes anisotropic diffusion of s with *diffusion tensor* $\langle \mathbf{u}\mathbf{u}^T \rangle$. The data term acts as a (grey value) source and is usually omitted. Obviously u and s are coupled via BCCE (Eq. (1.59)). However u is not s and when varying s, u may be kept constant.

1.6.2 Diffusion for Multiple Orientations

The single orientation model $\nabla^T g\mathbf{u} = 0$ (cf. Eq. (1.59)) above can be rewritten as

$$(\mathfrak{D}^T s)\mathbf{p} = 0, \tag{1.62}$$

where $\mathbf{p}$ is the parameter vector and $\mathfrak{D}$ is an operator vector applied to the data s. In Eq. (1.59) we have $\mathbf{p} = \mathbf{u}$ and $\mathfrak{D} = \nabla$. For any linear model that can be written in the form Eq. (1.62), we can construct a diffusion-like reconstruction scheme as done for the single orientation model above via the cost function (cf. Eq. (1.60))

$$E(s) = \int_\Omega (s - r)^2 \, d\mathbf{x} + \alpha \int_\Omega \langle((\mathfrak{D}^T s)\mathbf{p})^2\rangle \, d\mathbf{x}. \tag{1.63}$$

As above we derive a gradient descent scheme

$$\partial_t s = (r - s) - \alpha\bar{\mathfrak{D}}^T \langle \mathbf{p}\mathbf{p}^T \rangle \mathfrak{D} s, \tag{1.64}$$

where $\bar{\mathfrak{D}}$ is a mirrored version of $\mathfrak{D}$. The expression $\langle \mathbf{p}\mathbf{p}^T \rangle$ again is a square, symmetric, positive definite matrix. Analog to the approach in Sect. 1.2.2 we can approximate it via the extended structure tensor $\mathbf{J}_\mathfrak{D}$

$$\mathbf{J}_\mathfrak{D}(\mathbf{x}) = \int w(\mathbf{x} - \mathbf{x}')(\mathfrak{D} s(\mathbf{x}'))(\mathfrak{D}^T s(\mathbf{x}')) \, d\mathbf{x}' \tag{1.65}$$

belonging to the model $(\mathfrak{D}^T s)\mathbf{p} = 0$ and exchange the eigenvalues μ_i of $\mathbf{J}_\mathfrak{D}$ via the edge stopping functions from Edge-Enhancing Diffusion (EED, Eq. (1.12)) or Orientation-Enhancing Diffusion (OED, Eq. (1.15)).

1.6.3 Example: Double-Orientation-Enhancing Diffusion

In this example, we construct a diffusion-like scheme for enhancement of transparently overlayed structures resulting in two local orientations in 2d data. The linear model describing this is (cf. [394], Eq. (11))

$$\partial_{xx} s p_1 + \partial_{xy} s p_2 + \partial_{yy} s p_3 = 0 \quad \text{or} \quad (\mathfrak{D}^T s)\mathbf{p} = 0, \tag{1.66}$$

where $\mathfrak{D} = (\partial_{xx}, \partial_{xy}, \partial_{yy})^T$ is an operator vector containing second order partial derivative operators, s is the image data and $\mathbf{p}$ is a parameter vector (containing mixed orientation parameters, that we do *not* need to disentangle, cf. [394]). This operator vector $\mathfrak{D}$ is now plugged into Eq. (1.64). Further $\langle \mathbf{p}\mathbf{p}^T \rangle$ in that equation is replaced by the extended structure tensor from Eq. (1.65) with exchanged eigenvalues. Discretization of $\mathfrak{D}$ is done using separable convolution filters

$$\partial_{xx} = \mathscr{L}_x * \mathscr{S}_y, \qquad \partial_{xy} = \mathscr{D}_x * \mathscr{D}_y, \qquad \partial_{yy} = \mathscr{S}_x * \mathscr{L}_y, \tag{1.67}$$

where lower indices give the application direction, $\mathscr{L}$ is a discrete 1d second order derivative, $\mathscr{S}$ is a 1d smoother and $\mathscr{D}$ is a 1d first order derivative filter.

For the calculation of $\mathfrak{D}$ and $\bar{\mathfrak{D}}$ in Eq. (1.64) we use the 3×3 scheme from [366]

$$\mathscr{L} = [1, -2, 1], \qquad \mathscr{D} = [0.5, 0, -0.5], \qquad \mathscr{S} = [0.21478, 0.57044, 0.21478]. \tag{1.68}$$

Structure tensor $\mathbf{J}_{\mathfrak{D}}$ (Eq. (1.65)) needed to approximate $\langle \mathbf{pp}^T \rangle$ in Eq. (1.64) is calculated using 5×5 filters optimized for accurate double orientation estimation [367]

$$\begin{aligned} \mathscr{L} &= [0.2068, 0.1729, -0.7593, 0.1729, 0.2068], \\ \mathscr{D} &= [0.06295, 0.3741, 0, -0.3741, -0.06295], \\ \mathscr{S} &= [0.01531, 0.2316, 0.5062, 0.2316, 0.01531]. \end{aligned} \tag{1.69}$$

We use an Euler-forward update scheme (cf. Eq. (1.45))

$$s^{t+\tau} = s^t - \tau \bar{\mathfrak{D}}^T \langle \mathbf{pp}^T \rangle \mathfrak{D} s^t, \tag{1.70}$$

where the upper index t indicates the diffusion time and τ is a (small) time step (cf. Eq. (1.61)). The initial data s^0 has to be the noisy (or otherwise corrupted) data r and the smoothing process has to be stopped manually or due to some criterion. In the simple tutorial experiment shown in Fig. 1.14 we run the smoothing process as long as the peak signal to noise ratio (PSNR, cf. [346])

$$PSNR = 20 \log \frac{255}{\|s - s_0\|_2} \tag{1.71}$$

rises. This is practicable because we have ground truth data s_0 (not to be confused with initial noisy image s^0 from Eq. (1.70)) available. Non-surprisingly anisotropic diffusion based on the standard single orientation model (BCCE, optical flow) either is stopped before crossed structures in Fig. 1.14 are denoised or severely corrupts these structures. Double-orientation diffusion delivers a result visually indistinguishable from the original, noise free data.

1.7 Conclusion

Diffusion is a concept defined in continuous space and time originating from physics. Consequently discrete diffusion methods have to be numerically consistent with diffusion. If they are not consistent, they are merely nonlinear regularization schemes. Nevertheless they can be excellent schemes. Being a diffusion is not a sign of high or low quality. It is just an observation.

Diffusion-like methods can be derived from statistics. The usual way is to design a prior term formulating a smoothness assumption on the underlying continuous undisturbed signal. Very similar schemes can be derived using a likelihood term and a constancy assumption on the underlying signal. The difference between the two

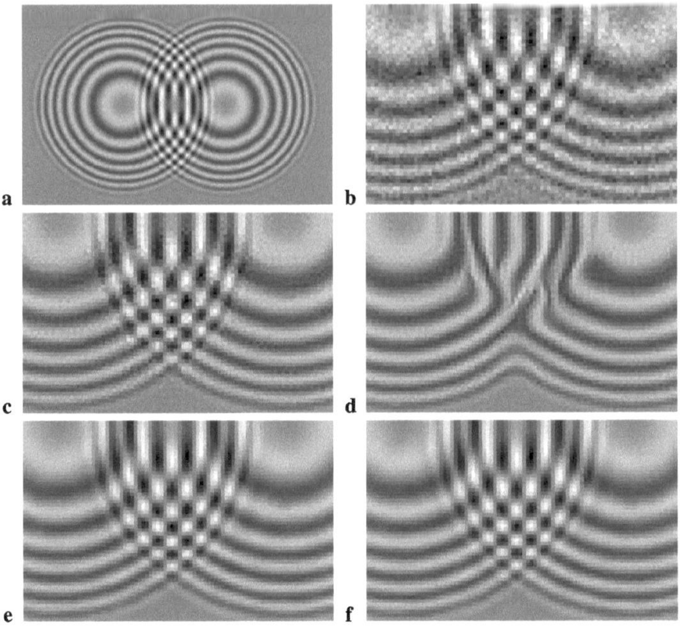

Fig. 1.14 Transparently overlaid rings: **a**: original, **b**: with added noise ($\sigma = 15$). Reconstructed with anisotropic diffusion, applying **c**: OED, **d**: EED. Reconstructed with double-orientation-enhancing diffusion, applying **e**: OED, **f**: EED

approaches is that a prior yields diffusion in a narrower sense, i.e. a regularization scheme ending in a flat signal if no data-term counterbalances it. A likelihood term is always based on the input data (like in mean-shift filtering) and does not yield a flat signal in the end—it is a data-term.

The advantage of a statistical formulation of diffusion is that all parameters, like diffusivities, can be learned from training data. In addition, originating from statistics the scheme applied is an optimization procedure and therefore we know in which sense the data becomes "better" if we change it by our scheme, i.e. we know a reason why to apply our algorithm and no other.

Extensions to diffusion come from the observation that derivatives are not the only operators making sense in an energy functional or Gibbs-distribution. Best suited operators for regularization of a given type of data can be derived in the isotropic nonlinear case. In the anisotropic nonlinear case filter selection can be recast into selection of a linear model the data is assumed to fulfill locally.

Chapter 2
Adaptive Filtering Using Channel Representations

Michael Felsberg

Abstract This review article gives an overview on adaptive filtering methods based on channel representations. The framework of channel representations and its relation to density estimation is introduced. The fast and accurate scheme of virtual shift decoding is introduced and applied in several variants of channel smoothing:

- channel smoothing with alpha-synthesis for improving stability of edge-enhancing filtering
- orientation adaptive channel smoothing with applications to coherence-enhancing filtering
- channel smoothing using graph-cuts for improving filtering results at corners
- channel-coded feature maps (CCFMs) which lead to a significant speed-up of channel averaging
- CCFM-based smoothing based on optimal parameters derived from a novel uncertainty relation

For each method, an algorithmic description and some examples of results are provided, together with discussions and references of the original papers. Cross connections to other articles in this volume are given where appropriate.

2.1 Introduction

Adaptive filtering is an important field of image processing that has been considered by many researchers during the past three decades. Many different variants of adaptivity have been discussed in the literature, but the most relevant instances are probably orientation and scale.

The concept of scale was first introduced systematically in terms of the concept of *linear scale space* [232, 261, 450], establishing a 3D space of spatial coordinates and a scale coordinate. Often identified with Gaussian low-pass filtering, a rigorous

---

M. Felsberg (✉)
Computer Vision Laboratory, Department of Electrical Engineering, Linköping University, 58183 Linköping, Sweden
e-mail: michael.felsberg@liu.se

L. Florack et al. (eds.), *Mathematical Methods for Signal and Image Analysis and Representation*, Computational Imaging and Vision 41,
DOI 10.1007/978-1-4471-2353-8_2,

analysis of underlying scale space axioms [442] has led to the discovery of the Poisson scale space [138] and more general α scale spaces [116].

In practice, discrete scale spaces are mostly sub-sampled with increasing scale parameter, leading to the concept of scale-pyramids [63, 188], multi-scale analysis and wavelet theory [92, 298], cf. Chap. 7. While pyramids and wavelets speedup the computation of linear operators and transforms, non-linear scale space methods are widely used, e.g. for image enhancement. Non-linear scale space is based on a non-stationary or anisotropic diffusivity function [334, 432], cf. also Chaps. 5 and 9.

More recently, non-linear methods have been introduced which are less directly connected to linear scale space and diffusion, but allow for faster processing and partially superior results [139, 346]. The former method is based on wavelets, whereas the latter one is based on the channel representation [189] and is called channel smoothing. Combining the channel representation with a systematic decimation of spatial resolution, similar to the pyramid approach, has been applied in blob-detection [160] and in channel-coded feature maps (CCFM) [241, 244], a density representation in spatio-featural domain, see also [137].

Non-linear filtering based on anisotropic diffusivity is closely related to orientation adaptive filtering [134]. Orientation adaptive filtering, often referred to steerable filters [171], goes actually back on the work [259]. In more recent work, it has been extended to orientation selective channel smoothing [136] and orientation scores and their processing [117].

This review paper focusses on the channel-based variants of adaptive filtering and it is structured as follows. The framework of channel representations and its relation to density estimation is introduced in Sect. 2.2. The fast and accurate scheme of virtual shift decoding is introduced in Sect. 2.3. Sections 2.4–2.8 summarize several variants of channel smoothing:

1. channel smoothing with alpha-synthesis for improving stability of edge-enhancing filtering
2. orientation adaptive channel smoothing with applications to coherence-enhancing filtering
3. channel smoothing using graph-cuts for improving filtering results at corners
4. channel-coded feature maps (CCFMs) which lead to a significant speed-up of channel averaging
5. CCFM-based smoothing based on optimal parameters derived from a novel uncertainty relation

For each method, an algorithmic description and some examples of results are provided, together with discussions of the original papers.

2.2 The Channel Representation

Channel coding, also called population coding [347, 460], is a biologically inspired data representation, where features are represented by weights assigned to ranges of feature values [189, 230], see Fig. 2.1. Similar feature representations can also be found in the visual cortex of the human brain, e.g. in the cortical columns.

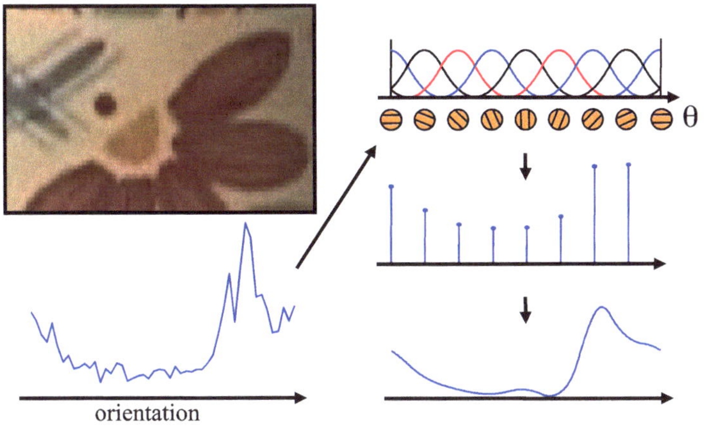

Fig. 2.1 Orientation distribution is encoded into channels, resulting in a (low-pass filtered) reconstruction using maximum entropy [242], see also Chap. 16. Figure courtesy by Erik Jonsson

The closer the current feature value f to the respective feature interval center n, the higher the channel weight c_n:

$$c_n(f) = k(f - n), \quad n \in \mathbb{N}, \tag{2.1}$$

where $k(\cdot)$ is a suitable kernel function and where f has been scaled such that it has a suitable range (note that we chose to place the channel centers at integers). By introducing z as a continuous feature coordinate, $k_n(z) = k(z - n)$, and $\delta_f(z) = \delta(z - f)$ denoting the Dirac-delta at f, the encoding can be written as a scalar product

$$c_n(f) = \langle \delta_f | k_n \rangle = \int \delta_f(z) k_n(z)\, dz \tag{2.2}$$

or as a sampled correlation in the feature-domain:

$$c_n = (\delta_f \star k)(n) = \int \delta_f(z') k(z' - z)\, dz' \bigg|_{z=n}. \tag{2.3}$$

From the weights of all channels the feature value can be decoded unambiguously by finding the mode, where the decoding depends on the kernel function. In what follows, we have been using quadratic B-splines:

$$B_2(z) = \begin{cases} (z + 3/2)^2/2 & -3/2 < z \le -1/2, \\ 3/4 - z^2 & -1/2 < z \le 1/2, \\ (z - 3/2)^2/2 & 1/2 < z < 3/2, \\ 0 & \text{otherwise.} \end{cases} \tag{2.4}$$

This results in the encoding scheme Algorithm 2.1, where N denotes the number of channels.

Algorithm 2.1 Channel Encoding Algorithm

Require: $f \in [1.5; N - 0.5]$

$\mathbf{c} \Leftarrow 0$
for all samples f **do**
  $i \Leftarrow \text{round}(f)$
  $g \Leftarrow f - i$
  $c_{i-1} \Leftarrow c_{i-1} + (g - 1/2)^2/2$
  $c_i \Leftarrow c_i + 3/4 - g^2$
  $c_{i+1} \Leftarrow c_{i+1} + (g + 1/2)^2/2$
end for

The features can be scalar valued or vector valued, e.g. grey-scales, color vectors, or orientations. In the case of scalar features the decoding from quadratic B-splines has been considered in detail in [139], see next section. For the case of non-interfering channel weights, a simplified scheme based on the quotient of linear combinations can be used:

$$M_n = c_{n-1} + c_n + c_{n+1}, \qquad n_0 = \arg\max M_n, \qquad \hat{f} = \frac{c_{n_0+1} - c_{n_0-1}}{M_{n_0}} + n_0, \tag{2.5}$$

where $\hat{f}$ is our estimate of the feature f that had been encoded in c_n.

2.3 Virtual Shift Decoding

The decoding scheme (2.5) is not sufficient in all applications due to its quantization effects, i.e., the decoding result is biased toward the channel centers. If avoiding quantization effects in the decoding is essential, the *virtual shift decoding* [139] can be applied. For its derivation the reader should refer to the original paper. The algorithm works as follows. Let $h = 2\sqrt{2} - 3$. In a first processing step, the channels are forward and backward filtered according to

$$c_n^+ = c_n + hc_{n-1}^+, \quad (n = 2, \dots, N), \tag{2.6}$$

$$c_n^- = h(c_{n+1}^- - c_n^+), \quad (n = N - 1, \dots, 1), \tag{2.7}$$

$$c_n' = 8c_n^-. \tag{2.8}$$

For finite domains, the boundary conditions are given as

$$c_1^+ = c_1, \tag{2.9}$$

$$c_N^- = \frac{h}{h^2 - 1} c_N^+. \tag{2.10}$$

For periodic domains, a DFT-based method should be applied instead of (2.6–2.10)

$$\text{DFT}_N(\mathbf{c}'') = 8(\text{DFT}_N([6\ 1\ 0\ \dots\ 0\ 1]_N))^{-1}\text{DFT}_N(\mathbf{c}), \tag{2.11}$$

where DFT_N denotes the N-point DFT along the channel index. The new channel vector $\mathbf{c}''$ must be extended at both ends according to

$$\mathbf{c}' = [c''_{N-1}\ c''_N\ \mathbf{c}''^T\ c''_1\ c''_2]^T$$

in order to process it further as if it was non-periodic. Sampling the continuous function

$$p(f) = \sum_{n=1}^{N} c'_n B_2(f-n) \tag{2.12}$$

results in the original channel vector $\mathbf{c}$ again, i.e., $p(f)$ is a proper interpolation of the channel vector.

In order to extract the modes f_n at channels n, define $\beta_n = f_n - n$ and solve

$$0 = \lambda\beta_n^2 + \mu\beta_n + \nu \quad \text{with} \tag{2.13}$$

$$\lambda = (c'_{n0-2} - 2c'_{n0-1} + 2c'_{n0+1} - c'_{n0+2})/2, \tag{2.14}$$

$$\mu = (-c'_{n0-2} + 2c'_{n0} - c'_{n0+2})/2, \tag{2.15}$$

$$\nu = (c'_{n0-2} + 6c'_{n0-1} - 6c'_{n0+1} - c'_{n0+2})/8 \tag{2.16}$$

such that the minimum of the error corresponds to

$$\beta_n = \frac{-\mu/2 + \sqrt{\mu^2/4 - \nu\lambda}}{\lambda}. \tag{2.17}$$

Solutions where $|\beta_n| > 1/2$ must be excluded, since they are located outside the unit interval around n. For valid solutions of β_n, the decoded mode is given by $f_n = n + \beta_n$. For periodic domains, the correct values are obtained after a modulo operation.

In order to extract the maximum likelihood estimate or to order the modes, the robust error needs to be known. It is given at channel n as

$$E(n) = \frac{23}{24} + \beta_n\nu + \beta_n^2\mu/2 + \beta_n^3\lambda/3 - \frac{c'_{n0-2} + 24c'_{n0-1} + 46c'_{n0} + 24c'_{n0-1} + c'_{n0-2}}{48}. \tag{2.18}$$

The virtual shift decoding algorithm is summarized in Algorithm 2.2.

Channel representations obviously need more memory than directly storing features, but this investment pays off in several ways which we will show in the subsequent sections.

Algorithm 2.2 Virtual Shift Decoding Algorithm

Require: $\mathbf{c}$ is non-negative and normalized

if periodic domain **then**
  $\mathbf{c} \Leftarrow \mathrm{IDFT}_N(8(\mathrm{DFT}_N([6\ 1\ 0\ \dots\ 0\ 1]_N))^{-1}\mathrm{DFT}_N(\mathbf{c}))$
  $\mathbf{c} \Leftarrow [c_{N-1}\ c_N\ \mathbf{c}^T\ c_1\ c_2]^T$
else
  $h \Leftarrow 2\sqrt{2} - 3$
  for $n = 2$ to N **do**
    $c_n \Leftarrow c_n + hc_{n-1}$
  end for
  $c_N \Leftarrow 8\frac{h}{h^2-1}c_N$
  for $n = N - 1$ to 1 **do**
    $c_n \Leftarrow h(c_{n+1} - 8c_n)$
  end for
end if
$\boldsymbol{\lambda} \Leftarrow \mathrm{conv}(\mathbf{c}, [-\frac{1}{2}\ 1\ 0\ -1\ \frac{1}{2}])$
$\boldsymbol{\mu} \Leftarrow \mathrm{conv}(\mathbf{c}, [-\frac{1}{2}\ 0\ 1\ 0\ -\frac{1}{2}])$
$\boldsymbol{\nu} \Leftarrow \mathrm{conv}(\mathbf{c}, [-\frac{1}{8}\ -\frac{3}{4}\ 0\ \frac{3}{4}\ \frac{1}{8}])$
$\boldsymbol{\beta} \Leftarrow (-\boldsymbol{\mu}/2 + \sqrt{\boldsymbol{\mu}.^2/4 - \boldsymbol{\nu}.\boldsymbol{\lambda}})./\boldsymbol{\lambda}$
$\boldsymbol{\gamma} \Leftarrow \mathrm{conv}(\mathbf{c}, [\frac{1}{48}\ \frac{1}{2}\ \frac{23}{24}\ \frac{1}{2}\ \frac{1}{48}])$
$\mathbf{f} \Leftarrow \boldsymbol{\beta} + [1\ 2\ \dots\ N]$
$\mathbf{E} \Leftarrow \frac{23}{24} + (-1 < 2\boldsymbol{\beta} < 1).(\boldsymbol{\beta}.\boldsymbol{\nu} + \boldsymbol{\beta}.^2.\boldsymbol{\mu}/2 + \boldsymbol{\beta}.^3.\boldsymbol{\lambda}/3 - \boldsymbol{\gamma})$

2.4 Channel Smoothing

The idea of channel smoothing is based on considering the feature f in the encoding (2.1) as a stochastic variable. It has been shown in [139] that the distribution p_f is approximated by c_n in expectation sense (see also Fig. 2.2):

$$E\{c_n(f)\} = (p_f \star k)(n) \tag{2.19}$$

such that $\hat{f}$ becomes a maximum-likelihood estimate of f.

If we assume that p_f is locally ergodic, we can estimate $\hat{f}$ from a local image region, which corresponds to a local averaging of the channel weights within a spatial neighborhood. The algorithm consisting of the three steps channel encoding, channel averaging, and channel decoding is called channel smoothing (see Algorithm 2.3 and Fig. 2.3) and has been shown to be superior to many other robust smoothing methods [139], cf. Chap. 1. Due to its structure, Algorithm 2.3 is very well suited for parallel implementations. A nearly identical scheme has been proposed independently, but later, in [329].

Due to the point-wise decoding in Algorithm 2.3, the positioning of region boundaries might violate the sampling theorem, resulting in unstable edge-pixels. To avoid this effect, a modification to the channel decoding has been proposed

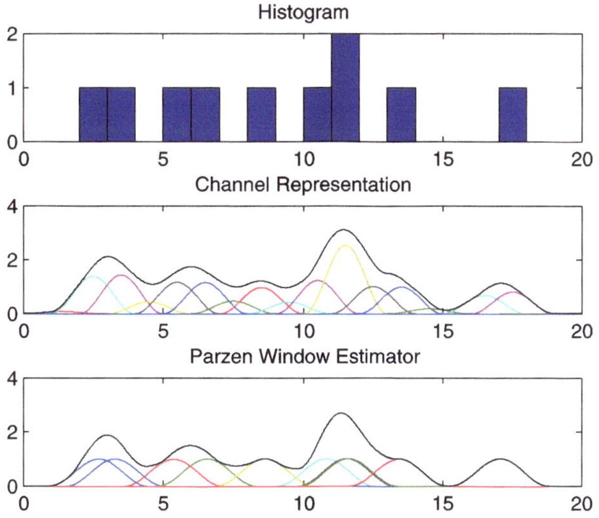

Fig. 2.2 Schematic comparison of density estimation using a histogram (*top*), a channel representation (*middle*), and a Parzen window estimator (*bottom*)

Algorithm 2.3 Channel Smoothing Algorithm

Require: $f \in [1.5; N - 0.5]$

1: **for all x do**

2: $\quad \mathbf{c}(\mathbf{x}) \Leftarrow \text{encode}(f(\mathbf{x}))$

3: **end for**

4: **for** $n = 1$ to N **do**

5: $\quad c_n \Leftarrow \text{conv2}(c_n, g_\sigma)$

6: **end for**

7: **for all x do**

8: $\quad [\mathbf{f}(\mathbf{x})\ \mathbf{E}(\mathbf{x})] \Leftarrow \text{decode}(\mathbf{c}(\mathbf{x}))$

9: $\quad i(\mathbf{x}) \Leftarrow \arg\max_n E_n(\mathbf{x})$

10: $\quad [\hat{f}(\mathbf{x})\ \hat{E}(\mathbf{x})] \Leftarrow [f_{i(\mathbf{x})}(\mathbf{x})\ E_{i(\mathbf{x})}(\mathbf{x})]$

11: **end for**

---

in [159], called α-synthesis, which creates smooth transitions between neighborhoods with different feature levels. Instead of extracting only the maximum in (2.5) or Algorithm 2.3, line 9, all local decodings are combined, weighted by the respective robust errors

$$\hat{f} = \frac{\sum_n f_n(\frac{23}{24} - E_n)^\alpha}{\sum_n(\frac{23}{24} - E_n)^\alpha}. \tag{2.20}$$

For the choice of α see [159]; we used $\alpha = 2$ throughout this paper. This method avoids aliasing artifacts at edges, but it does not avoid the rounding of corners, see Fig. 2.4.

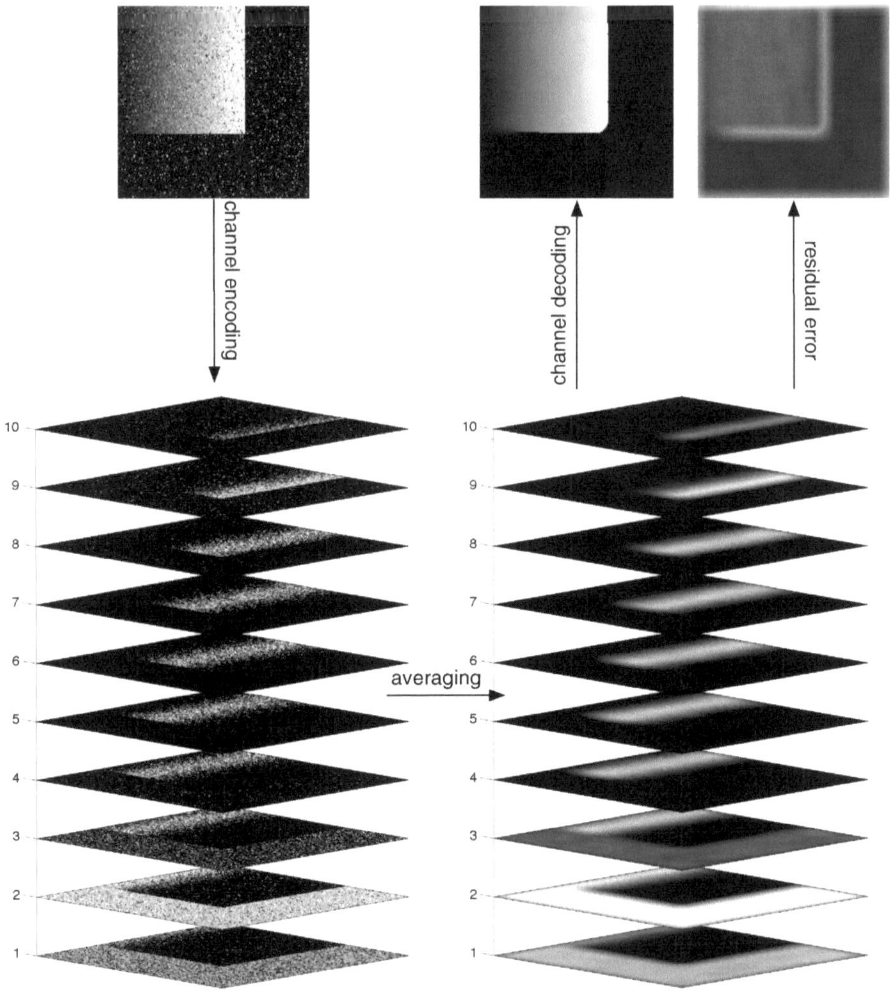

Fig. 2.3 The simple synthetic example (*top left*) smoothed by channel smoothing, using a Gaussian filter with $\sigma = 10$. No quantization effects are visible but note the rounding of the corner. *On the top right*: Robust error E of the decoding

Fig. 2.4 *From left to right*: Original noisy test image, result without alpha-synthesis (note the flip-overs at the edges), and result with alpha-synthesis

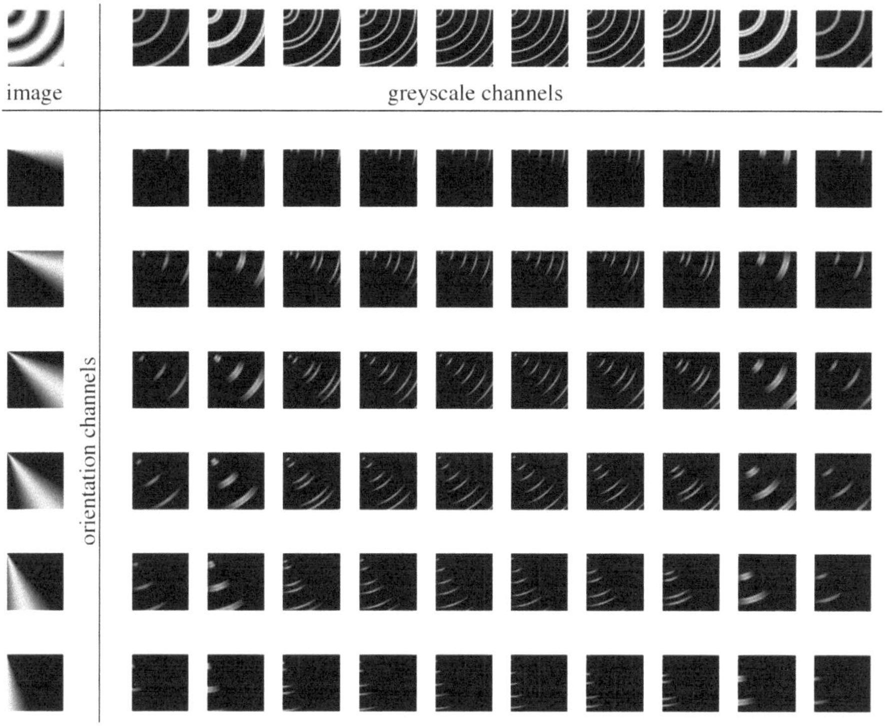

Fig. 2.5 Channel matrix using ten grey-scale channels and six orientation channels

2.5 Orientation Adaptive Channel Smoothing

The contents of this section is based on the publication [136], but the work is also closely related to Chaps. 3 and 10, as well as [295].

Channel smoothing as introduced in the previous section preserves edges, similar to edge-enhancing diffusion. Close to an edge, the filter support of a comparable linear filter is a semi-disc lying entirely on one side of the edge, thus giving a stable response everywhere except for points lying exactly on the edge. The latter problem has been solved by α-synthesis. For elongated, line-like structures, for instance in images of finger prints, the ordinary channel smoothing becomes however instable, as the majority of points are edge points. As a consequence, the resulting image suffers from 'flip-over' effects and contours become fragmented or displaced.

This problem is avoided if the channel smoothing is only performed along a 1D subspace, given by the local signal orientation θ_f, similar to the principle of adaptive filters [259] or coherence enhancing diffusion [439]. Hence, the channel smoothing operation must depend on the local signal orientation, which is itself represented using channels. The orientation information is typically of much slower variation than the intensity information and thus ordinary channel smoothing gives good results [139].

Algorithm 2.4 Orientation Adaptive Channel Smoothing Algorithm

Require: $f \in [1.5; N - 0.5]$
Require: $\theta_f \in [1.5; M - 0.5]$
1: **for all** $\mathbf{x}$ **do**
2: $\quad \mathbf{c}_f(\mathbf{x}) \Leftarrow \text{encode}(f(\mathbf{x}))$
3: $\quad \mathbf{c}_\theta(\mathbf{x}) \Leftarrow \text{encode}(\theta_f(\mathbf{x}))$
4: **end for**
5: **for** $m = 1$ to M **do**
6: $\quad c_{\theta,m} \Leftarrow \text{conv2}(c_{\theta,m}, g_\sigma)$
7: $\quad$ **for** $n = 1$ to N **do**
8: $\quad\quad c_{n,m} \Leftarrow \text{conv2}(c_{\theta,m} c_{f,n}, g_m)$
9: $\quad$ **end for**
10: **end for**
11: **for all** $\mathbf{x}$ **do**
12: $\quad$ **for** $n = 1$ to N **do**
13: $\quad\quad [c_{n,m}(\mathbf{x})]_m \Leftarrow \text{normalize}([c_{n,m}(\mathbf{x})]_m)$
14: $\quad\quad [\mathbf{f}(\mathbf{x})\ \mathbf{E}(\mathbf{x})] \Leftarrow \text{decode}([c_{n,m}(\mathbf{x})]_m)$
15: $\quad\quad i(\mathbf{x}) \Leftarrow \arg\max_m E_m(\mathbf{x})$
16: $\quad\quad c_{f,n}(\mathbf{x}) \Leftarrow \frac{23}{24} - E_{i(\mathbf{x})}(\mathbf{x})$
17: $\quad$ **end for**
18: $\quad \mathbf{c}_f(\mathbf{x}) \Leftarrow \text{normalize}(\mathbf{c}_f(\mathbf{x}))$
19: $\quad [\mathbf{f}(\mathbf{x})\ \mathbf{E}(\mathbf{x})] \Leftarrow \text{decode}(\mathbf{c}_f(\mathbf{x}))$
20: $\quad i(\mathbf{x}) \Leftarrow \arg\max_n E_n(\mathbf{x})$
21: $\quad [\hat{f}(\mathbf{x})\ \hat{E}(\mathbf{x})] \Leftarrow [f_{i(\mathbf{x})}(\mathbf{x})\ E_{i(\mathbf{x})}(\mathbf{x})]$
22: **end for**

The smoothed orientation channels and the grey-scale channels are then combined in an outer product, see Fig. 2.5. The channel matrix encodes explicitly which grey-scale is present at which orientation. Hence, it is straightforward to adapt the smoothing kernel to the local orientation: we simply use *differently oriented* anisotropic smoothing kernel g_m for each orientation channel.

This smoothed, extended channel representation is then decoded in three steps using the standard decoding: Decode the orientation channel for each grey-scale, use the obtained error estimates to build a new grey-scale channel vector, and decode the new channel vector. The complete algorithm is summarized in Algorithm 2.4.

The ratio behind the sketched method is as follows. For oriented structures the channel matrix has a clear 2D maximum. Generating the grey-scale channel vector in line 16 yields a vector with a maximum at the appropriate grey-scale. The effective filter kernel is dominated by the anisotropic kernel corresponding to the strongest orientation channel, i.e., the structure is smoothed along its orientation and maintained perpendicular to its orientation, see Fig. 2.6.

For unoriented structures, the orientation decoding is random, but the resulting grey-scale channel will still be correct as the grey-scales are identical for all orientations. The effective filter kernel is a combination of anisotropic kernels with random

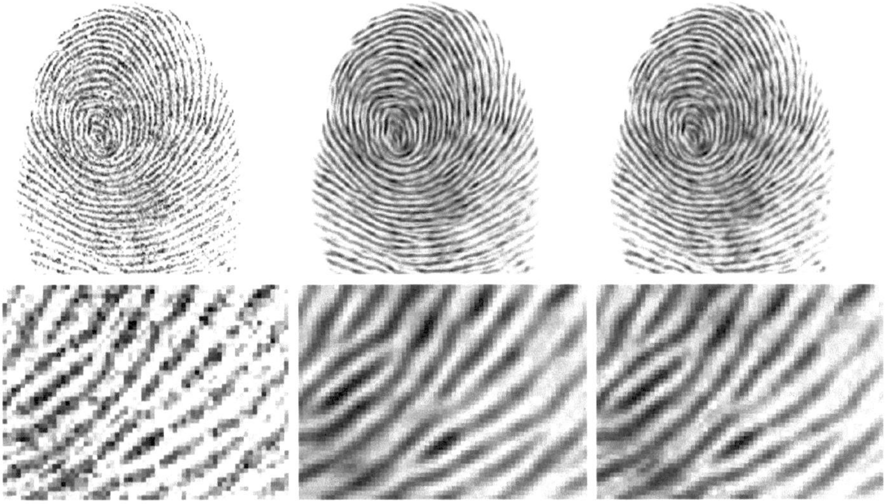

Fig. 2.6 Fingerprint experiment from [136]. *From left to right*: Original images, results from coherence enhancing diffusion [420], and results from channel smoothing. *Top*: Fingerprint at 300 dpi. *Bottom*: Zoomed detail. For further details and parameters, refer to [136]

orientation, thus resulting in an isotropic kernel. Hence, the filter output corresponds to isotropic smoothing.

A further aspect of adaptive filtering is the choice of the smoothing kernels, depending on the noise level [132] (for related work on noise level estimation, see also [140, 363]) and the noise distribution, e.g., multiplicative noise [372]. The selection of filter kernels is however out of the scope of this review and the interested reader is referred to the original publications.

2.6 Channel Smoothing Without Corner Rounding

The method described in this section is based on the publication [133], which proposes a method to avoid rounding of corners by restricting the smoothing to a generic domain where the respective channels are active. This means in practice: channel values should be averaged on bounded domains. Before looking into the issue of determining the active region of a channel, we have to modify the averaging step in channel smoothing in order to apply it to a bounded domain. Filtering of (uncertain) data f from a bounded domain is well modeled in terms of normalized convolution [258] of 0th order (normalized averaging):

$$\hat{f} = \frac{(a * (bf))}{(a * b)}, \tag{2.21}$$

where a denotes the applicability function, i.e., a suitable averaging kernel (typically a Gaussian function), b is the certainty function, which describes the bounded domain Ω:

$$b(\mathbf{x}) = \begin{cases} 1 & \mathbf{x} \in \Omega, \\ 0 & \mathbf{x} \notin \Omega, \end{cases} \tag{2.22}$$

and $*$ denotes the convolution operation.

The co-domain of normalized convolution is however unbounded, and therefore, we cannot apply it directly to the case of channels c_n that are active in a bounded region. To remain within the active region of each channel, we mask the result from normalized convolution to the *same bounded domain* of each channel by means of the certainty function b_n:

$$\hat{c}_n = b_n \frac{(a * (b_n c_n))}{(a * b_n)}. \tag{2.23}$$

What remains to be considered is the estimation of the active region, or equivalently, the certainty function for each channel. We have to find those regions where the image was sufficiently stationary in order to produce *similar* channel vectors $\mathbf{c}(\mathbf{x})$. Similar channel vectors have the same active components, where we classify the activity by a simple threshold θ. As stationarity does not make sense without spatial context, we require the active region to be as connected as possible.

For each channel n we formulate the following objective function:

$$E(b_n) = \sum_{\mathbf{x}} b_n(\mathbf{x})(\theta - c_n(\mathbf{x})) + \lambda \sum_{\{\mathbf{x},\mathbf{y}\} \in \mathscr{N}} |b_n(\mathbf{x}) - b_n(\mathbf{y})|, \tag{2.24}$$

where we use the following parameters throughout the remainder of this paper:

- $\mathscr{N}$ is the four-neighborhood
- $\lambda = 0.3$ is the penalty for discontinuities in b_n
- $\theta = \frac{1}{16}$ is the threshold for active channels

All parameters were chosen according to [133] (as is the width of the Gaussian filter $\sigma = 10$ and the number of channels $N = 10$). The interested reader is referred to [132, 161] for estimation of the channel averaging filter and the number of channels. The threshold θ can be derived from classical decision theory (see e.g. [406], Chap. 3) and depends also on the number of channels. The meta parameter λ depends on the neighborhood structure and the signal statistics. It should be at least one third of the maximum channel value (minus θ) to fill in one-pixel gaps in a contour with four-neighborhood. Too large values will remove structural details from the active region. For quadratic B-spline channels and $\theta = \frac{1}{16}$ this happens for $\lambda \geq \frac{11}{32}$.

A binary labeling problem as formulated in (2.24) (see also Chap. 14) is efficiently solved by graph-cut algorithms [52]. Using graph-cut for determining the activation of channels, we obtain the graph-cut channel smoothing algorithm as given in Algorithm 2.5. The synthetic example from Fig. 2.3 shows that graph-cut channel smoothing does not suffer from the drawback of multi-label graph-cut (coarse quantization), nor does it suffer from rounding of corners as pure channel smoothing does, see Fig. 2.7. The computational complexity of the proposed method is

Algorithm 2.5 Graph-Cut Channel Smoothing Algorithm

Require: $f \in [1.5; N - 0.5]$

```
for all x do
    c(x) ⇐ encode(f(x))
end for
for n = 1 to N do
    b_n ⇐ binary_graph_cut(c_n, 𝒩, λ, θ)
    c_n ⇐ b_n conv2(b_n c_n, g_σ)/conv2(b_n, g_σ)
end for
for all x do
    [f(x) E(x)] ⇐ decode(c(x))
    i(x) ⇐ arg max_n E_n(x)
    [f̂(x) Ê(x)] ⇐ [f_i(x)(x) E_i(x)(x)]
end for
```

somewhat higher than that of pure channel smoothing and it is dominated by the N binary graph-cut computations.

2.7 Channel-Coded Feature Maps

Channel-coded feature maps have been suggested in [243] and are related to the theory in Chap. 4 in the sense that spatio-featural densities are estimated using channel representations.

One major drawback of channel smoothing is the extensive use of memory if many feature channels are required. A high density of channels is only reasonable if the spatial support is large, which implies that the individual feature channels are heavily low-pass filtered along the spatial dimension. Therefore, the feature channels have a lower band limit and can be sub-sampled in the spatial domain without losing information. If the three steps of channel encoding, channel averaging, and sub-sampling are integrated into a single step, channel-coded feature maps (CCFMs) are generated. The advantage of CCFMs is a much higher number of channels, e.g. by combining several features as in Fig. 2.8, without increasing the memory requirements significantly. The CCFM encoding of a single feature point can be written as (cf. (2.1)):

$$c_{l,m,n}(f(x,y),x,y) = k_f(f(x,y)-n)k_x(x-l)k_y(y-m), \tag{2.25}$$

where k_f, k_x, k_y are the 1D kernels in feature domain and spatial domain. Note that x and y are scaled such that they suit the integer spatial channel centers l, m. Similar to (2.1), the encoding (2.25) of a set of feature points can be written as a scalar product in 3D function space or as a 3D correlation, where we use

$$\delta_f(x,y,z) = \delta(z - f(x,y)) \tag{2.26}$$

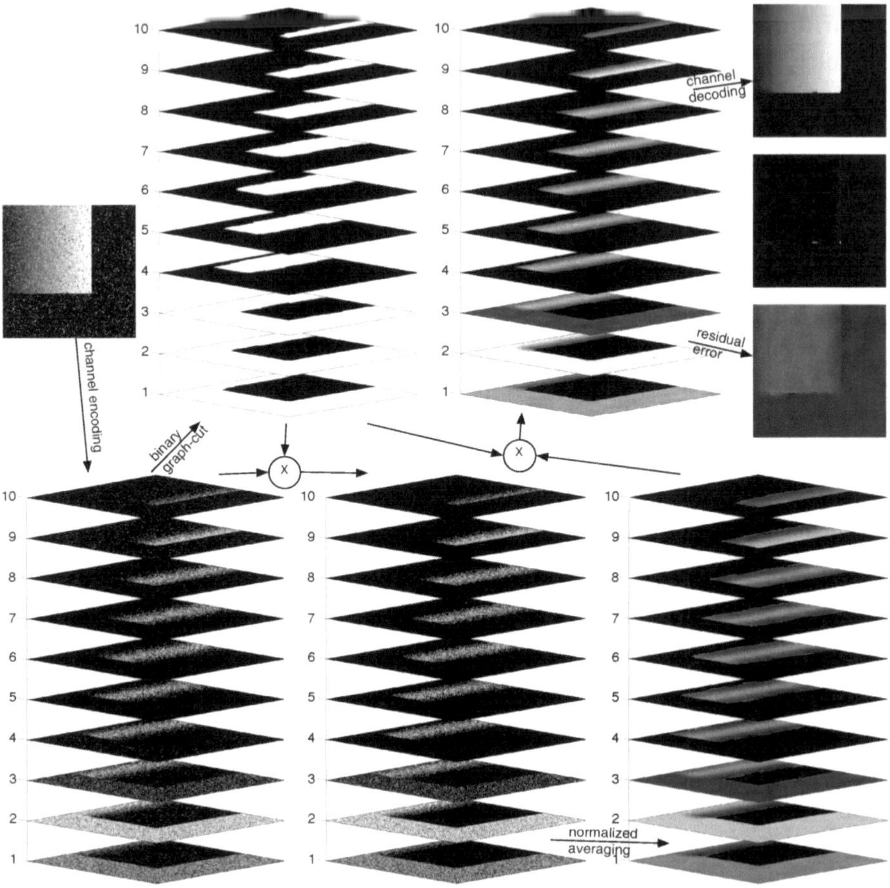

Fig. 2.7 Simple synthetic example from Fig. 2.3 smoothed by graph-cut channel smoothing, using a Gaussian filter with $\sigma = 10$. *Top right*: Result from graph-cut channel smoothing. *Below*: Actual absolute error compared to noise-free gradient image and reconstruction error. Note the absence of rounding of the corner

and $k_{f,n}(z) = k_f(z - n)$, $k_{x,l}(x) = k_x(x - l)$, $k_{y,m}(y) = k_y(y - m)$:

$$c_{l,m,n}(f) = \langle \delta_f | k_{f,n} k_{x,l} k_{y,m} \rangle = \iiint \delta_f(x, y, z) k_{f,n}(z) k_{x,l}(x) k_{y,m}(y) \, dz \, dy \, dx$$
$$= (\delta_f \star (k_f k_x k_y))(n, m, l). \tag{2.27}$$

The final formulation is the starting point of the CCFM scale space, see next section. CCFMs can be computed very efficiently using monopieces [241, 244], but here we restrict ourselves to the more basic Algorithm 2.6, where $\otimes$ denotes the outer (Kronecker) product. CCFMs and their derivatives can be used efficiently for robust visual tracking [243].

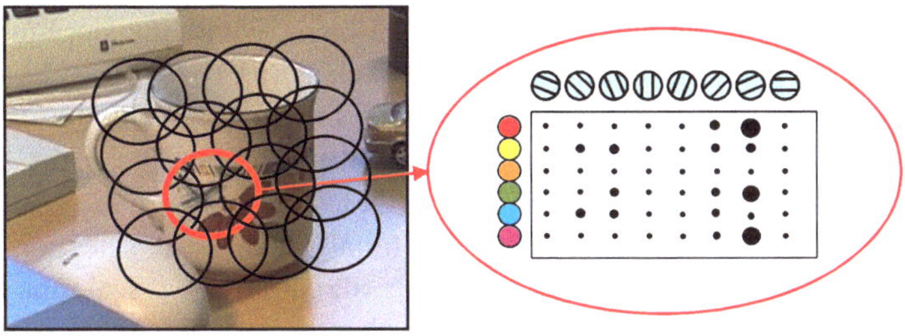

Fig. 2.8 Simultaneous encoding of orientation and color in a local image region. Figure taken from [241] courtesy Erik Jonsson

Algorithm 2.6 CCFM Algorithm

Require: $f \in [1.5; N - 0.5]$
Require: $\mathbf{x} = (x, y)^T \in [1.5; X - 0.5] \times [1.5; Y - 0.5]$
$\mathbf{C} \Leftarrow 0$
for all x do
  $\mathbf{c}_f \Leftarrow \text{encode}(f(\mathbf{x}))$
  $\mathbf{c}_x \Leftarrow \text{encode}(x)$
  $\mathbf{c}_y \Leftarrow \text{encode}(y)$
  $\mathbf{C} \Leftarrow \mathbf{C} + \mathbf{c}_f \otimes \mathbf{c}_x \otimes \mathbf{c}_y$
end for

2.8 CCFM Scale Space

This section summarizes some recent results [135] on the concept of CCFM scale space and spatial-featural uncertainties.

The starting point is to embed the image $f(x, y)$ as a 3D surface according to (2.26). One might try to generate a 3D α scale space [116] (Gaussian as a special case $\alpha = 1$ and all α-kernels are symmetric, i.e., correlation and convolution are the same):

$$F_s(x, y, z) = (k_s^{(\alpha)} \star \delta_f)(x, y, z). \tag{2.28}$$

However, the semi-group property of scale space implies that all dimensions (spatial dimensions and the feature dimension) become increasingly blurred. Despite the fact that this implies a rapidly growing loss of information with increasing scale and a singular zero scale, this procedure is insensible from a statistical perspective and does not comply with the notion of scale selection [126, 290].

Since the latter argument is not straightforward, we explain our rationale in some more detail. From the requirement that the dimensionless derivative attains its maximum at a position proportional to the wavelength of the signal [290] (Sect. 13.1), we conclude that the scale of a structure is proportional to its spatial scale (a trivial fact) and anti-proportional to its feature scale. The latter can be shown by looking at the Taylor expansion of a harmonic oscillation $A\sin(\omega x)$ in the origin: $A\omega x$. The steepness of a sinusoid $A\omega$ in the origin grows linearly with the amplitude and the frequency, i.e., it is anti-proportional to the wavelength $\lambda = \frac{2\pi}{\omega}$.

Alternatively, one can consider the energy of a harmonic oscillation. The energy is proportional to the square of the amplitude times the square of the frequency: $E \propto A^2\omega^2 \propto \frac{A^2}{\lambda^2}$. That means, if we apply a 3D lowpass filter to the spatio-featural domain, the energy decays with a power of four. Hence, scale selection would favor the highest possible frequencies in nearly all cases. If we scale the amplitude anti-proportionally to the spatial domain, the change of energy is balanced and will reflect intrinsic properties of the signal.

This relation is formalized in terms of a spatio-featural uncertainty relation, which is derived based on the group structure of spatio-featural transformations. We choose a methodology which is based on the isotropic model used in [264], although restricted to the 1D case. The higher-dimensional case generalizes straightforwardly. The group that we consider contains the shearing group and the translation group given as

$$x' = x + t_x, \tag{2.29}$$

$$f' = f + \tan(\phi)x + t_f. \tag{2.30}$$

The shearing transformation corresponds to the rotation of a Euclidean space and is obtained since the f-coordinate is a null-vector [264], i.e., $f \cdot f = 0$. The parametrization is chosen such that it reflects the fact that points move along the surface/curve with angle ϕ towards the ground plane. Using this definition we state the following

Theorem 2.1 *Let the spatio-featural domain be described by the isotropic model. The uncertainty product in the spatio-featural domain has a lower bound*

$$\exists k > 0: \quad (\Delta x)(\Delta f) \geq k \tag{2.31}$$

and the lower bound is given as

$$k = \frac{1}{2}\sigma_f \sigma_x, \tag{2.32}$$

where σ_f^2 *is the variance of the feature domain marginal distribution and* σ_x^2 *is the variance of the spatial domain distribution.*

The proof of this theorem is given in [135]. As a consequence of this theorem, optimal parameters for CCFM computation are derived. As an experimental validation, images have been reconstructed from these CCFMs showing that a good perceptual quality is maintained for a wide range of channels, see Fig. 2.9. The CCFM-smoothing algorithm is summarized in Algorithm 2.7, where the numbers of channels X, Y, and N are the optimized parameters mentioned before and the interpolation in line 3 generates a new channel vector from adjacent channel vectors by linear interpolation.

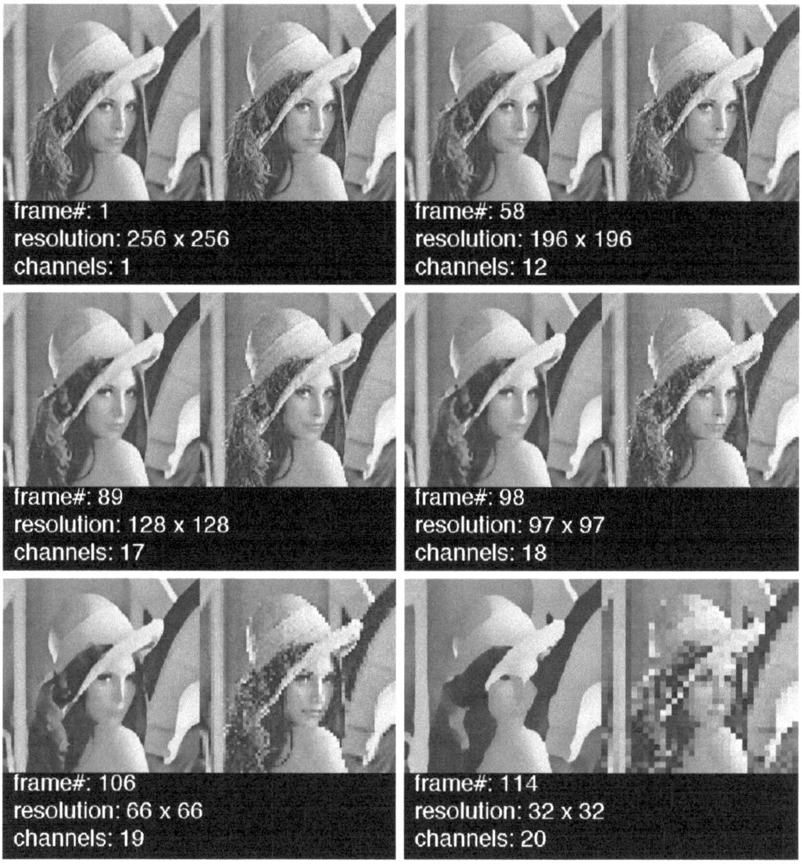

Fig. 2.9 Examples for CCFM-smoothing at different scales. The spatial and featural (denoted as *channels*) resolutions are given according to the spatio-featural uncertainty. The feature considered here is the greyscale

Algorithm 2.7 CCFM Smoothing Algorithm

Require: $f \in [1.5; N - 0.5]$
Require: $\mathbf{x} = (x, y)^T \in [1.5; X - 0.5] \times [1.5; Y - 0.5]$
$\mathbf{C} \Leftarrow \mathrm{CCFM}(x, y, f)$
for all x do
  $\mathbf{c}_f \Leftarrow \mathrm{interpolate}(\mathbf{C}, \mathbf{x})$
  $[\mathbf{f}(\mathbf{x})\ \mathbf{E}(\mathbf{x})] \Leftarrow \mathrm{decode}(\mathbf{c}_f)$
  $i(\mathbf{x}) \Leftarrow \arg\max_n E_n(\mathbf{x})$
  $[\hat{f}(\mathbf{x})\ \hat{E}(\mathbf{x})] \Leftarrow [f_{i(\mathbf{x})}(\mathbf{x})\ E_{i(\mathbf{x})}(\mathbf{x})]$
end for

2.9 Conclusion

In this review paper, we have given a compact and concise overview over the field of channel-based filtering. Research on this topic is still in progress and more results on the efficient computation and other types of features are to be expected in the near future. Code for most of the presented work is available at the author's website.

Acknowledgements The author would like to thank P.-E. Forssén for various discussions about the paper, in particular on alpha-synthesis. The research leading to these results has received funding from the European Community's Seventh Framework Programme (FP7/2007-2013) under grant agreements no 215078 (DIPLECS) and 247947 (GARNICS) as well as the VR project 2009-4282.

Chapter 3
3D-Coherence-Enhancing Diffusion Filtering for Matrix Fields

Bernhard Burgeth, Luis Pizarro, Stephan Didas, and Joachim Weickert

Abstract Coherence-enhancing diffusion filtering is a striking application of the structure tensor concept in image processing. The technique deals with the problem of completion of interrupted lines and enhancement of flow-like features in images. The completion of line-like structures is also a major concern in diffusion tensor magnetic resonance imaging (DT-MRI). This medical image acquisition technique outputs a 3D matrix field of symmetric (3×3)-matrices, and it helps to visualize, for example, the nerve fibers in brain tissue. As any physical measurement DT-MRI is subjected to errors causing faulty representations of the tissue corrupted by noise and with visually interrupted lines or fibers.

In this paper we address that problem by proposing a coherence-enhancing diffusion filtering methodology for matrix fields. The approach is based on a generic structure tensor concept for matrix fields that relies on the operator-algebraic properties of symmetric matrices, rather than their channel-wise treatment of earlier proposals.

Numerical experiments with artificial and real DT-MRI data confirm the gap-closing and flow-enhancing qualities of the technique presented.

B. Burgeth (✉) · J. Weickert
Mathematical Image Analysis Group, Faculty of Mathematics and Computer Science, Saarland University, 66041 Saarbruecken, Germany
e-mail: burgeth@mia.uni-saarland.de

J. Weickert
e-mail: weickert@mia.uni-saarland.de

L. Pizarro
Department of Computing, Imperial College London, 180 Queen's Gate, London SW7 2AZ, UK
e-mail: l.pizarro@imperial.ac.uk

S. Didas
Abteilung Bildverarbeitung, Fraunhofer-Institut für Techno- und Wirtschaftsmathematik, Fraunhofer-Platz 1, 67663 Kaiserslautern, Germany
e-mail: stephan.didas@itwm.fraunhofer.de

L. Florack et al. (eds.), *Mathematical Methods for Signal and Image Analysis and Representation*, Computational Imaging and Vision 41,
DOI 10.1007/978-1-4471-2353-8_3,

3.1 Introduction

An important task in diffusion tensor magnetic resonance imaging (DT-MRI) is the so-called fiber tracking [438]. In parts of the brain, e.g. the corpus callosum, nerve fibers form bundles with a coherent structure. In principle the matrix field produced by DT-MRI allows for the application of fiber tracking techniques and hence for the accurate visualisation of the nerve fibers. There is a vast literature on various techniques to achieve this goal, see for example [40, 64, 84, 146, 306, 464, 467]. However, ubiquitous measurement errors during acquisition cause gaps and interrupted fibers in the final visualization. It would be desirable to have a method available that enhances flow-like patterns such as bundles of nerve fibers. For scalar and vector-valued images a method achieving this goal is already at our disposal: Coherence enhancing diffusion filtering [436]. It is governed by the equation

$$\begin{aligned} \partial_t u - \operatorname{div}(D \cdot \nabla u) &= 0 \quad \text{in } I \times \Omega, \\ \partial_n u &= 0 \quad \text{in } I \times \partial\Omega, \\ u(x,0) &= f(x) \quad \text{in } \Omega, \end{aligned} \tag{3.1}$$

where $\Omega \subset \mathbb{R}^d$ is the image domain and $I = [0, T[$ a potentially unbounded time interval. To the authors' best knowledge, no extension of this method to matrix-valued images, matrix fields for short, has been reported in the literature.

The essential ingredient in this equation is the *diffusion tensor* D of the scalar image u which steers the diffusion process: It amplifies diffusion along flow-like structures, and hinders diffusion perpendicular to those patterns. Postponing the detailed construction of D to Sect. 3.2, for now we only remark that it is a function of the structure tensor [162], which is given by

$$S_\rho(u(x)) := G_\rho * \left(\nabla u(x) \cdot (\nabla u(x))^\top\right) = \left(G_\rho * \left(\partial_{x_i} u(x) \cdot \partial_{x_j} u(x)\right)\right)_{i,j=1,\dots,d}.$$

Here $G_\rho *$ indicates a convolution with a Gaussian of standard deviation ρ, however, more general averaging procedures can be used. If $\nabla u(x) \neq 0$ the matrix

$$\left(\nabla u(x) \cdot (\nabla u(x))^\top\right)$$

has rank one, the eigenvector $\nabla u(x)$ belongs to the only non-zero eigenvalue $|\nabla u(x)|^2$. The eigenvalues represent the contrast in the directions of the eigenspaces. The averaging process then creates a matrix with full rank which contains valuable directional information. Note that the averaging of the structure tensor avoids cancelation of directional information. If one would average the gradients instead, neutralization of vectors with opposite sign would occur. In many applications it is advantageous to use a presmoothed image $u_\sigma := G_\sigma * u$ instead of u in order to reduce the influence of noise for better numerical results. The structure tensor is a classical tool in image processing to extract directional information from an image, for more details the reader is referred to [35] and the references therein, as well as to Chaps. 1 and 5. It is not straightforward to generalize both the structure and the diffusion tensor concept to the setting of matrix-valued images. To fix notation in this work matrix-valued images or matrix fields $M(x)$ are considered as mappings from $\mathbb{R}^d$ into the set $\mathrm{Sym}_n(\mathbb{R})$ of symmetric $n \times n$-matrices

$$M : x \mapsto M(x) = \left(m_{i,j}(x)\right)_{i,j=1,\ldots,n} \in \mathrm{Sym}_n(\mathbb{R}),$$

and denoted by capital letters while indexed lower case letters indicate their components. In [53, 437] di Zenso's approach to a structure tensor for multi channel images is generalized: Each channel considered as independent scalar image gives rise to a structure tensor, then these structure tensors are summed up to give the structure tensor of Weickert

$$J_\rho(U(x)) := \sum_{i,j=1}^{n} S_\rho(u_{i,j}(x)).$$

This construction has been refined to a customizable structure tensor in [381]. There the resulting structure tensor is a weighted sum of tensors of scalar quantities that are now not just the channels, but other meaningful scalar quantities derived from the matrix field. The weights are provided by the user, and depending on the choice of weights the emerging structure tensor has a sensitivity for certain features of the matrix field. A special constellation of the weights turns the customizable structure tensor into Weickert's structure tensor. It is important to mention that in case of a 3D matrix field of 3×3 symmetric matrices these concepts yield also a 3×3 structure tensor, the very same order as a 3D scalar image. Here we opt for a different approach: We assume an operator-algebraic view on symmetric matrices as finite dimensional instances of selfadjoint Hilbert space operators. The exploitation of the algebraic properties of matrices ensures proper interaction between the different matrix channels.

Promising proposals to generalize nonlinear regularization methods and related diffusion filters for scalar images to matrix fields have been made in [60, 61]. These approaches are based on a basic differential calculus for matrix fields, which will be useful in this context as well. Other approaches to tensor field regularisation have a more differential geometric background [71, 412] where the set of positive definite matrices is endowed with a Riemannian metric stemming form the DT-MRI field. A different approach to regularization of tensor fields is proposed in Chap. 5, which is based on the Beltrami framework, a local differential geometric approach. For a basic introduction to the Beltrami framework see also Chap. 1. Further investigations in connection with CED based on Lie group theory can be found in Chap. 8, as well as in [114]. Regarding the problem of completing lines and closing gaps, a very different approach to ours is considered in Chap. 10.

In this chapter we will present a general concept for a large size structure tensor that carries all the directional information of the matrix field. We will show how this information can be deduced from this large tensor by a reduction process. We will follow the presentation in [62].

The rest of this chapter is structured as follows: Sect. 3.2 is devoted to a brief review of coherence enhancing diffusion (CED) filtering of scalar images. Notions necessary to construct the diffusion tensor and a basic differential calculus for matrix fields necessary to construct the diffusion tensor is provided in Sect. 3.3. In Sect. 3.4 we present the structure tensor concept for matrix fields, study some of its properties by investigating the connection to already known structure tensors for matrix-valued data. Employing this structure tensor concept we introduce coherence enhancing diffusion for matrix fields in Sect. 3.5. We then describe in Sect. 3.6 how explicit schemes for two-dimensional data can be extended by a Sobel-type method to three-dimensional ones. The results of our experiments with matrix-valued coherence enhancing diffusion applied to real DT-MRI images are discussed in Sect. 3.7. Section 3.8 is made up by concluding remarks.

3.2 Coherence Enhancing Diffusion

The rationale behind the construction of the diffusion tensor D is as follows: The matrix $S_\rho(u)$ as the positive average of different symmetric positive semidefinite matrices has the very same property. Hence $S_\rho(u)$ has a orthonormal system $\{w_1, \ldots, w_d\}$ of eigenvectors corresponding to the non-negative eigenvalues $\mu_1 \geq \mu_2 \geq \cdots \geq \mu_d \geq 0$ indicating the contrast in each direction. In the line defined by w_d, the *coherence orientation*, the contrast is the least compared to other orientations, since w_d belongs to the smallest eigenvalue μ_d. The coherence or anisotropy of an image structure is essentially captured in the eigenvalue distribution of the structure tensor S_ρ. In [435] the quantity

$$\kappa := \sum_{i=1}^{d-1} \sum_{j=i+1}^{d} (\mu_i - \mu_j)^2$$

is proposed to measure coherence. Strongly differing eigenvalues result in a large value of κ, while similar values produce a small κ-value indicating a structure with isotropic character. The matrix D has the same eigenvectors as S_ρ, however, its eigenvectors λ_i are altered via the tensor map H according to

$$\lambda_i := H(\mu_i) := \alpha \quad \text{for } i = 1, \ldots, d-1$$

and

$$\lambda_d := H(\mu_d) := \begin{cases} \alpha & \text{if } \kappa = 0, \\ \alpha + (1-\alpha)\exp(-\frac{C}{\kappa}) & \text{else} \end{cases}$$

with a threshold $C > 0$.

In this chapter we show how coherence enhancing diffusion filtering can be extended to matrix fields. We follow the exposition in [62] and provide basic notions of a calculus for matrix fields in the next section.

3.3 Basic Differential Calculus for Matrix Fields

In this section we provide briefly the *basic definitions* for the formulation of a differential calculus for matrix fields. This material is instigated in [59] but for a more detailed exposition the reader is referred to [60].

1. *Functions of matrices.* The standard definition of a function h on $\mathrm{Sym}_n(\mathbb{R})$ is given by [229]:

$$h(U) = V^\top \operatorname{diag}(h(\lambda_1), \ldots, h(\lambda_n))V \in \mathrm{Sym}_n(\mathbb{R}),$$

if $U = V^\top \operatorname{diag}(\lambda_1, \ldots, \lambda_n)V$ is the spectral/eigen decomposition of the symmetric matrix U, and if $\lambda_1, \ldots, \lambda_n$ lie in the domain of definition of h. We encountered already an example of a function of a symmetric matrix; the diffusion tensor as a function of the structure tensor S_ρ with coherence κ under the tensor map H, $D = H(S_\rho)$.

2. *Partial derivatives.* Let $\omega \in \{x_1, \ldots, x_d, t\} \in \mathbb{R}^{d+1}$ stand for a spatial or temporal variable, $e_\omega \in \mathbb{R}^{d+1}$ a unit vector along the ω-direction, and $(x,t) = (x_1, \ldots, x_d, t)$. A partial derivative for a matrix field is naturally defined *componentwise* as the limit of a difference quotient:

$$\begin{aligned}\overline{\partial}_\omega U(x,t) &= \lim_{h\to 0} \frac{U((x,t)+h\cdot e_\omega) - U(x,t)}{h} \\ &= \left(\lim_{h\to 0} \frac{u_{ij}((x,t)+h\cdot e_\omega) - u_{ij}(x,t)}{h}\right)_{i,j} \\ &= (\partial_\omega u_{ij}(x,t))_{i,j}.\end{aligned}$$

The generalization to directional derivatives is straightforward, viz. by letting e_ω refer to any arbitrary unit vector. Higher order partial differential operators, such as the Laplacian, or other more sophisticated operators, find their natural counterparts in the matrix-valued framework in this way as well.

3. *Generalized gradient of a matrix field.* The gradient of a matrix field with sufficiently smooth component functions is defined via

$$\overline{\nabla} U(x) := (\overline{\partial}_{x_1} U(x), \ldots, \overline{\partial}_{x_d} U(x))^\top \in (\mathrm{Sym}_n(\mathbb{R}))^d.$$

Hence, the generalised gradient $\overline{\nabla}U(x)$ at a voxel x is regarded as an element of the module $(\mathrm{Sym}_n(\mathbb{R}))^d$ over $\mathrm{Sym}_n(\mathbb{R})$ in close analogy to the scalar setting where $\nabla u(x) \in \mathbb{R}^d$. In the sequel we will call a mapping from $\mathbb{R}^d$ into $(\mathrm{Sym}_n(\mathbb{R}))^d$ a *module field* rather than a vector field.

4. For the sake of completeness we include the formal definition of the *generalized structure tensor of a matrix field* here. We will discuss its derivation, properties and application in the next section. The novel structure tensor for a matrix field is given by

$$\begin{aligned}(\overline{\mathscr{S}}_L(U))(x) &:= (G_\rho * (\overline{\nabla}U(\cdot) \cdot (\overline{\nabla}U(\cdot))^\top))(x) \\ &= ((G_\rho * (\overline{\partial}_{x_i} U(\cdot) \cdot \overline{\partial}_{x_j} U(\cdot)))(x))_{i,j=1,\ldots,d}. \qquad (3.2)\end{aligned}$$

Table 3.1 Extensions of elements of scalar valued calculus (*middle*) to the matrix-valued setting (*right*)

Setting	Scalar valued	Matrix-valued
Function	$h: \begin{cases} \mathbb{R} \to \mathbb{R} \\ x \mapsto h(x) \end{cases}$	$h: \begin{cases} \mathrm{Sym}_n(\mathbb{R}) \longrightarrow \mathrm{Sym}_n(\mathbb{R}) \\ U \mapsto V^\top \mathrm{diag}(h(\lambda_1), \dots, h(\lambda_n))V \end{cases}$
Partial derivatives	$\partial_\omega u$, $\omega \in \{t, x_1, \dots, x_d\}$	$\overline{\partial}_\omega U := (\partial_\omega u_{ij})_{ij}$, $\omega \in \{t, x_1, \dots, x_d\}$
Gradient	$\nabla u(x) := (\partial_{x_1} u(x), \dots, \partial_{x_d} u(x))^\top$, $\nabla u(x) \in \mathbb{R}^d$	$\overline{\nabla} U(x) := (\overline{\partial}_{x_1} U(x), \dots, \overline{\partial}_{x_d} U(x))^\top$, $\overline{\nabla} U(x) \in (\mathrm{Sym}_n(\mathbb{R}))^d$
Structure tensor	$(G_\rho * (\overline{\nabla} u(\cdot) \cdot (\overline{\nabla} u(\cdot))^\top))(x)$	$(\overline{\mathscr{S}}_L(U))(x) :=$ $(G_\rho * (\overline{\nabla} U(\cdot) \cdot (\overline{\nabla} U(\cdot))^\top))(x)$
Product	$a \cdot b$	$A \bullet_J B := \frac{1}{2}(AB + BA)$

5. *Symmetric product of symmetric matrices*. The product of two symmetric matrices $A, B \in \mathrm{Sym}_n(\mathbb{R})$ is not symmetric unless the matrices commute. However, it is vital to our interests to have a symmetric matrix product at our disposal. There are numerous options to define a symmetric matrix product, however, we concentrate on a specific one known from algebra and called Jordan product:

$$A \bullet_J B = \frac{1}{2}(AB + BA) \quad \text{for } A, B \in \mathrm{Sym}_n(\mathbb{R}). \tag{3.3}$$

For commuting A and B we have $A \bullet_J B = A \cdot B$. This product is commutative and distributive but not associative. Most important, it does not preserve the positive semi-definiteness of its arguments [62].

We summarized the definitions from above and juxtapose them with their scalar counterparts in Table 3.1. The matrix field $U(x)$ is assumed to be diagonalizable with $U = (u_{ij})_{ij} = V^\top \mathrm{diag}(\lambda_1, \dots, \lambda_n)V$, where $V \in O(n)$, the set of all orthogonal $n \times n$-matrices, and $\lambda_1, \dots, \lambda_n \in \mathbb{R}$.

3.4 The Structure Tensor for Matrix Fields

In order to extract d-dimensional information we *reduce* $\overline{\mathscr{S}}_L(U) \in \mathrm{Sym}_{nd}(\mathbb{R})$ to a structure tensor $S(U) \in \mathrm{Sym}_n(\mathbb{R})$ in a generalized projection step employing the block operator matrix

$$\mathrm{tr}_A := \begin{pmatrix} \mathrm{tr}_A & \cdots & 0 \\ \vdots & \ddots & \vdots \\ 0 & \cdots & \mathrm{tr}_A \end{pmatrix} \tag{3.4}$$

containing the trace operation. We set $\mathrm{tr} := \mathrm{tr}_I$. This operator matrix acts on elements of the space $(\mathrm{Sym}_n(\mathbb{R}))^d$ as well as on block matrices via formal blockwise matrix multiplication.

$$\begin{pmatrix} \mathrm{tr}_A & \cdots & 0 \\ \vdots & \ddots & \vdots \\ 0 & \cdots & \mathrm{tr}_A \end{pmatrix} \begin{pmatrix} M_{11} & \cdots & M_{1n} \\ \vdots & \ddots & \vdots \\ M_{n1} & \cdots & M_{nn} \end{pmatrix} = \begin{pmatrix} \mathrm{tr}_A(M_{11}) & \cdots & \mathrm{tr}_A(M_{1n}) \\ \vdots & \ddots & \vdots \\ \mathrm{tr}_A(M_{n1}) & \cdots & \mathrm{tr}_A(M_{nn}) \end{pmatrix},$$

provided that the square blocks M_{ij} are compatible with tr_A, that means here, have the same size as A. The reason for choosing tr_A as reduction operators is their homogeneity:

$$\mathrm{tr}_A(tM) = t\,\mathrm{tr}_A(M) \quad \text{for all } t \in \mathbb{R}.$$

The subsequent result stated in [62] gives a first insight into the role of this reduction operation and its connection to other structure tensors:

Proposition (Weickert's Tensor as an Elementary Reduction of $\overline{\mathscr{S}}_L$) *Let $U(x) \in \mathrm{Sym}_n(\mathbb{R})$ be a d-dimensional matrix-field. Then the Weickert tensor J_ρ is a reduced version of $\overline{\mathscr{S}}_L$,*

$$\mathrm{Tr}\,\overline{S_L}(U) = J_\rho(U) \in \mathrm{Sym}_n(\mathbb{R}).$$

The reduction operation is accompanied by an extension operation defined via the *Kronecker product*:

Definition The I_n-extension operation is the mapping from $\mathrm{Sym}_d(\mathbb{R})$ to $\mathrm{Sym}_{nd}(\mathbb{R})$ given by the Kronecker product $\otimes$:

$$\begin{pmatrix} v_{11} & \cdots & v_{1d} \\ \vdots & \ddots & \vdots \\ v_{d1} & \cdots & v_{dd} \end{pmatrix} \longmapsto \begin{pmatrix} v_{11} & \cdots & v_{1d} \\ \vdots & \ddots & \vdots \\ v_{d1} & \cdots & v_{dd} \end{pmatrix} \otimes \begin{pmatrix} I_n & \cdots & I_n \\ \vdots & \ddots & \vdots \\ I_n & \cdots & I_n \end{pmatrix}$$
$$:= \begin{pmatrix} v_{11} I_n & \cdots & v_{1d} I_n \\ \vdots & \ddots & \vdots \\ v_{d1} I_n & \cdots & v_{dd} I_n \end{pmatrix}.$$

If the $d \times d$-matrix $(v_{ij})_{ij}$ is Kronecker-multiplied with

$$\begin{pmatrix} C & \cdots & 0 \\ \vdots & \ddots & \vdots \\ 0 & \cdots & C \end{pmatrix} \begin{pmatrix} I_n & \cdots & I_n \\ \vdots & \ddots & \vdots \\ I_n & \cdots & I_n \end{pmatrix} = \begin{pmatrix} C & \cdots & C \\ \vdots & \ddots & \vdots \\ C & \cdots & C \end{pmatrix}$$

we speak of a C-extension.

3.4.1 A Novel Diffusion Tensor $\overline{D}$ for Matrix Fields

Now it is possible to give an analog $\overline{D}$ to the diffusion tensor D in the framework of matrix fields. We proceed in four steps:

1. The matrix field $\mathbb{R}^d \ni x \mapsto U(x)$ provides us with an module field of generalized gradients $\overline{\nabla} U(x)$ from which we construct the generalized structure tensor $\overline{\mathscr{S}}_L U(x)$ possibly with a certain integration scale ρ. This step corresponds exactly to the scalar case.
2. We infer reliable d-dimensional directional information by reducing $\overline{\mathscr{S}}_L U(x)$ with tr_A with the help of the block operator matrix given in (3.4) leading to a symmetric $(d \times d)$-matrix S, for example $S = J_\rho$ if $A = I_n$.

$$S := \begin{pmatrix} \mathrm{tr}_A & \cdots & 0 \\ \vdots & \ddots & \vdots \\ 0 & \cdots & \mathrm{tr}_A \end{pmatrix} \overline{\mathscr{S}}_L U(x).$$

3. The symmetric $(d \times d)$-matrix S is spectrally decomposed, and the tensor map H is applied to S yielding the diffusion tensor D,

$$D := H(S).$$

4. Finally we enlarge the $(d \times d)$-matrix D to a $(nd \times nd)$-matrix $\overline{D}$ by the extension operation:

$$\overline{D} = D \otimes \left(\begin{pmatrix} C & \cdots & 0 \\ \vdots & \ddots & \vdots \\ 0 & \cdots & C \end{pmatrix} \begin{pmatrix} I_n & \cdots & I_n \\ \vdots & \ddots & \vdots \\ I_n & \cdots & I_n \end{pmatrix} \right).$$

This last step gives another possibility to steer the filter process by the choice of the matrix C. However, this is the subject of current research. For this work we restricted ourselves to $C = I_n$.

3.5 Coherence-Enhancing Diffusion Filtering for Matrix Fields

Now we have gathered the necessary ingredients to formulate the matrix-valued equivalent to the scalar coherence enhancing diffusion as expressed in Eq. (3.1).

$$\overline{\partial}_t U - \sum_{i=1}^{d} \overline{\partial}_{x_i} (\overline{D} \bullet \overline{\nabla} U) = 0 \quad \text{in } I \times \Omega,$$

$$U(x,0) = F(x) \quad \text{in } \Omega.$$

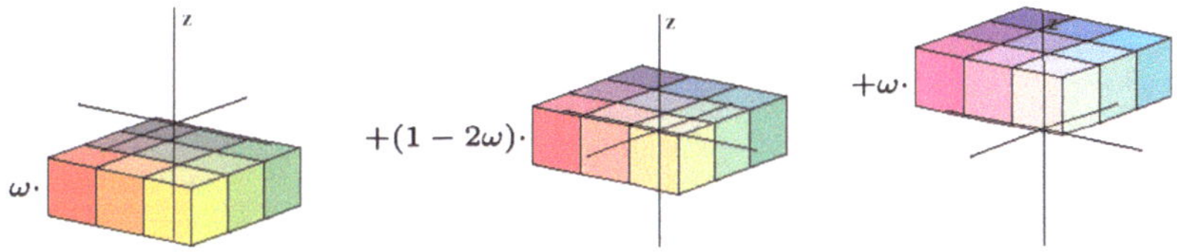

Fig. 3.1 Construction of a 3D-stencil specific for the z-direction. Each layer represents a discrete approximation of an essentially 2D divergence term

Note that the Jordan-multiplication in $\overline{D} \cdot \overline{\nabla} U$ is understood in the blockwise sense of partitioned matrices. Moreover, we translated the divergence differential operator acting on a vector-valued function $u = (u_1, \ldots, u_d)$, $\operatorname{div} u = \sum_{i=1}^{d} \partial_{x_i} u$, into its matrix-valued counterpart acting on a module field $W \in \mathrm{Sym}_n(\mathbb{R})^d$ by

$$\overline{\operatorname{div}} W = \sum_{i=1}^{d} \overline{\partial}_{x_i} W.$$

3.6 Numerical Issues

In the scalar, two-dimensional case an explicit scheme can be found in [433]. We used a matrix-valued version employing the calculus framework for matrix fields as presented before. This provides us with a matrix-valued solution scheme for matrix fields defined over a two-dimensional image domain Ω.

We derive a three-dimensional scheme by employing a Sobel-type construction using the 2D discrete divergence approximation. The idea is based on the following decomposition:

$$\begin{aligned}
&\operatorname{div}\left(\begin{pmatrix} 2a & b & c \\ b & 2d & e \\ c & e & 2f \end{pmatrix} \nabla u\right) \\
&\quad = \operatorname{div}\left(\begin{pmatrix} a & b & 0 \\ b & d & 0 \\ 0 & 0 & 0 \end{pmatrix} \nabla u\right) + \operatorname{div}\left(\begin{pmatrix} a & 0 & c \\ 0 & 0 & 0 \\ c & 0 & f \end{pmatrix} \nabla u\right) \\
&\qquad + \operatorname{div}\left(\begin{pmatrix} 0 & 0 & 0 \\ 0 & d & e \\ 0 & e & f \end{pmatrix} \nabla u\right).
\end{aligned} \tag{3.5}$$

Each term on the right hand side of (3.5) can be approximated by a two dimensional discretisation where we employ additionally a Sobel-type construction with a weight $\omega \in [0, \frac{1}{3}]$ as indicated in Fig. 3.1. For each of the three summands associated with the z-, y- and x-directions a $(3 \times 3 \times 3)$-stencil is obtained which, when added give the final stencil. We used the matrix-valued version of this stencil. Note that this construction can be applied to more sophisticated two-dimensional stencils.

Fig. 3.2 (**a**) *Top left*: Artificial data set of ellipsoids indicating a crossing. (**b**) *Top right*: Effect of coherence-enhancing filtering (CED-filtering) if preference is given to the direction $(1, -1)$. (**c**) *Bottom left*: Effect of CED-filtering if preference is given to the direction $(1, 1)$. (**d**) *Bottom right*: Effect of CED-filtering if no directional priority is established

3.7 Experiments

We use two data sets in our numerical experiments: The artificial matrix fields of (3×3)-symmetric matrices exhibit various coherent structures ranging from simple line-like to curved features, Figs. 3.2, 3.3, and 3.4. Important is the fact that these structures are not complete but interrupted. We will use these data to demonstrate the gap-closing and enhancing properties of our technique. The other matrix field stems from a 2D slice extracted from a 3D DT-MRI data set of size of a $128 \times 128 \times 30$ of a human head.

The data are represented as ellipsoids via the level sets of the quadratic form $\{x^\top A^{-2} x = const. : x \in \mathbb{R}^3\}$ associated with a matrix $A \in \mathrm{Sym}^+(3)$. By using A^{-2} the length of the semi-axes of the ellipsoid correspond directly with the three

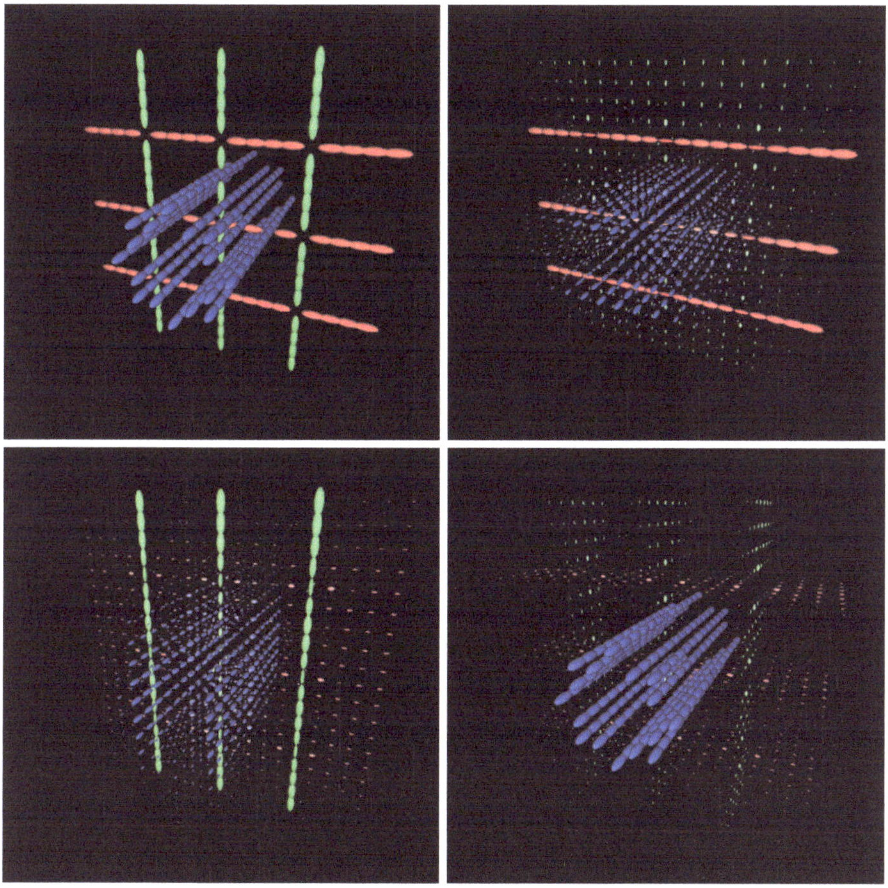

Fig. 3.3 (**a**) *Top left*: Original matrix field with grid-like structure. (**b**) *Top right*: Result of CED-filtering with preference on the horizontal x-direction. (**c**) *Bottom left*: Result of CED-filtering with preference on the horizontal x-direction. (**d**) *Bottom right*: The same but with preference on the z-direction

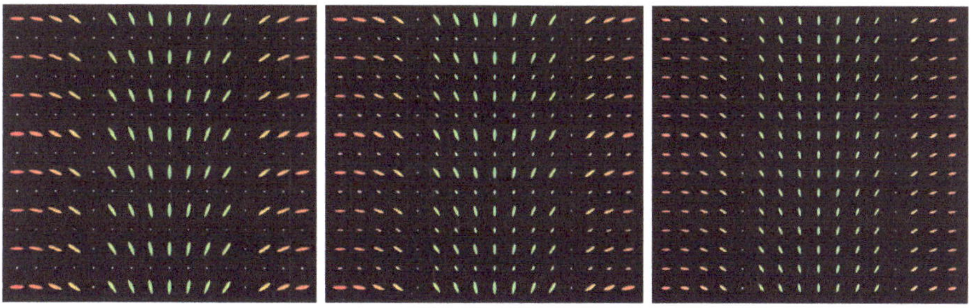

Fig. 3.4 (**a**) *Left*: Artificial incomplete coherent structure. (**b**) *Middle*: After CED-filtering with stopping time $t = 0.3$. (**c**) *Right*: After CED-filtering with stopping time $t = 3$

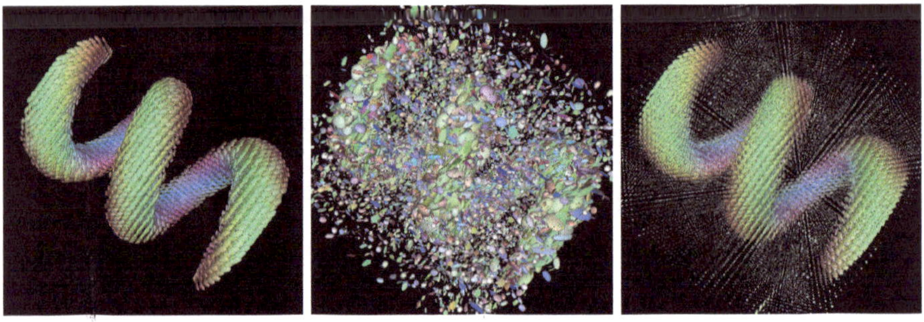

Fig. 3.5 (**a**) *Left*: Synthetic 3D-data set. (**b**) *Middle*: Polluted with truncated Gaussian noise in the eigenvalues while the orthogonal matrices result from three Euler matrices with uniformly distributed angles. (**c**) *Right*: After CED-filtering with stopping time $t = 2$ and $\omega = \frac{1}{4}$

eigenvalues of the matrix. We have added random positive definite matrices to the data to demonstrate the denoising capabilities of our coherence enhancing filtering, CED-filtering for short. The eigenvectors of this noise were obtained by choosing Gaussian-distributed numbers with standard deviation $\sigma = 1000.0$ and taking the absolute value for positive definiteness. The high standard deviation can be explained by the fact that in real-world data the typical eigenvalues are in the order of magnitude of 1000. The eigenvectors of the artificial noise result in choosing three uniformly distributed angles and rotating the matrix by these angles around the coordinate axes. The resulting data is shown in Fig. 3.5.

The artificial data set displayed in Fig. 3.2 imitates a crossing of nerve fibers. Depending on the choice of the reduction matrix A in tr_A either the diagonal directed downward,

$$A = \begin{pmatrix} 1 & -1 \\ -1 & 1 \end{pmatrix},$$

or the one directed upward,

$$A = \begin{pmatrix} 1 & 1 \\ 1 & 1 \end{pmatrix},$$

is given preference in the CED-filtering results. If no priority is set, $A = I$, a homogeneous structure is developing in the center, as it is expected due to the high symmetry of the image, see Fig. 3.2(d). Directional preferences can be conveyed to the diffusion process in 3D as well by selecting, for example, the x-direction via the specification

$$A = \begin{pmatrix} 1 & 0 & 0 \\ 0 & 0 & 0 \\ 0 & 0 & 0 \end{pmatrix}.$$

This and other examples are shown in Fig. 3.3 depicting a grid-like synthetic structure.

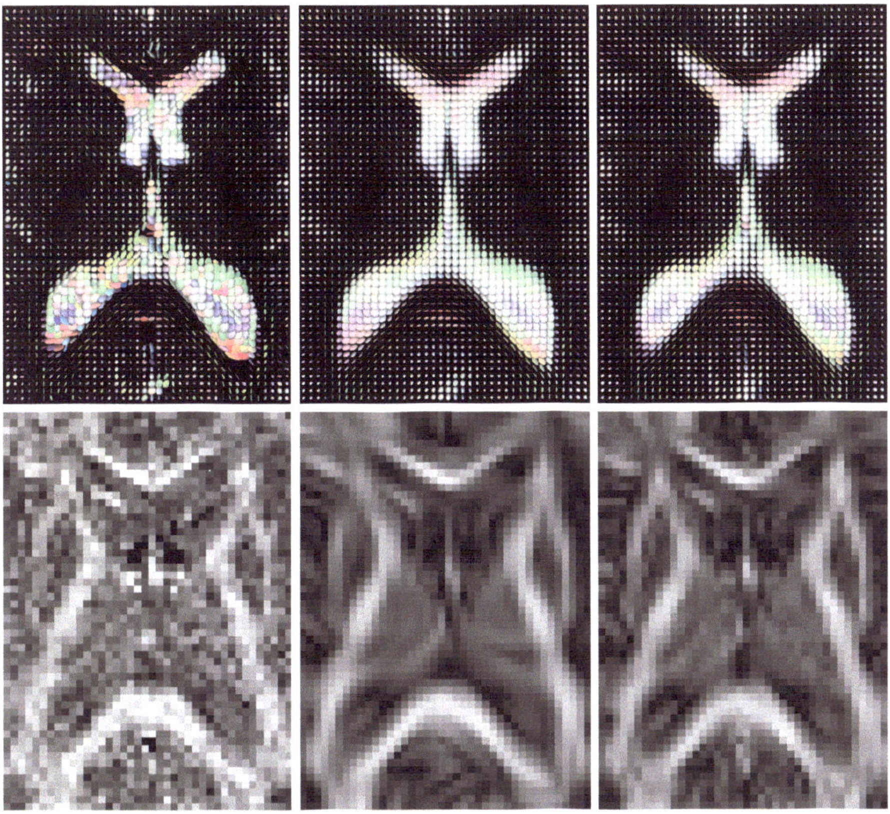

Fig. 3.6 (**a**) *Left*: 2D slice of 3D DT-MRI data and corresponding FA-map below. (**b**) *Middle*: After 2D-CED filtering with $t = 2$ and FA-map. (**c**) *Right*: After 3D-CED filtering with $t = 2$ and FA-map

The experiment depicted in Fig. 3.4 demonstrates that even areas with no information may constitute a coherent structure. A matrix field with a visually diverging structure CED-filtered without directional preferences. The (almost) empty lines in x-direction are getting filled while the two lines in y-direction remain untouched by the filtering. The explanation is that in x-direction we have changes in the shape and orientation of the ellipsoids in the vicinity of the empty lines, hence, the gap-closing quality of CED-filtering is coming into effect. However, proceeding in y direction no changes of the surrounding ellipsoids is discernable, rendering the CED-filtering idle in this direction.

A impression of the denoising capabilities can be obtained from the results in Fig. 3.5. The helix-shaped data is heavily polluted by random matrices/ellipsoids, and this noisy version is then CED-filtered without directional preferences, that is, with $A = I$.

We applied CED-filtering without directional preference to real DT-MRI data as well in order to investigate its usefulness as a pre-processing step. A comparison

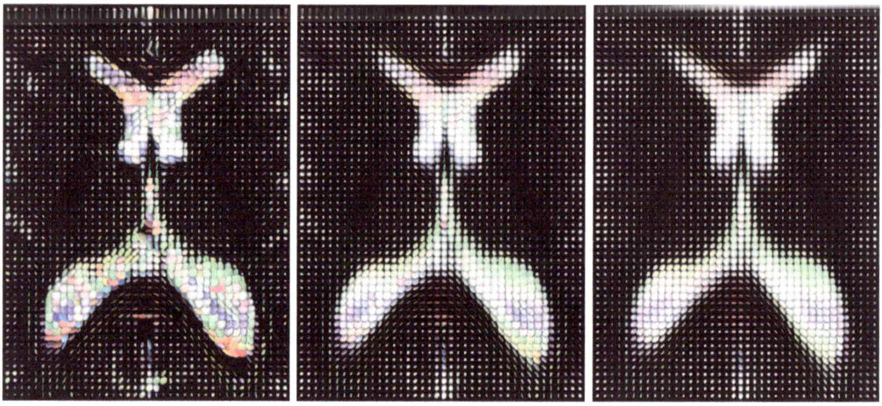

Fig. 3.7 (**a**) *Left*: 2D slice of 3D DT-MRI data. (**b**) *Middle*: After 3D-CED filtering with $t = 2$. (**c**) *Right*: After 3D-CED filtering with $t = 5$

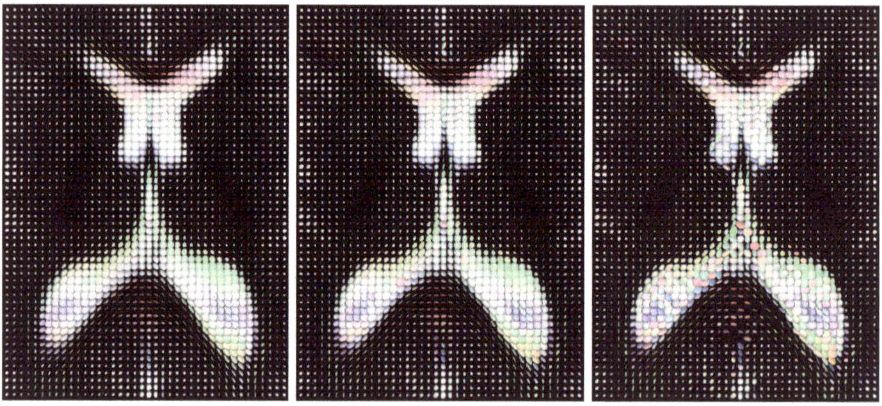

Fig. 3.8 (**a**) *Left*: 3D-CED-filtering with $t = 2$ and $\omega = 0$. (**b**) *Middle*: Filtering with $t = 2$ and $\omega = \frac{1}{4}$. (**c**) *Right*: Filtering with $t = 2$ and $\omega = \frac{1}{3}$

of two-dimensional and three-dimensional matrix-valued CED filtering is depicted in Fig. 3.6. The 3D-version is slightly more detailed than the 2D-variant after a diffusion time $t = 2$, as it incorporates information from adjacent layers. This higher degree of detail of the 3D-variant is confirmed when the corresponding FA-maps are considered. The stencil weight was set to $\omega = \frac{1}{4}$ giving the central slices in the three-dimensional stencil a clear predominance.

In Fig. 3.7 we display the results of 3D CED-diffusion after $t = 2$ and $t = 5$. The results confirm the regularising effect and the enhancement of coherent structures such as the fiber bundles below the Corpus Callosum.

Finally, for the sake of completeness, we study the influence of variations in $\omega \in [0, \frac{1}{3}]$ on the CED-diffusion process. The weights $\omega = 0$ and $\omega = \frac{1}{4}$ produce quite similar results. However, the choice $\omega = \frac{1}{4}$, where only the corners of the

$(3 \times 3 \times 3)$-stencil have non-zero weights, results in a very slow diffusion process, see Fig. 3.8.

3.8 Conclusion

Based on an operator-algebraic view on matrices we described in this chapter a structure tensor concept for matrix fields that is inspired by the classical structure tensor for scalar images. It allows us to develop a directionally selective coherence-enhancing diffusion filtering of matrix fields by employing a generic differential calculus framework for matrices. The matrix-valued CED-filtering exhibits similar behavior as its scalar counterpart. Current work encompasses the investigation of further opportunities to steer the filtering process, e.g. in the extension step, and its relation to other customizable tensor concepts for matrix fields. Future research will focus on further applications of the extended structure tensor concepts in image processing for matrix fields.

Acknowledgements We are grateful to Anna Vilanova i Bartrolí (Eindhoven University of Technology) and Carola van Pul (Maxima Medical Center, Eindhoven) for providing us with the DT-MRI data set and for discussing questions concerning data conversion. The original helix data is by courtesy of Gordon Kindlmann (University of Chicago).

Chapter 4
Structural Adaptive Smoothing: Principles and Applications in Imaging

Jörg Polzehl and Karsten Tabelow

Abstract Structural adaptive smoothing provides a new concept of edge-preserving non-parametric smoothing methods. In imaging it employs qualitative assumption on the underlying homogeneity structure of the image. The chapter describes the main principles of the approach and discusses applications ranging from image denoising to the analysis of functional and diffusion weighted Magnetic Resonance experiments.

4.1 Introduction

Images are often characterized by qualitative properties of their spatial structure, e.g. spatially extended regions of homogeneity that are separated by discontinuities. Images or image data with such a property are the target of the methods considered in this chapter. Alternative geometric characterizations using orientation or channels in feature space are discussed Chaps. 2, 8 and 9 in this book as well as elsewhere [114, 139, 166], and could be combined with the approach pursued here.

The methods summarized under the term structural adaptive smoothing try to employ a qualitative assumption on the spatial structure of the data to simultaneously describe the structure and efficiently estimate parameters like image intensities. Structural adaptive smoothing generalizes several concepts in non-parametric regression.

These include kernel smoothing and local polynomials, see e.g. [130, 386, 427] or [51], the filter proposed by Lee [279], bilateral smoothing [409] and scale space methods, see e.g. [70, 334]. Relations probably exist to diffusion methods in the Beltrami framework.

Our approach provides an alternative to non-linear diffusion methods, see e.g. Chap. 1 and [433], and generalizes linear diffusion in a different way. Information on the error distribution and qualitative assumptions on the underlying structure are

J. Polzehl (✉) · K. Tabelow
Weierstrass Institute for Applied Analysis and Stochastics, Mohrenstr. 39, 10117 Berlin, Germany
e-mail: polzehl@wias-berlin.de

K. Tabelow
e-mail: tabelow@wias-berlin.de

L. Florack et al. (eds.), *Mathematical Methods for Signal and Image Analysis and Representation*, Computational Imaging and Vision 41,
DOI 10.1007/978-1-4471-2353-8_4,

effectively coded by the definition of statistical penalties. The methods are designed to provide intrinsic balance between variability and bias of the reconstruction results. In contrast to diffusion methods this leads to a meaningful limit for increasing bandwidth, or correspondingly diffusion time. An approach derived from a related idea is described in Chap. 12.

A first attempt to use the idea of structural adaptive smoothing was proposed in [339] under the name adaptive weights smoothing. This was generalized and refined especially in [340] providing a theory for the case of one-parameter exponential families. Several extensions have been made to cover locally smooth images [341], color images [342] and special applications like functional Magnetic Resonance Imaging (fMRI) [343, 396] and Diffusion Tensor Imaging (DTI) [344, 397].

The next section introduces the general approach. We then illustrate how this translates to different imaging modalities and problems like image denoising in 2D and 3D, signal detection in fMRI and smoothing in DTI. As a summary we give some information on implementations and numerical complexity of the algorithms.

4.2 Structural Adaptive Smoothing

Within this chapter we assume the following data structure. We denote by $x_1, \ldots, x_n \in \mathscr{X} \subseteq \mathbb{R}^p$ the experimental design. In imaging x_i usually will be a point on a p dimensional grid, although this assumption is not necessary for the approach. At each design point x_i we assume to observe a scalar or vector $Y_i \in \mathscr{Y} \subseteq \mathbb{R}^q$.

We assume that the observed values Y_i follow a probability distribution $P_{\theta(x_i)}$ from a family $\mathscr{P} = \{P_\theta; \theta \in \Theta\}$ and that we are interested in estimating θ or some function $g(\theta)$ as a function of x. Traditional methods in non-parametric regression allow for varying parameters but usually assume that $\theta(x)$ is a smooth function in x. This is violated for image data characterized by strong discontinuities.

Instead we try to describe the image by its local homogeneity structure. We assume that there exists a partitioning

$$\mathscr{X} = \bigcup_{m=1}^{M} \mathscr{X}_m \tag{4.1}$$

such that

$$\theta(x) \approx \theta(x_i) \quad \Leftrightarrow \quad \exists m : x \in \mathscr{X}_m \wedge x_i \in \mathscr{X}_m$$

i.e. that θ is approximately constant on each $\mathscr{X}_m$. This assumption is very weak in the sense that the number M of partitions may be large and that there are no restrictions on the form of the sets $\mathscr{X}_m$. Nevertheless the assumption will prove helpful if there exists a partitioning with $M \ll n$ and where the $\mathscr{X}_m$ have some spatial extent. This structural assumption is used within an iterative procedure.

We do not directly enforce the partitioning (4.1). Instead, for each design point x_i, we describe a set $U(x_i)$ containing x_i by a weighting scheme

$$W(x_i) = (w_1(x_i), \dots, w_n(x_i)) = (w_{i1}, \dots, w_{in}).$$

A positive weight w_{ij} will be assigned if the estimates of θ_j and θ_i are not significantly different. In this case x_j would be contained in $U(x_i)$. Within the iteration process $U(x_i)$ can be imagined as being a subset or an estimate of a set $\mathscr{X}_m$ containing x_i from the assumed partition. $U(x_i)$ and $U(x_j)$ will usually not coincide at any stage of the process even if the structural assumption is valid exactly and x_i and x_j belong to the same set $\mathscr{X}_m$.

We try to determine the structure, e.g. sets of similar parameters, and estimate the parameters in an iterative procedure. We start at each design point with an initial estimate, if possible solely obtained from the observation at this point, and initialize a bandwidth h such that a ball of radius h just contains some neighboring points. We will formalize this later. We now alternate the following steps. At all design points x_i and for all design points x_j within a ball of radius h we assign a positive weight w_{ij} if $\hat{\theta}(x_j)$ belongs to a confidence region for $\theta(x_i)$ and a zero weight for all other points, thereby creating a new weighting scheme $W_i = W(x_i)$. This means we employ information from the estimates to learn on the underlying structure. We then use the generated weighting scheme to obtain a new estimate $\hat{\theta}(x_i)$ by weighted local likelihood or minimizing a weighted risk. Before continuing we synchronize and increase the bandwidth h thereby allowing for more positive weights and for a decrease in variability of the estimates. We stop iterating when a prespecified bandwidth, corresponding to a maximal possible variance reduction, is reached.

We now more formally describe how we generate the weighting schemes. Let x_i be fixed and $\hat{\theta}(x_i)$ be obtained employing a weighting scheme $W^{(k-1)}(x_i) = W_i^{(k-1)} = (w_{i1}^{(k-1)}, \dots, w_{in}^{(k-1)})$. We consider all x_j such that $\|x_i - x_j\|^2/h^2 \leq 1$. New weights $w_{ij}^{(k)}$ are then generated as the product of two terms

$$w_{ij}^{(k)} = K_{\mathrm{loc}}(l_{ij}) K_{\mathrm{st}}(s_{ij}^{(k-1)}),$$

where

$$l_{ij} = \|x_i - x_j\|^2/h^2 \quad \text{and} \quad s_{ij}^{(k-1)} = \frac{N_i}{\lambda} T(\hat{\theta}(x_j), \hat{\theta}(x_i)) \quad \text{with } N_i = \sum_{j=1}^{n} w_{ij}^{(k-1)}$$

are two penalties measuring the spatial distance between the two design points and the difference between the two estimates. Both terms depend on kernel functions K_{loc} and K_{st} respectively. The second term should reflect both the difference of the estimated parameters and the variability of the parameter estimate at point x_i. Such a statistics $T(\hat{\theta}(x_j), \hat{\theta}(x_i))$ can often be derived as the Kullback-Leibler distance $\mathscr{K}(\hat{\theta}_j, \hat{\theta}_i)$ distance between the probability distributions $P_{\hat{\theta}_i}$ and $P_{\hat{\theta}_j}$. As an alternative, if $\hat{\theta}(x_i)$ is obtained by minimization of a risk $R(Y, W_i; \theta)$ we may define

$$T(\hat{\theta}(x_j), \hat{\theta}(x_i)) = 2(R(Y, W_i; \hat{\theta}(x_j)) - R(Y, W_i; \hat{\theta}(x_i)))$$

which in case of a logarithmic likelihood corresponds to using likelihood profiles.

Given a weighting scheme W_i we estimate parameters by either weighted (local) likelihood

$$\hat{\theta}(x_i) = \arg\max_{\theta} l(Y, W_i; \theta) = \arg\max_{\theta} \sum_{j=1}^{n} w_{ij} p(Y_j; \theta)$$

or weighted (local) risk minimization, e.g. least squares,

$$\hat{\theta}(x_i) = \arg\min_{\theta} R(Y, W_i; \theta) = \arg\min_{\theta} \sum_{j=1}^{n} w_{ij} \|Y_j - f(\theta)\|^2,$$

where $f : \Theta \mapsto \mathbb{R}^q$ is a suitable function on the parameter space. A formal description of the algorithm is then given as

- Initialization: Set $k = 0$, $W_i^{(0)}$ such that $w_{ij}^{(0)} = \delta_{ij}$, $\hat{\theta}^{(0)}(x_i)$ defined as a weighted likelihood or least squares estimate, $h^{(0)} = 1$.
- Adaptation: $\forall i, j$ define

$$w_{ij}^{(k)} = K_{\text{loc}}(l_{ij}^{(k)}) K_{\text{st}}(s_{ij}^{(k-1)}).$$

- Estimation: $\forall i$ define

$$\hat{\theta}^{(k)}(x_i) = \arg\max_{\theta} l(Y, W_i^{(k)}; \theta) \quad \Big(\text{or } \arg\min_{\theta} R(Y, W_i^{(k)}; \theta)\Big).$$

- Iterate: Stop if $k \geq k^*$, else select $h^{(k+1)}$ such that $\sum_j K_{\text{loc}}(l_{ij}^{(k+1)}) = c_h \sum_j K_{\text{loc}}(l_{ij}^{(k)})$ ($c_h = 1.25$), set $k := k + 1$, and continue with adaptation.

The proposed procedure involves several parameters. The most important one is the scale parameter λ in the statistical penalty s_{ij}. The special case $\lambda = \infty$ simply leads to a kernel estimate with bandwidth $h_{\max} = h^{(k^*)}$. We propose to choose λ as the smallest value satisfying a propagation condition (4.2) [340]. This condition requires that, if the local assumption is valid globally, i.e. $\theta(x) \equiv \theta$ does not depend on x, then with high probability and for all k the estimate coincides at every point with the nonadaptive estimate. More formally we request that in this case for each iteration k

$$\mathbf{E} \sum_{i=1}^{n} |\hat{\theta}^{(k)}(x_i) - \check{\theta}^{(k)}(x_i)| < \alpha \mathbf{E} \sum_{i=1}^{n} |\check{\theta}^{(k)}(x_i) - \theta| \tag{4.2}$$

for a specified constant $\alpha > 0$. Here

$$\check{\theta}^{(k)}(x_i) = \sum_j K_{\text{loc}}(l_{ij}^{(k)}) Y_j \Big/ \sum_j K_{\text{loc}}(l_{ij}^{(k)})$$

Table 4.1 Statistical model, corresponding statistical penalty and R-package

Model	Penalty s_{ij}	R-package
1D-, 2D- and 3D- regression models	$\frac{N_i}{2\lambda\sigma^2}(\hat\theta_i - \hat\theta_j)^2$	aws/adimpro
1D-, 2D- and 3D- exponential families	$\frac{N_i}{\lambda}\mathscr{K}(\hat\theta_j, \hat\theta_i)$	aws
1D-, 2D- local polynomial regression	$\frac{N_i}{\lambda}(R(Y, W_i; \hat\theta_j) - R(Y, W_i; \hat\theta_i))$	aws/adimpro
1D-, 2D- and 3D- Gaussian models with parametric mean-variance model $g(x, \theta, \eta)$ (η-global parameter)	$\frac{N_i}{2\lambda\hat\sigma^2(x_i)}(\hat\theta_i - \hat\theta_j)^2$ $\sigma(x_i) = g(x_i, \theta_i, \eta)$	aws
Color images with constant and linear parametric mean-variance model and spatial correlation	$\frac{N_i}{\lambda C(h,g)}(\hat\theta_i - \hat\theta_j)^T \hat\Sigma_i^{-1}(\hat\theta_i - \hat\theta_j)$	adimpro
Functional MR (smoothing of SPM's)	$\lambda^{-1}(\mathrm{Var}\,\hat\theta_i)^{-1}(\hat\theta_i - \hat\theta_j)^2$	fmri
Diffusion tensor imaging (DTI)	$\frac{N_i}{\lambda}[R(Y, W_i; \hat\theta_j) - R(Y, W_i; \hat\theta_i)]$	dti

denotes the non-adaptive kernel estimate employing the bandwidth $h^{(k)}$ from step k. The value λ provided by this condition usually does not depend on the unknown model parameter θ and can therefore be found by simulation in a global parametric situation. This enables us to select default values for λ depending on the specified family of the probability distribution $\mathscr{P} = (P_\theta, \theta \in \Theta)$ and the chosen statistics T. Default values for λ in the examples below are selected for a value of $\alpha = 0.2$.

The second parameter of interest is the maximal bandwidth $h_{\max}$ which controls both numerical complexity of the algorithm and smoothness within homogeneous regions.

Additionally we specify a number of parameters and kernel functions that have less influence on the resulting estimates. As a default the kernel functions are chosen as $K_{\mathrm{loc}}(x) = (1 - x^2)_+$ and $K_{\mathrm{st}}(x) = \min(1, 2(1 - x))_+$. If the design is on a grid, e.g. for images, the initial bandwidth $h^{(0)}$ is chosen as the distance between neighboring pixel.

Applications usually require an appropriate description of the statistical model, the structural assumption and a corresponding definition of the statistical penalty s_{ij}. Table 4.1 provides an overview of currently implemented models and the corresponding software packages for the R environment for statistical computing [352].

4.3 Image Denoising

The algorithm described in the last section is essentially dimension free. It can be easily applied to reconstruct 2D and 3D images. We illustrate this using a 3D-MR image of a head.

We apply adaptive weights smoothing assuming a model with additive Gaussian errors

$$Y_i = \theta(x_i) + \varepsilon_i,$$

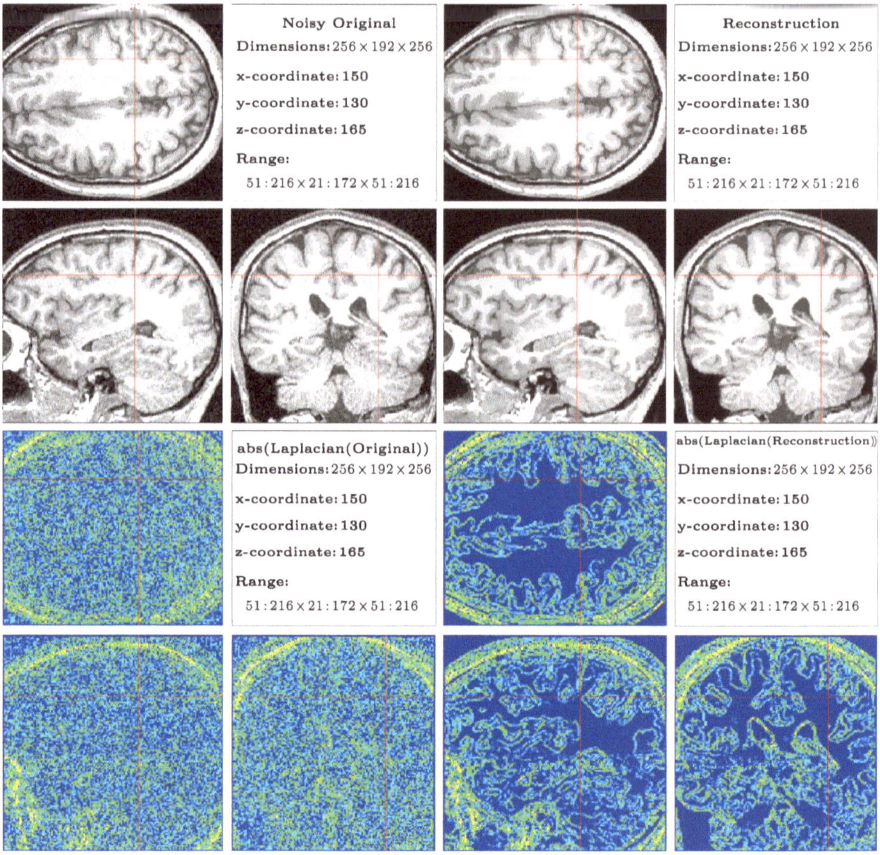

Fig. 4.1 *Top*: Original (*left*) and reconstruction (*right*) of an 3D-MR-image. *Bottom*: Corresponding absolute values of a Laplacian edge filter

to describe the gray value in voxel x_i. The statistical penalty used is $s_{ij} = \frac{N_i}{2\lambda\sigma^2}(\hat{\theta}_i - \hat{\theta}_j)^2$. The error variance σ^2 is estimated from the image. We employ a maximal bandwidth $h_{\max} = 6$. The value of $\lambda = 3.45$ fulfills the propagation condition for $\alpha = 0.1$. Special interest in this example is in detection and/or enhancement of tissue borders. We illustrate the results in Fig. 4.1. Additionally to the image we provide the results in terms of absolute values of a Laplacian filter which illustrates the gain in edge detection.

Within this example we essentially assumed that the image intensity is locally constant. This assumption may be too rigid and can be replaced, at the cost of sensitivity to discontinuities, by assuming the image to consist of locally smooth regions. The Propagation-Separation approach from [340] assumes that within a homogeneous region containing $x_i = (i_h, i_v)$, i.e. for $x_j \in U(x_i)$, the gray value or color Y_{j_h,j_v} can be modeled as

$$Y_{j_h,j_v} = \theta(x_i)^\top \Psi(j_h - i_h, j_v - i_v) + \varepsilon_{j_h,j_v},$$

Fig. 4.2 *From left to right*: Noisy image, local constant reconstruction and local quadratic reconstruction ($h_{\max} = 12$)

where the components of $\Psi(\delta_h, \delta_v)$ contain values of basis functions

$$\psi_{m_1,m_2}(\delta_h, \delta_v) = (\delta_h)^{m_1}(\delta_v)^{m_2}$$

for integers $m_1, m_2 \geq 0$, $m_1 + m_2 \leq p$ and some polynomial order p. For a given local model $W(x_i)$ estimates of $\theta(x_i)$ are obtained by local Least Squares as

$$\tilde{\theta}(x_i) = B_i^{-1} \sum_j w_{ij} \Psi(j_h - i_h, j_v - i_v) Y_{j_h, j_v},$$

with

$$B_i = \sum_j w_{ij} \Psi(j_h - i_h, j_v - i_v) \Psi(j_h - i_h, j_v - i_v)^\top.$$

The parameters $\theta(x_i)$ are defined with respect to a system of basis functions centered in x_i. Parameter estimates $\hat{\theta}(x_j, x_i)$ employing the local model $W(x_j)$ with basis functions centered at x_i can be obtained by a linear transformation from $\hat{\theta}(x_j)$. In iteration k a statistical penalty can now be defined as

$$s_{ij}^{(k)} = \frac{1}{\lambda 2\sigma^2} \big(\hat{\theta}^{(k-1)}(x_i) - \hat{\theta}^{(k-1)}(x_j, x_i)\big)^\top B_i \big(\hat{\theta}^{(k-1)}(x_i) - \hat{\theta}^{(k-1)}(x_j, x_i)\big).$$

For a more detailed description and discussion of the resulting algorithm see [341].

Figure 4.2 illustrates results obtained by local constant and a quadratic structural adaptive smoothing for a piecewise smooth image. The local constant reconstruction gives a cartoon-like impression which is due to the use of an, for this image, inappropriate structural assumption.

In digital color images the information in each pixel consists of a vector of three values. Each value is a intensity in one channel of a three dimensional color space, usually the RGB space.

If the image was recorded under bad light conditions, employing a high sensitivity of the sensor, such images can carry a substantial noise. This noise is usually spatially correlated, i.e. colored. Additionally we observe a correlation between the noise components in the three RGB channels.

An appropriate model to describe such a situation is given by

$$Y_{i_h,i_v} = \theta(x_i) + \varepsilon_{i_h,i_v},$$

where the components of $x_i = (i_h, i_v)$ are the horizontal and vertical image coordinates. Y_{i_h,i_v}, $\theta(x_i)$ and ε_{i_h,i_v} take values in $\mathbb{R}^3$. The errors follow a distribution with $\mathbf{E}\varepsilon_{i_h,i_v} = 0$, $\operatorname{Var}\varepsilon_{i_h,i_v} = \Sigma$ and $\mathbf{E}\varepsilon^c_{i_h,i_v}\varepsilon^c_{i_h+1,i_v} = \mathbf{E}\varepsilon^c_{i_h,i_v}\varepsilon^c_{i_h,i_v+1} = \rho\sigma_c^2$ for each color channel c. The covariance matrix Σ may vary with the value of θ_{i_h,i_v}.

Structural adaptive smoothing can be applied in this situation with a statistical penalty

$$s_{ij}^{(k)} = \frac{N_i^{(k-1)}}{2\lambda C(g,h)} \left(\hat{\theta}_i^{(k-1)} - \hat{\theta}_j^{(k-1)}\right)^\top \Sigma^{-1} \left(\hat{\theta}_i^{(k-1)} - \hat{\theta}_j^{(k-1)}\right),$$

where $C(g,h)$ is a correction term for spatial correlation, see [342].

Figure 4.3 illustrates the effect of structural adaptive smoothing for color images. The lower row provides details for the regions marked in the original and the reconstructed image as well as an image of sum of weights N_i that illustrates the adaptivity of the approach.

4.4 Signal Detection in Functional MRI

Functional Magnetic Resonance Imaging (fMRI) is nowadays a standard tool for in-vivo examination of human brain function with plenty of applications both in research as well as in clinical practice such as diagnosis and treatment of brain lesions. Data obtained in human fMRI consists of time series of three dimensional data sets of the brain. The interscan interval is usually in the order of seconds, while the spatial resolution is commonly in the millimeter range [272, 273] with recent studies entering the sub-millimeter domain [74, 255, 268]. An interesting fact about fMRI is, that the blood oxygenation serves as a natural contrast making the method non-invasive [321, 322]. This effect is known as the BOLD-effect and can be used for example to localize cognitive functions within the brain. When performing a cognitive task, the MR signal in some voxels is increased due to the higher oxygenation level at the active site. Other voxels remain in their resting state. The increase of the signal can be described by the hemodynamic response function, which has been extensively studied in the past years. The BOLD-effect leads to the creation of various typical experimental designs, mainly block- or event-related. In recent years it has also been proposed that even the resting state pattern of the brain contains valuable information about the working brain and hence has attracted much interest [300].

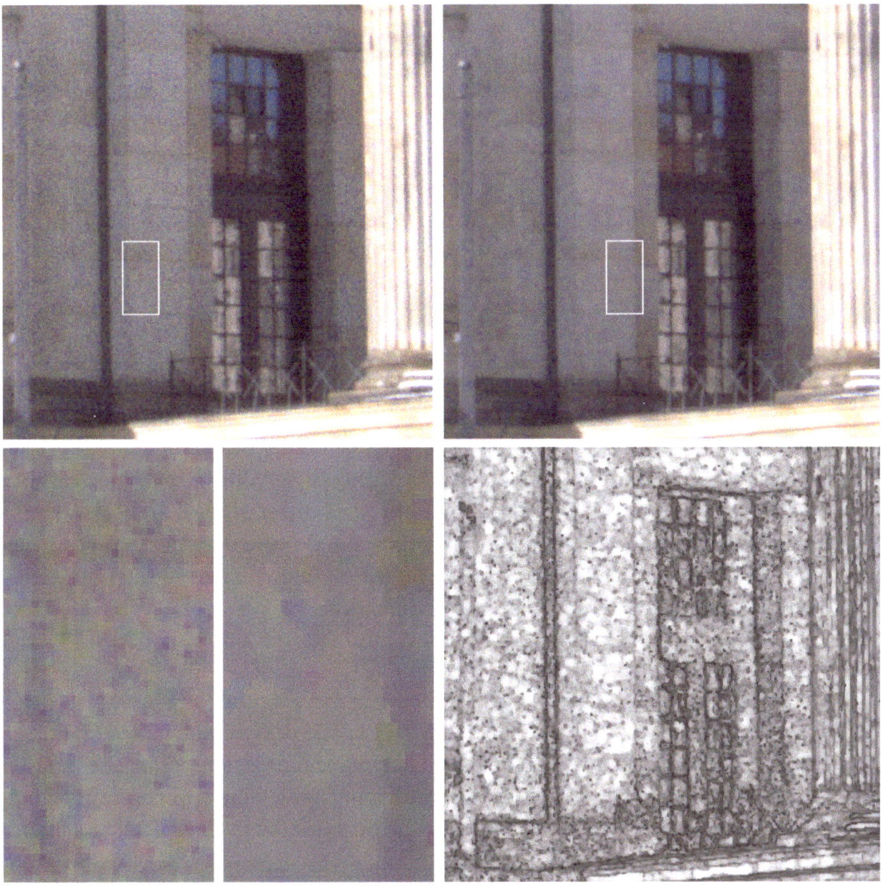

Fig. 4.3 Noisy image (*upper left*) and reconstruction by the proposed algorithm (bandwidth $h_{\max} = 6$, *upper right*). *The bottom row* shows details from both images and the sum of weights N_i in each voxel

Many different methods exist to analyze fMRI data depending on the experiment or the focus of the scientific questions, ranging from correlation analysis, ICA, spatiotemporal to fully Bayesian models, see [275] for an overview. Since from the knowledge of the design of an experiment the expected BOLD response is known, the linear model for fMRI data has perhaps become the most widely used approach.

Any analysis method however has to consider the fact, that fMRI data suffers from significant noise. Signal detection in fMRI data inherently involves a severe multiple test problem. In common experiments decisions have to be made at more than 100000 voxels, leading either to high thresholds (low sensitivity) or a high number of false positives (low specitivity). Analysis methods in fRMI use the fact, that activated areas have a spatial extent of several voxels. Spatial correlation, as introduced by smoothing, significantly reduces the number of independent tests.

Under the hypothesis of no activation and in a linear model based analysis spatial smoothing of fMRI data results in statistical parametric maps (SPM) that form random t- or F-fields. Results on excursion sets of random fields [4, 451] can be therefore used to define suitable thresholds for signal detection, see e.g. [452, 455]. However, common non-adaptive smoothing methods involve a significant loss of information on spatial structure and shape of activated areas. Several algorithms based on different methodology, from noise reduction with anisotropic diffusion processes, Bayesian approaches using spatial priors, region growing methods as established for image segmentation, and others, have been suggested to circumvent this.

Recently we proposed the use of structural adaptive smoothing [396], to avoid the loss of spatial information. The algorithm was developed in the context of the linear model for the BOLD-fMRI data [172, 453] but can easily be translated to other contexts. Our approach for smoothing fMRI data is mainly based on the observation that the structures of interest are defined by areas in which the parameter values corresponding to the BOLD signal are similar and differ significantly from zero. The common non-adaptive filtering approaches smooth the data cube at each time step separately, without making use of the information contained in the time series. We therefore suggested to first evaluate the linear model

$$Y_i = X\beta_i + \varepsilon_i$$

for the time series $Y_i = (Y_{it})_{t=1...T}$ at each voxel i. The design matrix X contains the expected BOLD response evaluated at scan acquisition times and nuisance parameters such as a slowly varying drift. After performing some appropriate prewhitening procedure, the error vector $\varepsilon_i = (\varepsilon_{it})_{t=1...T}$ can be assumed to have zero expectation and to be approximately uncorrelated in time.

We obtain fields of least squares estimates $\hat{\beta}_i$ for the parameter value β_i and, what is most important, its error variance $\operatorname{Var}\hat{\beta}_i$. Equipped with these the development of a specific structural adaptive smoothing algorithm as outlined in previous sections is canonical. First, we define a structural assumption of spatial homogeneity, which should be valid for the field of the true parameter β_i. In non-activated areas the parameter value is assumed to be zero. This serves as the null hypothesis and allows to again use Random Field Theory [4, 451] for signal detection, see [396]. In areas which are activated during the scan the parameter values differ from zero and are similar, provided that the BOLD %-changes are similar. Hence, our structural assumption is a local constant model for the BOLD-parameter. Activated areas may consist of more than one region with similar parameters. Based on this assumption, we use an iterative smoothing algorithm for the statistical parametric map (SPM) that is based on pairwise tests of homogeneity. The result is a smoothed SPM where the shape and borders of the activation structure are preserved. As a consequence, in contrast to other non-adaptive smoothing methods, the procedure does reduce noise while preserving the resolution of the scan as required by many modern applications.

The main parameter of our procedure is the maximum achievable variance reduction or equivalently a maximum achievable smoothness. Both can be specified by selecting a maximum bandwidth. Oversmoothing is avoided in the algorithm by construction as long as differences between the parameter values of two homogeneity regions are statistically significant. The largest homogeneous region is expected to be the non-activation area, where parameter values do not significantly differ from zero. Therefore we can choose the maximum bandwidth larger than in non-adaptive smoothing and achieve a larger amount of variance reduction without blurring. This has the effect of lowering the thresholds for signal detection, since the smoothness in non-activation areas, which determine the threshold under the hypothesis of no signal, is directly proportional to the bandwidth.

As statistical penalty we use

$$s_{ij}^{(k)} = \frac{1}{\lambda \operatorname{Var} \hat{\beta}_i^{(k-1)}} (\hat{\beta}_i^{(k-1)} - \hat{\beta}_j^{(k-1)})^2,$$

where $\hat{\beta}_i^{(k-1)}$ is the estimated BOLD-parameter from the previous iteration step. Its variance $\operatorname{Var} \hat{\beta}_i^{(k-1)}$ is estimated from the spatially smoothed residuals of the time series. From the final estimates for $k = k^\star$ a random t-field $\hat{\beta}_i^{(k^\star)} / (\operatorname{Var} \hat{\beta}_i^{(k^\star)})^{1/2}$ can be constructed such that again Random Field Theory can be applied for signal detection [396].

We now consider an application of this algorithm and compare it with the signal detection using no smoothing and Gaussian filtering. Experiments were performed on healthy volunteers and approved by the Institutional Review Board of Weill Cornell Medical College. Data was acquired on a 3.0 T General Electric (Milwaukee, WI) Signa Excite MRI scanner, using two-dimensional gradient echo echo planar imaging pulse sequences (GE-EPI) on an eight-channel head receive-only coil. A somatosensory motor task was performed by one male subject. For functional MRI, a GE-EPI sequence with TE/TR $= 40/2000$ ms was used and 20 axial slices of 4 mm thickness were acquired. We used a field-of-view of 24 cm with a matrix size of 128×128, yielding voxel dimensions of 1.88 mm, respectively. A task was performed in three blocks of 60 s duration; each block consisted of 30 s task and 30 s rest. The first 4 scans before these block were discarded, yielding in total 105 scans. The task consisted of bimanual tapping of the thumb against all fingers of the same hand, one by one and in quick succession. In Fig. 4.4 the effect of smoothing and structural adaptive smoothing in particular is demonstrated. While without smoothing only very few active voxels can be detected, smoothing in general leads to better detection results. However the significant and inherent blurring with Gaussian filtering can be avoided using structural adaptive smoothing. It has been recently shown, that this procedure is especially helpful in high resolution scans [399]. Structural adaptive smoothing in fMRI is capable to fully use high resolution, to correctly locate activation at tumor borders for pre-surgical planning, and to extract information on spatial structure and shape of the activation areas. For successful applications see e.g. [398, 399, 426].

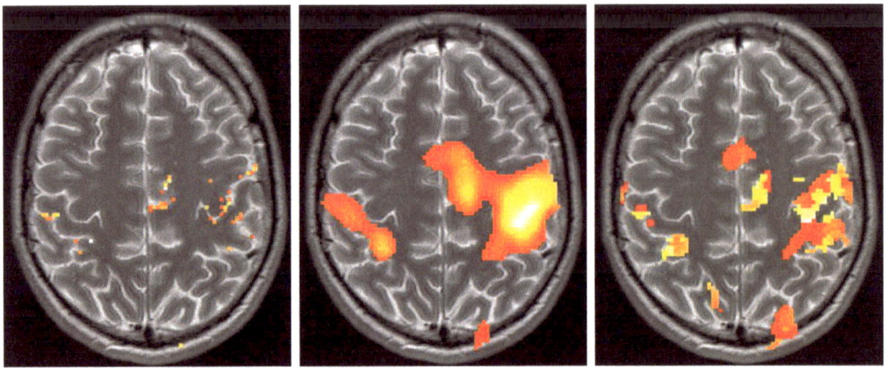

Fig. 4.4 Signal detection in an fMRI experiment: without smoothing (*left*), with Gaussian filtering (*center*) and structural adaptive smoothing (*right*)

4.5 Diffusion Tensor Imaging

Since the early times of nuclear magnetic resonance, it has been known that this phenomenon is sensitive to, and thus can be used to measure, diffusion of molecules in complex systems [67]. The basic principles of magnetic resonance diffusion weighted imaging (DWI) were introduced in the 1980's [276, 309, 402]. Since then, DWI has evolved into a versatile tool for in-vivo examination of tissues in the human brain and spinal cord, leading to a plethora of clinical and neuroscience applications. The broad interest in this technique grows from the fact that DWI probes microscopic structures well beyond typical image resolutions through water molecule displacement, which can be used in particular to characterize the integrity of neuronal tissue in the central nervous system.

Diffusion in neuronal tissue is usually not isotropic but depends on the particular microscopic structure of the tissue. Different diffusion directions can be probed by application of corresponding bipolar magnetic field diffusion gradients [392]. Compared to the non-diffusion weighted images S_0 the diffusion weighted images $S_{\mathbf{b}}$ for gradients in direction $\mathbf{b}$ are exponentially attenuated

$$S_{\mathbf{b}} = S_0 \exp(-bD(\mathbf{b})) \tag{4.3}$$

with the apparent diffusion coefficient $D(\mathbf{b})$ depending on the tested direction $\mathbf{b}$ and the "b-value" b depending on the magnetic field gradient parameters. In DTI [26, 27] this information is reduced to a three dimensional Gaussian distribution model for diffusion. Within this model, diffusion is completely characterized by the diffusion tensor $\mathtt{D}$, a symmetric positive definite 3×3 matrix with six independent components. Equation (4.3) thus generalizes to

$$S_{\mathbf{b}} = S_0 \exp(-b \cdot \mathbf{b}^\top \mathtt{D} \mathbf{b}).$$

The tensor itself is not invariant against rotations of the observation frame. Hence, only rotationally invariant quantities derived from the tensor contain physically meaningful measures. They are mainly based on the eigenvalues μ_i $(i = 1, 2, 3)$ of the tensor D with $\mu_i > 0$ for positive definite tensors. Mainly used are the trace, corresponding to the mean diffusivity $\langle\mu\rangle$, and the fractional (FA) or geodetic anisotropy (GA) measuring the anisotropy of the tensor:

$$\langle\mu\rangle = \frac{1}{3}\sum_{i=1}^{3}\mu_i,$$

$$FA = \sqrt{\frac{3}{2}}\sqrt{\sum_{i=1}^{3}(\mu_i - \langle\mu\rangle)^2 \bigg/ \sum_{i=1}^{3}\mu_i^2},$$

$$GA = \sqrt{\sum_{i=1}^{3}\left(\log(\mu_i) - \frac{1}{3}\sum_{i=1}^{3}\log(\mu_i)\right)^2}.$$

The diffusion tensor can be visualized as an ellipsoid with the length of main axis corresponding to the eigenvalues and the eigenvectors to the direction in space. Furthermore, the eigenvector for the largest eigenvalue directs in the main fiber direction. Evaluating the components of the vector as three components in a color space like RGB lead to color-coded directional maps with high diagnostic value due to its high contrast for interesting structures.

The diffusion tensor model describes diffusion completely if the microscopic diffusion properties within a voxel are homogeneous. In the presence of partial volume effects, like crossing or bifurcating fibers, the Gaussian model is only an approximation. Such effects are addressed in High Angular Resolution Diffusion Imaging (HARDI) [165, 416] see also Chaps. 8 and 9 within this book. For HARDI more sophisticated models exist, e.g. Q-ball [104, 415], higher order tensors [324], multi-tensor models [417] and tensor distribution functions [238, 283]. In this chapter, we restrict ourselves to the Gaussian diffusion tensor model for anisotropic diffusion, as used in DTI.

The diffusion tensor can be estimated by non-linear regression minimizing the risk:

$$\mathbf{R}(S_{\cdot,i}, \theta, \mathrm{D}) = \sum_{\mathbf{b}} \frac{(S_{\mathbf{b},i} - \theta\exp(-b \cdot \mathbf{b}^\top \mathrm{D}\mathbf{b}))^2}{\sigma_{\mathbf{b},i}^2} \tag{4.4}$$

with respect to the non-diffusion weighted parameter θ and the diffusion tensor D with the variability $\sigma_{\mathbf{b},i}^2$ of the diffusion weighted images.

DTI suffers from significant noise which may render subsequent analysis or medical decisions more difficult. This is especially important in low signal-to-noise applications, such as high-resolution DTI or DTI with high b-values [78, 240, 458]. It has been shown that noise may induce a systematically biased assessment of features. For example, a well known phenomenon is the biased estimation of anisotropy indices in the presence of noise [24, 198]. At high noise levels, in addition to the common random errors, the order of the diffusion eigenvectors is subject to a sorting bias. Noise reduction is therefore essential. Several approaches have been proposed for smoothing diffusion tensor data. They include common methods such as Gaussian smoothing [444], anisotropic kernel estimates [280], and methods based on non-linear diffusion [106, 330, 334, 433] or splines [221].

Procedures proposed within this book include coherence enhancing diffusion for matrix fields, cf. Chap. 3, and tensor regularization methods, cf. Chap. 5. Smoothing of tensor data requires to choose a Riemannian [148, 332, 462, 463] or log-Euclidian metric in the tensor space [14, 146]. We see some conceptional advantages in smoothing the diffusion weighted images instead of the tensor estimates. Estimating the tensor by Eq. (4.4) from noisy data leads, with a certain probability, to results outside the tensor space. This requires some kind of regularization. Reducing the noise level in the diffusion weighted images allows for a reduction of this probability in case of an underlying non-degenerate tensor. In case of high noise level in the diffusion weighted images, both the Rician distribution and the non-linearity of Eq. (4.3) lead to a bias in the tensor estimate. This bias can be reduced by smoothing the diffusion weighted images, but is not addressed if smoothing is performed in the tensor space itself. A correction for Rician bias [28, 194, 344] can be incorporated.

We therefore developed a structural adaptive smoothing algorithm for DWI data in the context of the diffusion tensor model [397] with extensions to include Rician bias correction and non-linear tensor estimation. Our underlying structural assumption is that for every voxel there is a neighborhood of this voxel in which the diffusion tensor is nearly constant. This assumption reflects the fact that the structures of interest are regions with a homogeneous fractional anisotropy, a homogeneous diffusivity, and a locally constant direction field. The shape of this neighborhood can be quite different for different voxels and cannot be described by few simple characteristics like bandwidth or principal directions.

The algorithm involves the statistical penalty

$$s_{ij}^{(k)} = \frac{N_i^{(k-1)}}{\lambda}\left[\mathbf{R}\big(\hat{\zeta}_{\cdot,i}^{(k-1)}, \hat{\theta}_{0,j}^{(k-1)}, \hat{\mathrm{D}}_j^{(k-1)}\big) - \mathbf{R}\big(\hat{\zeta}_{\cdot,i}^{(k-1)}, \hat{\theta}_{0,i}^{(k-1)}, \hat{\mathrm{D}}_i^{(k-1)}\big)\right]$$

based on previous estimates for the diffusion tensor and its variability. The corresponding weighting schemes are then directly applied to the diffusion weighted images, from which new estimates for the tensors with lower variability can be estimated.

In contrast to non-linear diffusion methods [332, 462, 463], cf. also Chap. 3, or non-adaptive smoothing [14, 146] this algorithm takes the variability of the tensor estimates into account and effectively uses the estimated underlying local structure to restrict the averaging process.

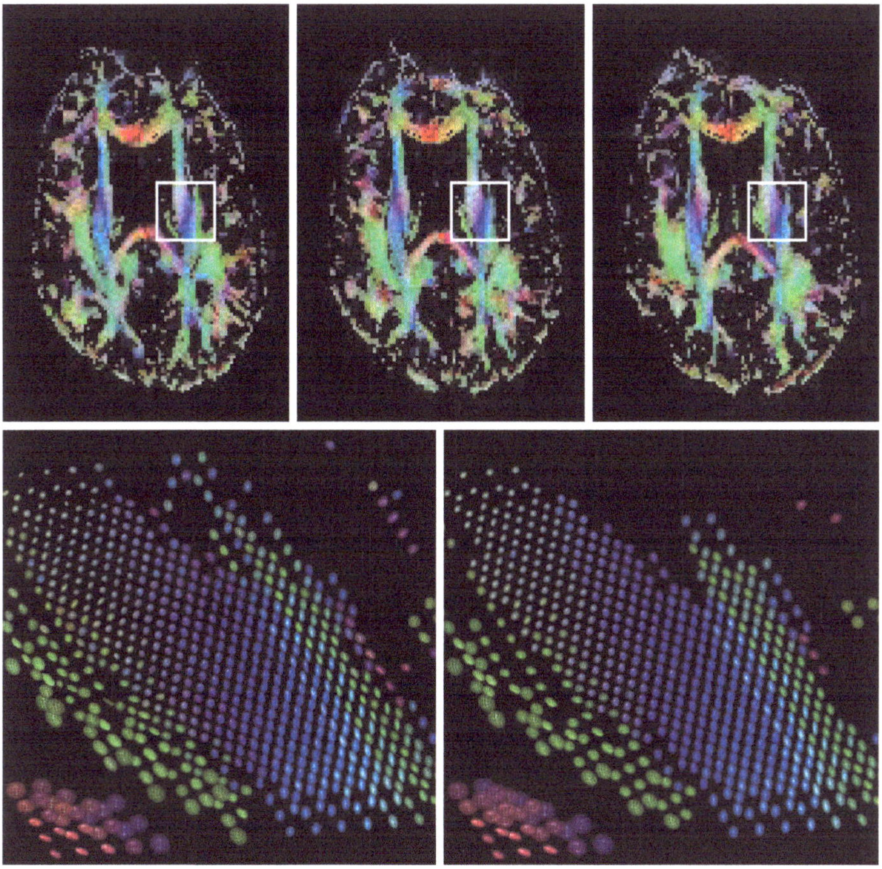

Fig. 4.5 Real DWI data example: *The upper row* shows the estimated color-coded directional map weighted with FA for the slices 22-24 of the CIBC-data set [76]. *White square* marks the extent of the region specified for *the lower row*. There, the noisy (*left*) and smoothed (bandwidth 4, *right*) tensors are shown. The structural adaptive smoothing apparently leads to a homogenization of the regions without blurring the structural borders

In Fig. 4.5 we demonstrate the effect of this procedure on experimental data. We use a DWI data set [76] made available by the NIH/NCRR Center for Integrative Biomedical Computing, P41-RR12553. This data set contains twelve diffusion weighted volumes and one non-diffusion-weighted ($b = 0$) reference volume. The data has a spatial resolution of 1.5 mm on each axis. The front of the head is at the top of the image. The scan goes from the top of the head down to about the middle of the brain, below the corpus callosum, but above the eyes.

The DTI data was collected on a 3 Tesla MRI scanner in the W.M. Keck Laboratory for Functional Brain Imaging and Behavior by Dr. Andrew Alexander, Departments of Medical Physics and Psychiatry, University of Wisconsin, Madison, funding: NIH RO1 EB002012.

Table 4.2 Computing time for three DWI data sets. The first column corresponds to the CIBC data set [76], the second to a data set kindly made available by A. Anwander, and a third data set kindly made available by H.U. Voss

Dimensions	101 × 146 × 38	72 × 100 × 50	146 × 193 × 47
# gradients	13	201	150
# voxel in mask	211235	255257	863134
CPU-time for sdpar	41 s	27 s	126 s
CPU-time for dtiTensor	16 s	118 s	156 s
CPU-time for dti.smooth ($h_{\max} = 2$)	220 s	667 s	893 s
CPU-time for dti.smooth ($h_{\max} = 4$)	415 s	1256 s	2140 s
CPU-time for dtiIndices	3.1 s	3.7 s	11 s
Mean N_i ($h_{\max} = 2/4$)	3.67/16.5	1.20/1.87	1.55/4.32

4.6 Implementations

The structural adaptive smoothing procedures described in this chapter have been mainly implemented as packages for the R environment for statistical computing [352].

Basic algorithms for denoising 1D- 2D and 3D structures using one parameter exponential family models and local polynomial models are implemented in the package *aws*.

The image processing of two dimensional color images has become omnipresent in the past years due to the availability of digital cameras. With the package *adimpro* we provide basic image processing functions, from reading/writing and color space transformations to structural adaptive smoothing.

The package *fmri* provides functions for reading and writing medical imaging formats and performing an analysis of BOLD-fMRI on the basis of the linear model. This includes the structural adaptive smoothing procedure described in the previous sections as well as signal detection based on Random Field Theory.

The package *dti* implements the structural adaptive smoothing in the context of the diffusion tensor model for diffusion weighted data. Table 4.2 compares the computing times and mean sum of weights for different data sets. The package also features extended visualization functions in two and three dimensions based on the R-interface to OpenGL provided by the R-package *rgl* [5].

4.7 Conclusion

Structural adaptive smoothing as described in this chapter can be very helpful in many contexts to remove noise without blurring interesting structures. While the main idea of structural adaptation in the Propagation-Separation approach is common for all applications, the resulting algorithms may be quite different and depend on the specific properties of the data and the application.

Acknowledgements This work is supported by the DFG Research Center MATHEON. It was made possible in part by software and a dataset from the NIH/NCRR Center for Integrative Biomedical Computing, P41-RR12553. The authors would like to thank A. Anwander at the Max Planck Institute for Human Cognitive and Brain Sciences (Leipzig, Germany) and H.U. Voss at the Citigroup Biomedical Imaging Center, Weill Cornell Medical College for providing diffusion-weighted and functional MR datasets. Furthermore, the authors would like to thank H.U. Voss for numerous intense and helpful discussions on Magnetic Resonance Imaging and related issues.

Chapter 5
SPD Tensors Regularization via Iwasawa Decomposition

Yaniv Gur, Ofer Pasternak, and Nir Sochen

Abstract In this chapter we propose a novel framework for regularization of symmetric positive-definite (SPD) tensors (e.g., diffusion tensors). This framework is based on a local differential geometric approach where the manifold of symmetric positive-definite (SPD) matrices, P_n, is parameterized via the Iwasawa coordinate system. The distances on P_n are measured here in terms of a natural $GL(n)$-invariant metric. Via the mathematical concept of fibre bundles, we describe the tensor-valued image as a section where the metric over the section is induced by the metric over P_n. Then, a functional over the sections accompanied by a suitable data fitting term is defined. The variation of this functional with respect to the Iwasawa coordinates leads to a set of $\frac{1}{2}n(n+1)$ coupled equations of motion for these coordinates. Then, by means of the gradient descent method, these equations of motion define a Beltrami flow over P_n. It turns out that the local coordinate approach via the Iwasawa coordinate system results in very simple numerics. Regularization results of structure tensors and diffusion tensors as well as results of fibres tractography for DTI are presented.

5.1 Introduction

Tensors are nowadays an important tool for image analysis as well as a source for physical information (e.g., [336, 433]). In this chapter we focus on symmetric positive-definite tensors. We suggest a general solution for the regularization prob-

Y. Gur (✉)
SCI Institute, University of Utah, 72 S. Central Campus Dr., Salt Lake City, UT 84112, USA
e-mail: yanivg@sci.utah.edu

O. Pasternak
Department of Psychiatry, Brigham and Women's Hospital Harvard Medical School,
1249 Boylston Street, Boston, MA 02215, USA
e-mail: ofer@bwh.harvard.edu

N. Sochen
Department of Applied Mathematics, Tel Aviv University, Ramat Aviv, Tel Aviv 69978, Israel
e-mail: sochen@post.tau.ac.il

L. Florack et al. (eds.), *Mathematical Methods for Signal and Image Analysis and Representation*, Computational Imaging and Vision 41,
DOI 10.1007/978-1-4471-2353-8_5,

lem of tensor-valued images of this type, and we demonstrate the proposed framework on DTI datasets and structure tensors. While in Chap. 3 an algebraic approach for tensors regularization is presented, this framework is based on Riemannian geometry which is the mathematical basis for various works on manipulating of diffusion tensors [29, 149, 311]. A completely different approach of tensor regularization for DTI smoothing is presented in Chap. 4. An alternative to the full tensor regularization approach we present in this chapter, is the method described in Chap. 10. This method applies to orientational information and hence can be used to regularize the white-matter fiber directions extracted from the estimated diffusion tensors.

Acquisition of high-resolution DT images generally leads to noisy images. In order to extract important features such as axon fibre bundles and to obtain smooth fibre tracts, the noise has to be removed. Therefore, regularization of Diffusion Tensor MRI data has been attracting much attention over the last few years (e.g., [72, 84, 131, 146, 312, 331, 332, 413, 428] and the book [438]). In DTI each image voxel is described in terms of a 3×3 symmetric positive-definite (SPD) diffusion tensor. The main challenge here is to perform a fast and efficient regularization process that preserves the properties of the tensors (i.e., symmetry and positive-definiteness), and respects the structure of the image.

In this chapter we tackle the tensors regularization problem by introducing a $GL(n)$-invariant Riemannian framework where the manifold of SPD tensors, P_n, is parameterized in terms of the *local* coordinates. Then, P_n is turned into a Riemannian manifold by the definition of natural $GL(n)$-invariant metric in terms of these coordinates. The local coordinate approach has two main advantages:

1. The numerics is fast and simple since it is free from matrix operations. In this method there are only operations between scalars to calculate (e.g., finite difference).
2. It is $GL(n)$-invariant (i.e., invariant under the action of any real invertible matrix of order n).

A proper choice of coordinate system to parameterize P_n is important both analytically and numerically. We show that the *Iwasawa* decomposition of SPD matrices yields a natural coordinate system. The choice of this coordinate system simplifies the analytical as well as the numerical calculations.

Based on the scale space representation discussed in Chap. 9, we define a product space via the mathematical notion of fibre bundles. The product space is composed of two spaces: the image domain (the base manifold), Ω, and the feature space of $n \times n$ symmetric positive-definite (SPD) matrices, P_n (the fibre). Using this concept a DT image may be described by the fibre bundle $\Omega \times P_3$ where Ω is in principal a two- or three-dimensional Euclidean space (DT slice or volume, respectively). A unique assignment of a feature space element (e.g., tensor) to each point of the base manifold (the image domain) is a *section* in the fibre bundle. By means of the Beltrami framework we define a functional over sections which is composed of a regularizer term and a data fitting term (fidelity term). The variation of this functional with respect to each one of the Iwasawa coordinates yields a set of $\frac{1}{2}n(n+1)$ coupled equations of motion (i.e., $n = 3$ for DTI). These equations of motion lead to a Beltrami flow on P_n via the gradient-descent method. Regularization of DTI data sets is performed by extracting the Iwasawa coordinates from the data and then solving this set of equations for these coordinates.

This chapter is organized as follows: In Sect. 5.2 we give a short introduction to fibre bundles and explain how a tensor-valued image is described via this concept. In Sect. 5.3 we discuss the basics of the Riemannian geometry of P_n. In this section we define a $GL(n)$-invariant metric that turns P_n into a Riemannian symmetric space. The metric over the sections in the fibre bundle is then defined in terms of this metric. In Sect. 5.4 we introduce a functional over sections by means of the Beltrami framework. Using calculus of variations we derive the equations of motion with respect to the coordinates on the section. The parametrization of P_n via the Iwasawa coordinates is discussed in Sect. 5.5. We show that for P_3 in particular we get six coupled Beltrami equations for the six Iwasawa coordinates. Finally, in Sect. 5.6 we present regularization of real volumetric DTI datasets as well as structure tensors.

5.2 Images as Fibred Space

Generally, an image may be described locally as the product of two spaces: The image domain (two- or three-dimensional usually) and the feature space which is composed of objects such as intensity, RGB values, orientation fields, etc. A product of two-spaces may be described mathematically via the concept of a fibre bundle. A fibre bundle is characterized by the total space E, the base manifold B, a structural group G (a group of homeomorphisms of the fibre F onto itself) together with a continuous surjective map $\pi : E \mapsto B$ which is called a projection. Since the dimension of E is higher than the dimension of B, the projection sends many points in E to one point in B. Locally, the total space E is described as product of the base manifold and the fibre: $E = B \times F$. Globally, this may not be the case since global features of the total space are not observed locally (e.g., the Möbius strip where the twist cannot be observed locally). When the identification of the total space with the direct product of the base space and the fibre is possible, this is known as a trivial bundle.

In computer vision the image domain is identified with the base manifold, and the feature space is identified with the fibre (see Fig. 5.1). The total space in this language is the *spatial-feature manifold*. For example, using this concept, a gray-level image is described locally by the product $\mathbb{R}^2 \times \mathbb{R}^+$. In this case the gray-levels are the fibre. Since an image selects just one gray-level for each pixel, we need the mathematical notion of a *section*.

A section of the bundle is a mapping $f : B \mapsto E$ such that $\pi(f(p)) = p$ for any $p \in B$. In our language, a given image is a section in a fibre bundle. For every pixel in the image domain we have a unique assignment of a feature space object. In this chapter we refer the section as the *image manifold*.

If we go back to DTI then we have a tensor-valued image where for each pixel there is unique assignment of a three-dimensional SPD matrix. Thus, a DT image is a section in the fibre bundle with the map $f : \mathbb{R}^m \mapsto \mathbb{R}^m \times P_3$, where $\mathbb{R}^m$ is in practice the two- or three-dimensional Euclidean image domain (DTI slice or

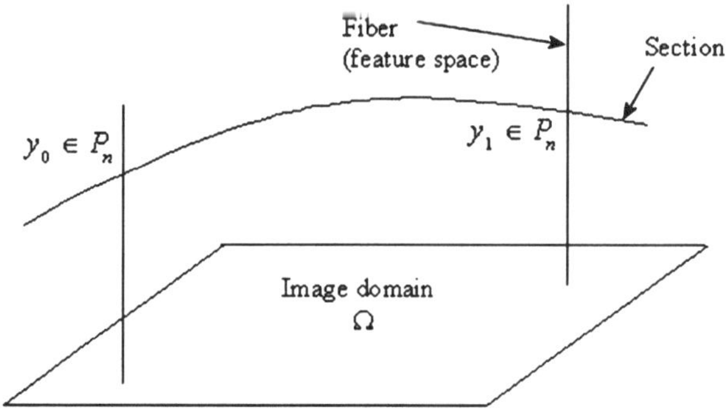

Fig. 5.1 Description of an image as a section of a fibre bundle. The base manifold B is the image domain Ω, the fibre is the feature space F (e.g., intensity values, color, DT). Here the image domain is Euclidean. Locally the bundle is described as a direct product $E = B \times F$ where E is the total space. In DTI the fibre is taken to be the space of 3×3 SPD matrices, P_3. A DT image is composed of a particular selection of an SPD matrix to each point of the base manifold. Thus, it is a section in the fibre bundle. The section is referred to here as the image manifold

volume, respectively) and P_3, which is the fibre, denotes the space of 3×3 SPD matrices. Thus, the spatial-feature manifold is described by the product $\mathbb{R}^m \times P_3$. The structural group in this case is simply $GL(3, \mathbb{R})$ which is the group of real invertible matrices of order 3. The action of $GL(3, \mathbb{R})$ on P_3 is given by $Y[g] = g^T Y g$ for $Y \in P_3$ and $g \in GL(3, \mathbb{R})$. This description is extendable for any n with the fibre-bundle $\mathbb{R}^m \times P_n$. In this case the structural group is $GL(n, \mathbb{R})$.

5.3 The Riemannian Geometry of P_n

For regularization of volumetric DTI data sets we take the image domain to be a three-dimensional Euclidean space. In order to write the metric over the spatial-feature manifold, the metric over the feature space P_n has to be defined. In this section we briefly review the important facts about the Riemannian geometry of P_n. Detailed discussions on this topic can be found in [245, 271].

The space of SPD matrices is a symmetric space of non-compact type [405]. Also, it was given in [271] as an example of a Riemannian manifold of non-positive curvature. It may be identified with an open cone in R^m where $m = n(n+1)/2$, i.e., for any $V, W \in P_n$ and for any positive scalar $c > 0$ we have $V + W \in P_n$ and $cV, cW \in P_n$. Since P_n is a connected manifold, according to the Hopf-Rinow theorem any geodesic segment can be extended indefinitely. Consequently, any two points in this space may be joined by a minimizing geodesic where its length is the geodesic distance between the two points. Moreover, it has been shown in [246] that there is precisely one geodesic connecting any two points on P_n. The geodesic distance between any two points $A, B \in P_n$ is given by [271]

$$d(A, B) = \left(\sum_{i=1}^{n} \log^2 \lambda_i \right)^{1/2}, \tag{5.1}$$

where λ_i are the eigenvalues of the matrix $A^{-1}B$. This geodesic distance is the so-called Riemannian distance which was used in [332] as a distance measure between two SPD tensors. With respect to this distance function, P_n is a geodesically complete Riemannian space.

Without getting into details, the tangent space at every point $Y \in P_n$ may be identified with the vector space of $n \times n$ symmetric matrices, SYM_n. Thus, the Riemannian metric at the point Y is defined in terms of the scalar product on SYM_n as in [271]

$$ds^2_{P_n} = \text{tr}((Y^{-1}dY)^2), \tag{5.2}$$

where $Y^{-1} = (y_{ij})^{-1}_{1\leq i,j\leq n}$ and $dY = (dy_{ij})_{1\leq i,j\leq n}$. This metric is by definition positive definite [271, 405]. Also, by defining the action of $g \in GL(n)$ on $Y \in P_n$ as $Y[g] = g^T Y g$, it can be easily shown that this metric is invariant under the action of $GL(n)$: Let $W = Y[g]$ where the differential is given by $dW = dY[g]$, then, upon plugging everything in $ds^2_{P_n}$ it follows that

$$ds^2_{P_n} = \text{tr}((Y^{-1}dY)^2) = \text{tr}((gW^{-1}g^T g^{-T} dWg^{-1})^2) = \text{tr}((W^{-1}dW)^2). \tag{5.3}$$

It can be easily verified that this metric is also invariant with respect to the inversion map $Y \mapsto Y^{-1}$. Thus, this map is an involutive isometry on P_n with respect to this metric. Therefore, this metric turns P_n into a Riemannian symmetric space. Another type of Riemannian symmetric spaces, namely compact Lie groups has been discussed by two of the authors in [196]. A bi-invariant metric of the same form as Eq. (5.2) has been used to formulate a coordinate-free regularization framework for Lie groups.

After the metric over P_n is introduced, the metric over the spatial-feature manifold $M = \mathbb{R}^3 \times P_n$ is simply given by $ds^2_M = \sum_{i=1}^3 dx_i^2 + \text{tr}((Y^{-1}dY)^2)$. Later we choose the coordinates to parameterize P_n.

5.4 Calculus of Variations in Fibre Bundles

Using functional analysis on sections in the fibre bundle we are able to derive a flow on the section. This will be done via the Beltrami framework (see [251, 383] and references therein). We now briefly review this framework's main ideas.

5.4.1 The Beltrami Framework

Denote by (Σ, γ) the image manifold and its metric and by (M, h) the embedding spatial-feature manifold (the fibre bundle) and its metric, then, the section of interest is expressed locally by the map $X : \Sigma \to M$. A functional over the space of sections is given in local coordinates by the following expression

$$S(X) = \int_{\Sigma} d^n x \sqrt{\gamma} \gamma^{\mu\nu} \frac{\partial X^i}{\partial x^\mu} \frac{\partial X^j}{\partial x^\nu} h_{ij}(X). \tag{5.4}$$

The case where $n = 2$ is known in the literature as the Polyakov action [338]. In this chapter we treat volumetric images (e.g., DTI volume) and hence $n = 3$. The integration is taken over the two- or three-dimensional image manifold where γ is the determinant of the image metric, $(\gamma^{\mu\nu})$ denotes the inverse of the image manifold's metric tensor and (h_{ij}) is the embedding space metric tensor (the spatial-feature fibre-bundle manifold). The coordinates in the spatial-feature space are denoted by X^i. The values of μ and ν range from 1 to $\dim \Sigma$ and the values of the i and j indices range from 1 to $\dim M$.

Using calculus of variations with respect to the embedding coordinates X^i, we obtain the Euler-Lagrange equations for the following action:

$$\begin{aligned} &\frac{1}{\sqrt{\gamma}} \partial_\mu (\sqrt{\gamma} \gamma^{\mu\nu} \partial_\nu X^i) + \Gamma^i_{jk} \gamma^{\mu\nu} \partial_\mu X^j \partial_\nu X^k \\ &\quad = \Delta_\gamma X^i + \Gamma^i_{jk} \langle \nabla X^j, \nabla X^k \rangle_\gamma = 0, \end{aligned} \tag{5.5}$$

where Δ_γ is the Laplace-Beltrami operator that is the generalization of the Laplacian to manifolds with metric γ. The solutions to these Euler-Lagrange equations are known as *harmonic maps*. By the gradient descent method we obtain a set of PDEs (i.e., Beltrami equations) with respect to the embedding coordinates. However, since the coordinates of the image domain are fixed, the interesting equations are for the coordinates of the fibre (P_n in our case). Hence,

$$X^i_t = \frac{1}{\sqrt{\gamma}} \partial_\mu (\sqrt{\gamma} \gamma^{\mu\nu} \partial_\nu X^i) + \Gamma^i_{jk} \gamma^{\mu\nu} \partial_\mu X^j \partial_\nu X^k, \tag{5.6}$$

where $i = 1, \ldots, (\dim(M) - \dim(\Sigma))$ and where Γ^i_{jk} are the Christoffel symbols. When the embedding space is Euclidean all the Christoffel symbols vanish. The Christoffel symbols are calculated with respect to (h_{ij}) as follows:

$$\Gamma^i_{jk} = \frac{1}{2} h^{il} (\partial_j h_{lk} + \partial_k h_{jl} - \partial_l h_{jk}). \tag{5.7}$$

The Beltrami equations are solved together at each iteration using standard finite-differences schemes. The components of the induced metric and the Christoffel symbols are evaluated at each iteration. An important result of this framework is that no constraint on the positive definiteness of the matrices is needed. Indeed the elipticity of the Laplace-Beltrami operator that generates the flow implies the validity of the extremum principle. This in turns mean that *the positive definiteness of the initial condition is a necessary and sufficient condition for the flow to stay on* P_n.

5.4.1.1 The Induced Metric

The metric over the image manifold (or section) is induced from the fibre bundle's metric. In this way the flow depends on the geometry of the data and not only on the geometry of the image domain. Moreover, the induced metric is a dynamic variable which changes along the flow.

We assume an isometric embedding, e.g., infinitesimal distances on the spatial-feature manifold, M, are equal infinitesimal distances on the image manifold, Σ. This assumption yields the pullback metric induced by X: $\gamma = X^*h$. Its components are given by

$$\gamma_{\mu\nu}(x) = \frac{\partial X^i}{\partial x_\mu}\frac{\partial X^j}{\partial x_\nu} h_{ij}(X). \tag{5.8}$$

Thus, the induced metric is actually calculated via the chain rule. Consequently, for a DTI volume where $M = \mathbb{R}^3 \times P_3$ we may write $dY = \sum_{i=1}^{3} \frac{\partial Y}{\partial x_i} dx_i$ where $Y \in P_3$. Then, we plug this expression into the metric on M to get the components of the induced metric:

$$\gamma_{\mu\nu} = \delta_{\mu\nu} + \beta\,\mathrm{tr}(Y^{-1}\partial_\mu Y Y^{-1}\partial_\nu Y), \tag{5.9}$$

where $\mu, \nu = 1, \ldots, 3$ are the indices of the local coordinates on the image manifold. The β parameter is introduced here in order to determine the ratio between the feature space and the image domain distances. It also determines the nature of the flow. In the limit $\beta \to 0$ the Laplace-Beltrami operator reduces to the usual Laplacian and hence we obtain a linear flow. Small β values lead to a multi-channel total variation (TV) norm [43] while large β values lead to potential surfaces [388].

In matrix form the induced metric is given by

$$(\gamma_{\mu\nu}) = \begin{pmatrix} 1+\beta\,\mathrm{tr}((Y^{-1}Y_{x_1})^2) & \beta\,\mathrm{tr}((Y^{-1}Y_{x_1})(Y^{-1}Y_{x_2})) & \beta\,\mathrm{tr}((Y^{-1}Y_{x_1})(Y^{-1}Y_{x_3})) \\ \beta\,\mathrm{tr}((Y^{-1}Y_{x_1})(Y^{-1}Y_{x_2})) & 1+\beta\,\mathrm{tr}((Y^{-1}Y_{x_2})^2) & \beta\,\mathrm{tr}((Y^{-1}Y_{x_2})(Y^{-1}Y_{x_3})) \\ \beta\,\mathrm{tr}((Y^{-1}Y_{x_1})(Y^{-1}Y_{x_3})) & \beta\,\mathrm{tr}((Y^{-1}Y_{x_2})(Y^{-1}Y_{x_3})) & 1+\beta\,\mathrm{tr}((Y^{-1}Y_{x_3})^2) \end{pmatrix}. \tag{5.10}$$

5.5 Iwasawa Coordinate-Based Formalism

In Sect. 5.4.1.1 we have defined the induced metric. Now we would like to express this metric in terms of the coordinates on P_3, explicitly.

The P_3 space is identified with an open cone in R^6. The 6 different entries of the symmetric matrix may be identified with the Cartesian coordinates for R^6. Therefore, it is straightforward to parameterize P_3 using these coordinates. However, taking the trace of $(Y^{-1}dY)^2$ where $Y = (y_{ij})_{1\le i,j\le 3}$ and $dY = (dy_{ij})_{1\le i,j\le 3}$, one gets long and cumbersome expressions with complicated terms involving the y_i's in the denominator. Hence, the metric tensor is cumbersome and there are 78 Christoffel symbols associated with it. As a result, the numerical implementation is problematic.

An important issue in analysis on manifolds is the right choice of coordinate system for the problem. We may try to choose a different coordinate system in order to simplify the analytical calculations as well as the numerics.

Another candidate is the coordinate system associated with the polar decomposition. However, as we have mentioned in the introduction this coordinate system is problematic because of the non-uniqueness of the polar decomposition. Luckily, there is a another set of coordinates called Iwasawa coordinates which corresponds to the Iwasawa decomposition of symmetric matrices. There are partial, as well as, full Iwasawa decompositions [245, 405]. The full Iwasawa decomposition is unique and obtained by applying repeatedly the partial Iwasawa decomposition to the matrices $Y \in P_n$. Surprisingly, the Iwasawa coordinates turn out to be the natural parametrization on P_n from analytical as well as from numerical considerations. Here we use the term "Iwasawa decomposition" to describe the full Iwasawa decomposition.

5.5.1 The Iwasawa Decomposition

Iwasawa has proved that every connected semi-simple Lie group G admits a unique representation as a product $G = KAN$ of an orthogonal subgroup K, an Abelian subgroup A and a nilpotent subgroup N [360]. In particular, for every invertible real matrix of order n, $G = GL(n, \mathbb{R})$, K is the orthogonal group $K = O(n)$, A is a positive diagonal matrix and N is a strictly upper-triangular matrix with ones on its diagonal.

However, the space of SPD matrices, P_n, is not a Lie-group but it is identified with the quotient space $GL(n, \mathbb{R})/O(n)$. Therefore, any $Y \in P_n$ may be identified with an Iwasawa decomposition of the form $Y = AN$. In this case the operation between A and N is not an ordinary matrix multiplication. The homeomorphism $GL(n, \mathbb{R})/O(n) \to P_n$ is given by the operation $g^T g$, $\forall g \in GL(n, \mathbb{R})$. Hence, any $Y \in P_n$ may be decomposed uniquely as follows: $Y = N^T AN$ [245, 405].

Calculating the metric over P_n with respect to the Iwasawa coordinates is done by a substitution of the Iwasawa decomposition into $ds^2_{P_n}$. This yields

$$ds^2_{P_n} = \mathrm{tr}((Y^{-1}dY)^2) = \mathrm{tr}((A^{-1}dA)^2) + 2\,\mathrm{tr}(A^{-1}A[dNN^{-1}]), \tag{5.11}$$

where the second term plays an important role here. It is the source for the coupling between the Iwasawa coordinates.

Let us now derive the metric tensor for the spatial-feature manifolds we explore in this chapter.

5.5.2 Structure Tensors

Structure Tensor (ST) is a source of information on image texture [131]. It is very useful for corner detection, texture segmentation [162, 354], and motion analysis [37].

Generally, for a two-dimensional gray-level image the structure tensor takes the form:

$$Y = \nabla I \nabla I^T = \begin{pmatrix} I_{x_1}^2 & I_{x_1} I_{x_2} \\ I_{x_1} I_{x_2} & I_{x_2}^2 \end{pmatrix}. \tag{5.12}$$

The two eigenvalues of this tensor are: $\lambda_1 = 0$, $\lambda_2 = |\nabla I|^2$, hence, it is positive semi-definite. To avoid noise effects, the image $I(x)$ is first convolved with a Gaussian kernel k_σ: $I_\sigma(x) = (k_\sigma * I)$. Then, each tensor is convolved componentwise with a Gaussian kernel k_ρ [433], and the structure tensor takes the form:

$$Y_\rho(\nabla I_\sigma) = k_\rho * \nabla I_\sigma (\nabla I_\sigma)^T. \tag{5.13}$$

This convolved tensor is symmetric and positive-definite, and its eigenvalues are:

$$\begin{aligned} \lambda_1 &= \frac{1}{2}\left(y_{11} + y_{22} + \sqrt{(y_{11} - y_{22})^2 + 4y_{12}^2}\right), \\ \lambda_2 &= \frac{1}{2}\left(y_{11} + y_{22} - \sqrt{(y_{11} - y_{22})^2 + 4y_{12}^2}\right), \end{aligned} \tag{5.14}$$

where y_{ij} are the components of the convolved tensor.

A 2×2 structure tensor may be parameterized using three Iwasawa coordinates. The Iwasawa decomposition in this case is simply given by

$$Y = \begin{pmatrix} 1 & 0 \\ w_3 & 1 \end{pmatrix} \begin{pmatrix} w_1 & 0 \\ 0 & w_2 \end{pmatrix} \begin{pmatrix} 1 & w_3 \\ 0 & 1 \end{pmatrix}, \quad Y \in P_2, \tag{5.15}$$

where $w_1, \ldots, w_3$ are identified with the Iwasawa coordinates.

The metric tensor of the spatial-feature manifold $M = \mathbb{R}^2 \times P_2$ is calculated directly using Eq. (5.8) and is given by

$$(h_{ij}) = \begin{pmatrix} 1 & 0 & 0 & 0 & 0 \\ 0 & 1 & 0 & 0 & 0 \\ 0 & 0 & \frac{1}{w_1^2} & 0 & 0 \\ 0 & 0 & 0 & \frac{1}{w_2^2} & 0 \\ 0 & 0 & 0 & 0 & \frac{2w_1}{w_2} \end{pmatrix}, \tag{5.16}$$

where,

$$ds_M^2 = (dx_1\, dx_2\, dw_1\, dw_2\, dw_3)(h_{ij}) \begin{pmatrix} dx_1 \\ dx_2 \\ dw_1 \\ dw_2 \\ dw_3 \end{pmatrix}. \tag{5.17}$$

With respect to this metric tensor we calculate the Christoffel symbols and the induced metric. Finally, regularization is performed by solving a system of three coupled Beltrami equations for the Iwasawa coordinates with initial conditions $X^i(t = 0) = w_i^0$, $i = 1, \ldots, 3$.

5.5.3 Diffusion Tensors

The diffusion process of water molecules in the brain may be modeled by a diffusion tensor. The diffusion tensor is a 3×3 SPD matrix. Let us write the Iwasawa decomposition for a SPD matrix:

$$
\begin{aligned}
Y &= A[N] = N^T A N \\
&= \begin{pmatrix} 1 & 0 & 0 \\ w_4 & 1 & 0 \\ w_5 & w_6 & 1 \end{pmatrix} \begin{pmatrix} w_1 & 0 & 0 \\ 0 & w_2 & 0 \\ 0 & 0 & w_3 \end{pmatrix} \begin{pmatrix} 1 & w_4 & w_5 \\ 0 & 1 & w_6 \\ 0 & 0 & 1 \end{pmatrix},
\end{aligned}
\tag{5.18}
$$

where in this case we identify $w_1, \ldots, w_6$ with the Iwasawa coordinates.

The metric tensor for the Iwasawa decomposition is obtained by using Eq. (5.11)

$$
(h_{ij}) = \begin{pmatrix}
1 & 0 & 0 & 0 & 0 & 0 & 0 & 0 & 0 \\
0 & 1 & 0 & 0 & 0 & 0 & 0 & 0 & 0 \\
0 & 0 & 1 & 0 & 0 & 0 & 0 & 0 & 0 \\
0 & 0 & 0 & \frac{1}{w_1^2} & 0 & 0 & 0 & 0 & 0 \\
0 & 0 & 0 & 0 & \frac{1}{w_2^2} & 0 & 0 & 0 & 0 \\
0 & 0 & 0 & 0 & 0 & \frac{1}{w_3^2} & 0 & 0 & 0 \\
0 & 0 & 0 & 0 & 0 & 0 & \frac{2w_1(w_3+w_2w_6^2)}{w_2w_3} & -\frac{2w_1w_6}{w_3} & 0 \\
0 & 0 & 0 & 0 & 0 & 0 & -\frac{2w_1w_6}{w_3} & \frac{2w_1}{w_3} & 0 \\
0 & 0 & 0 & 0 & 0 & 0 & 0 & 0 & \frac{2w_2}{w_3}
\end{pmatrix}.
\tag{5.19}
$$

The positive-definiteness of Y implies the constraint $w_1, w_2, w_3 > 0$. Hence, the components of the metric tensor cannot be singular.

There are only 26 non-trivial Christoffel symbols associated with this metric tensor. With respect to this metric tensor, the induced metric is calculated easily using Eq. (5.8) and the following set of six coupled Beltrami equations for the Iwasawa coordinates is constructed:

$$
\begin{aligned}
X_t^i &= \Delta_\gamma X^i + \Gamma^i_{jk} \langle \nabla X^j, \nabla X^k \rangle_\gamma = 0, \\
X^i_{t=0} &= w_i(t=0), \quad i = 1, \ldots, 6.
\end{aligned}
\tag{5.20}
$$

5.5.4 DTI Data Fitting Term

The coupled equations in (5.6) correspond to the *smoothing* term. However, to keep the regularized data fitting best to the original one we add a physical data fitting term to the functional in Eq. (5.4). This results in the following cost functional:

$$L(X) = \frac{\alpha}{2}\int_{\Sigma} d^3x \sum_{k=1}^{N}\left(e^{-bq_k^T D q_k} - \hat{E}(q_k)\right)^2 + \frac{1}{2}\int_{\Sigma} d^3x \sqrt{\gamma}\gamma^{\mu\nu}\partial_\mu X^i \partial_\nu X^j h_{ij}(X), \tag{5.21}$$

where $i = 1,\dots,6$. The term $e^{-bq_k^T D q_k}$ is the signal attenuation model [26] where D is the diffusion tensor, q_k is the k'th applied magnetic field gradient direction and b is a parameter of the measurement. The measured (noisy) signal is denoted by $\hat{E}(q_k)$.

The variation of this functional with respect to the Iwasawa coordinates yields the following equations of motion

$$-\alpha b \frac{1}{\sqrt{\gamma}} \sum_{k=1}^{N} (e^{-bq_k^T D q_k} - \hat{E}(q_k)) e^{-bq_k^T D q_k}\left(q_k^T \frac{\partial D}{\partial X^i} q_k\right) + \frac{1}{\sqrt{\gamma}}\partial_\mu(\sqrt{\gamma}\gamma^{\mu\nu}\partial_\nu X^i) + \Gamma^i_{jk}\gamma^{\mu\nu}\partial_\mu X^j \partial_\nu X^k = 0. \tag{5.22}$$

Then, the gradient descent equations are given by

$$X^i_t = -\alpha b \frac{1}{\sqrt{\gamma}} \sum_{k=1}^{N} (e^{-bq_k^T D q_k} - \hat{E}(q_k)) e^{-bq_k^T D q_k}\left(q_k^T \frac{\partial D}{\partial X^i} q_k\right) + \frac{1}{\sqrt{\gamma}}\partial_\mu(\sqrt{\gamma}\gamma^{\mu\nu}\partial_\nu X^i) + \Gamma^i_{jk}\gamma^{\mu\nu}\partial_\mu X^j \partial_\nu X^k. \tag{5.23}$$

The derivatives of D with respect to the Iwasawa coordinates are calculated as follows:

$$D = \begin{pmatrix} 1 & 0 & 0 \\ w_4 & 1 & 0 \\ w_5 & w_6 & 1 \end{pmatrix}\begin{pmatrix} w_1 & 0 & 0 \\ 0 & w_2 & 0 \\ 0 & 0 & w_3 \end{pmatrix}\begin{pmatrix} 1 & w_4 & w_5 \\ 0 & 1 & w_6 \\ 0 & 0 & 1 \end{pmatrix} = \begin{pmatrix} w_1 & w_1 w_4 & w_1 w_5 \\ w_1 w_4 & w_2 + w_1 w_4^2 & w_1 w_4 w_5 + w_2 w_6 \\ w_1 w_5 & w_1 w_4 w_5 + w_2 w_6 & w_3 + w_1 w_5^2 + w_2 w_6^2 \end{pmatrix}. \tag{5.24}$$

Thus, for example,

$$\frac{\partial D}{\partial w_1} = \begin{pmatrix} 1 & w_4 & w_5 \\ w_4 & w_4^2 & w_4 w_5 \\ w_5 & w_4 w_5 & w_5^2 \end{pmatrix}. \tag{5.25}$$

5.6 Experimental Results

5.6.1 Structure Tensor Regularization

Experimental results of the structure tensor regularization are presented in Fig. 5.2. The structure tensor has been extracted from a given image, as well as, from its noisy version (Gaussian noise with $\sigma = 0.01$). Then we applied our algorithm directly to the structure tensors while working on their Iwasawa coordinates. Thus, regularization is achieved by solving a set of three coupled Beltrami equations with respect to these coordinates. The regularization results are shown for 150 iterations. Most of the noise has been removed while the edges that has been extracted by taking the trace of the tensor, have been preserved. In Fig. 5.3 we present regularization of various Brodatz textures. In this case we compare the linear and the nonlinear case. In the nonlinear case, the sharp edges are preserved as well as the borders between the different textures. Some edges that are less sharp, as in the central texture, were smoothed. As expected, in the linear case blurring occurs.

5.6.2 DTI Regularization

The regularization framework was tested on in vivo, human noisy data that was acquired on a GE Signa 3T. A DW-EPI sequence was used with the following parameters: $\Delta/\delta = 31/25$ ms, a 22 cm FOV, matrix size of 128×128 and 1.7 mm slice thickness with 72 slices covering the entire brain. The TE was 88 ms and the gradient strength was 4 G/cm, resulting in a b value of 1000 s/mm^2 measured in 33 non-collinear gradient directions. In addition, a single non-weighted b_0 image was acquired. The sequence was repeated 4 times. In order to avoid intrinsic pulsative brain motion artifacts, the sequence was gated to the cardiac cycle with effective TR of 30 R-R intervals. The MRI protocol was approved by the local IRB committee, and informed consent was obtained from a healthy volunteer.

5.6.3 Data Analysis

The collected volumes were first corrected for patient motion by rigid-body spatial transformation (using SPM2) accompanied with gradient-orientation compensation. A noisy tensor field was then generated by applying DTI on a single repetition of the DW-EPI sequence. Then, it was regularized by the proposed framework (50 iterations with fidelity weight $\alpha = 1$ and $\beta = 1$). Finally, the regularized tensors dataset was compared with the initial noisy data set. Also, it was compared with the mean of the four repetitions data set. In order to use the noisy tensor field as an initial guess for the Beltrami framework minimization, the attenuation signal of any voxel with negative eigenvalue was replaced with the mean signal of its positive neighbors

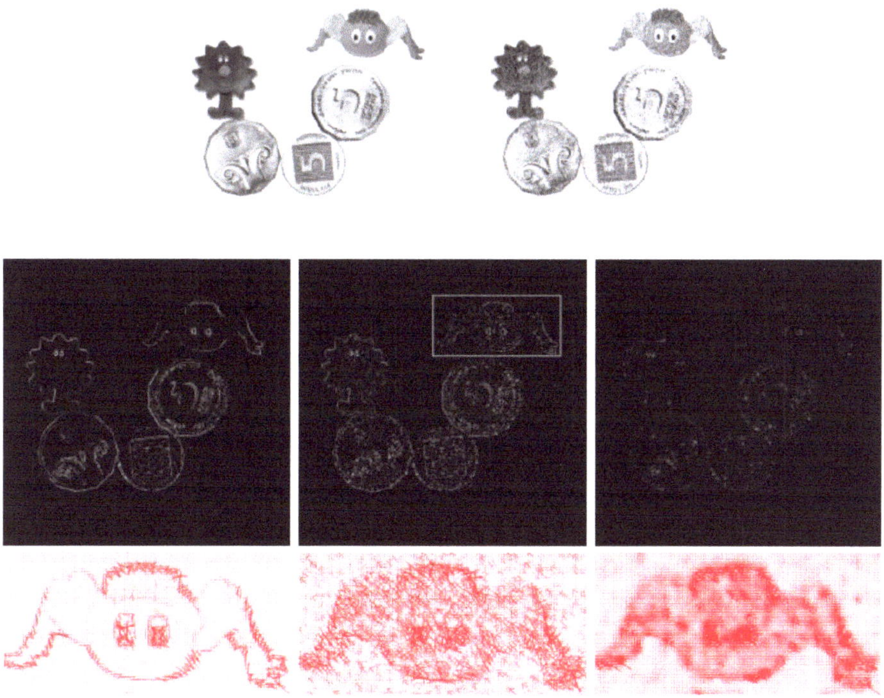

Fig. 5.2 Regularization of structure tensor. Original image and noisy image (*top row*, *left to right*). The trace of the structure tensor for the original image, the noisy image and the regularized image (*middle row*, *left to right*). The structure tensors of the marked region in ellipses representation for the three images. The axes of the ellipses are proportional to the eigenvalues of the matrix whereas the orientations of the ellipses are determined by the eigenvectors

(if there were any, otherwise it was omitted). Visualization of the tensor fields was generated by spectral decomposition of the three different tensor fields for the rendering of fractional anisotropy (FA) maps [25], RGB color schemes [327], and three dimensional tractography objects. The latter were acquired by brute-force streamline tractography [314] overlayed on SPGR anatomical images, co-registered with a b_0 diffusion image. A FA threshold of 0.2 was used for fibre initialization and termination.

5.6.4 Comparison Between the Three Cases

The high voxel resolution yielded low-grade tensor images for the single repetition data set (Fig. 5.4, left column): The noise is easily visible in all parts of the image, including in deep white matter structures such as the genu of the corpus callosum, and especially in lateral brain areas where it is difficult to recognize small fibre bun-

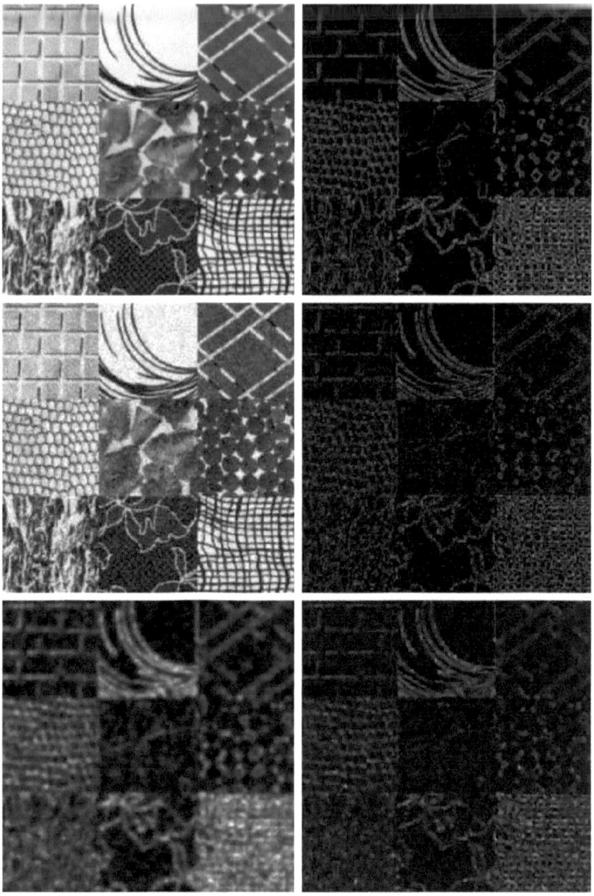

Fig. 5.3 Regularization of Brodatz textures. Original image and the trace of the structure tensor (*top row*, *left to right*). *Middle row*: Gaussian noise of variance 0.01 was added and the structure tensors of the noisy image were extracted. *Bottom row*: Linear diffusion with $\beta = 0$, 150 iterations (*left*) and nonlinear diffusion with $\beta = 5$, 150 iterations (*right*)

dles. This noise effect is less noticeable in the 4-repetitions data set (Fig. 5.4, right column). In this case, finer fibres at the intersection with gray matter become visible. Therefore, we may expect that the outcome of the regularization process are images that are more similar to the 4-repetitions images. This is indeed the case (Fig. 5.4, middle column): The noise level of the regularized images have been improved significantly comparing to the original, single repetition images. Anatomically expected homogeneous fibre areas appear smoother with much less gray level value changes, and the images become similar to the 4-repetitions images. While the FA maps demonstrate that the eigenvalues have been regularized, we can further see in the color-coded schemes (Fig. 5.5) that the Beltrami regularization has also regularized the principal orientations of the tensors. The colors better represent different fibre bundles, and demonstrate again that the regularized images are not only smoother but also preserve the edges between these bundles.

Very similar structures appear in the 4-repetitions color schemes, reassuring the assumption of piece-wise smooth organization of fibres, and supporting the regularized results. The difference between the regularized and the 4-repetitions images is

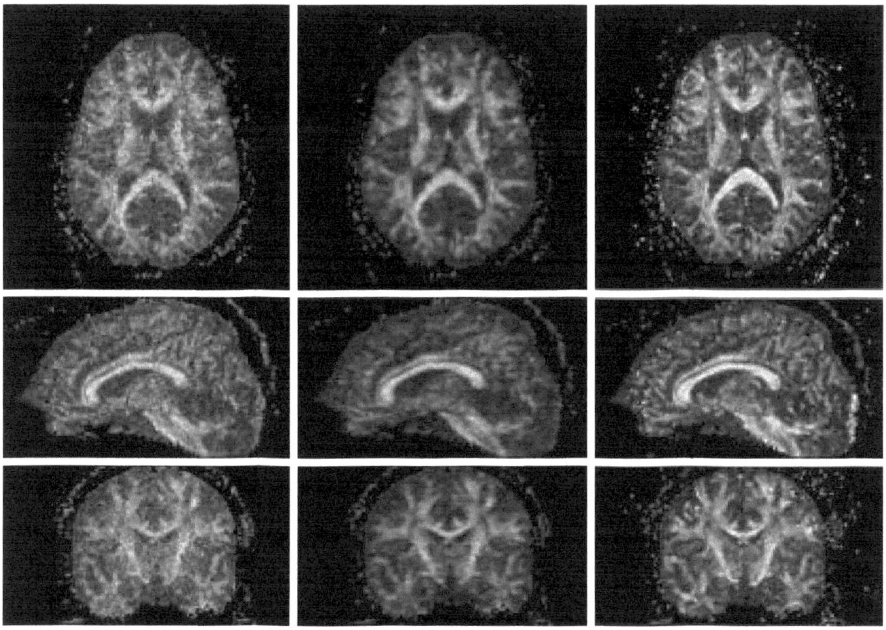

(a) Noisy (b) Beltrami (c) 4-repetitions

Fig. 5.4 FA images. Presented are mid-Axial (*top row*), mid-Sagittal (*middle row*) and mid-Coronal (*bottom row*) slices. The Beltrami regularized maps (*middle column*) show significantly lower noise level compared to the original noisy images (*left column*). The regularized images resemble the high SNR images obtained by 4-repetitions (*right column*)

the noise level where the regularized image seems to have a better one. However, the image edges of the 4-repetitions images are sharper.

Further visualization and validation of the regularized results is obtained from tractography images. Identical ROIs were chosen for the three data sets to acquire streamline tractography of the cingulum. The ROIs were placed on two coronal slices, posterior and anterior to the mid-coronal slice, on anatomical landmarks identified with the cingulum. The streamlines generated had to pass through both ROIs. The cingulum is a fibre bundle passing between parts of the limbic system and is wrapped around the corpus callosum. It connects the anterior part of the cingulate gyrus (aCG), which is in proximity to the genu of the corpus callosum (gcc), passing superior to the body of the corpus callosum (bcc) and ends at the posterior part of the cingulate gyrus (pCG) next to the splenium of the corpus callosum (scc). From the posterior part of the cingulate gyrus it projects to the amygdala and hippocampus complex.

In Fig. 5.6 we can see the trajectories obtained for the three data sets. The three images were able to delineate the complete tract from the anterior to the posterior cingulate gyrus. Interestingly, the noisy data set had also delineated the projection to the amygdala and hippocampus, where the 4-repetitions delineation was terminated at the posterior cingulate gyrus. As for the regularized data set, it had also obtained

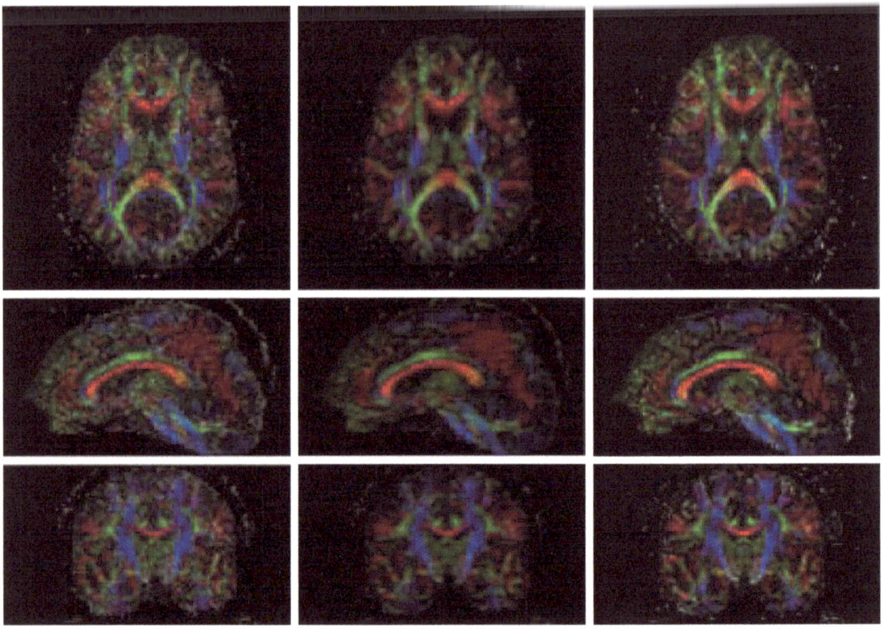

(a) Noisy (b) Beltrami (c) 4-repetitions

Fig. 5.5 Colorcoded images. The same slices as in Fig. 5.4 coded by the tensor's principal orientation. Orientations were regularized as well, while important edges between different fibre bundles were preserved

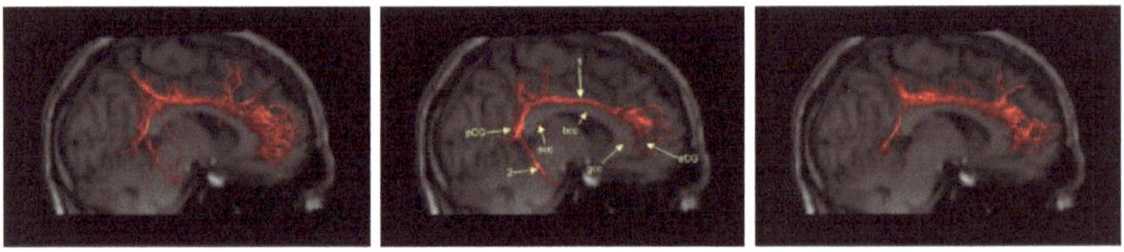

(a) Noisy (b) Beltrami (c) 4-repetitions

Fig. 5.6 The cingulum. In the three images (the noisy, regularized and the 4-repetitions) the cingulum between the aCG to the pCG has been delineated. In the regularized image the projection to the amygdala and hippocampus was also obtained (marked by 2). The regularized tract has much less spurious fibres and is smoother than the noisy tract

the complete expected trajectory. Further comparing the three images reveals that the regularized data set provided a smooth and robust shape of fibres, where the noisy data set presents many spurious fibres that seem as noise. Observing a zoomed perspective of the same images (Fig. 5.7), reveals much finer differences between the images. Here it is easy to notice that the regularized data set provided much more continuous and smooth trajectories than in the noisy data set. In the noisy data set the effect of noise is seen as sharp turns along the fibre as well as discontinuities

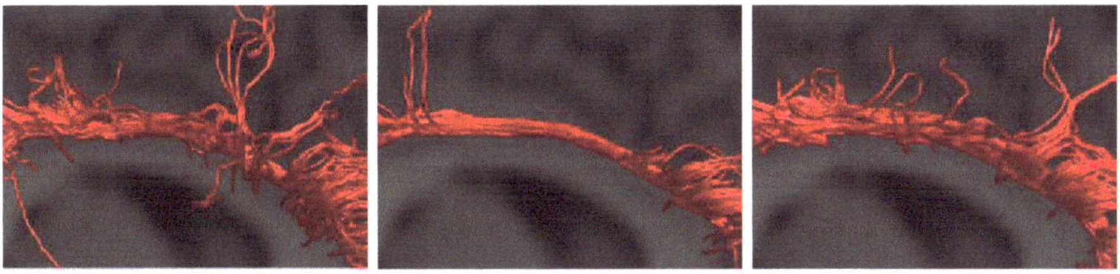

(a) Noisy (b) Beltrami (c) 4-repetitions

Fig. 5.7 Cingulum zoomed. Area 1 in Fig. 5.6(b) is zoomed. This reveals that the regularized fibre orientations change smoothly along neighboring voxels, unlike sharp edges and discontinuities in the noisy dataset

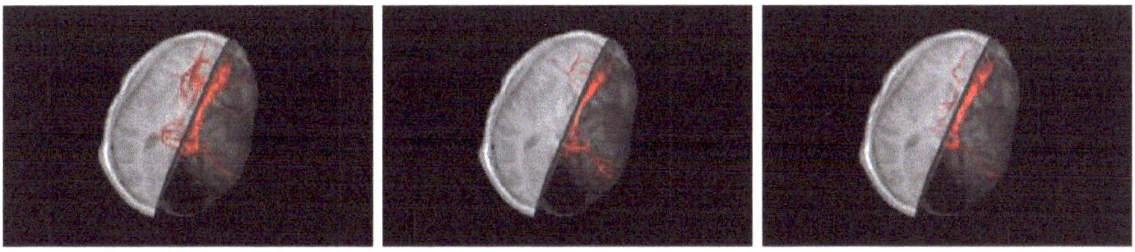

(a) Noisy (b) Beltrami (c) 4-repetitions

Fig. 5.8 Deviation to the corpus callosum. Since the cingulum is very close to the corpus callosum, small deviations caused by noise may switch the trajectory to the corpus callosum trajectory. Any fibre that crosses to the opposite hemisphere is false positive. The regularized image reduced dramatically the number of false positive fibres obtained in the original data. Similar results were obtained in the 4-repetitions data

along the tract. Comparing the regularized result with the 4-repetitions result shows that the regularized trajectories seems smoother than the 4-repetitions ones.

Since in most of its trajectory the cingulum is very close to the corpus callosum, small orientational noise is expected to cause deviations that might cross to the neighboring bundle and follow it to the other hemisphere. The two-ROIs method used here is supposed to eliminate most of these fibres. However, it is limited to the area that is between these ROIs. In areas posterior and anterior to the chosen ROIs there is still the possibility of noise causing deviation on the corpus callosum. Observing the trajectory from a different viewing angle (Fig. 5.8) demonstrates clearly some of the false positives fibres that cross to the other hemisphere. This visualization shows us that indeed in the noisy data set there are many fibres that deviate on the corpus callosum tract, where both in the regularized and the 4-repetitions data sets the number of such fibres is dramatically decreased.

As a conclusion from the tractography of the cingulum fibre bundle, the regularization process provides smoother, more robust, and less false positive fibres than in the original noisy data set. Also, it preserves the anatomical shape of the fibre and increases its separation from other fibre bundles.

5.7 Conclusion

In this chapter we have presented a novel geometric framework for regularization of data in P_n, the space of SPD matrices of order n. The basis of this framework is the description of P_n as a Riemannian manifold. We have shown that parametrization of P_n in terms of the *Iwasawa* coordinates simplifies the analytical and the numerical calculations. Then, regularization in this framework is achieved by solving a set of coupled Beltrami equations for these coordinates. We have exemplified this framework on real DTI datasets and structure tensors.

Chapter 6
Sparse Representation of Video Data by Adaptive Tetrahedralizations

Laurent Demaret, Armin Iske, and Wahid Khachabi

Abstract Natural videos are composed of a superposition of moving objects, usually resulting from anisotropic motions into different directions. By discretization with respect to time, a video may be regarded as a sequence of consecutive natural still images. Alternatively, when considering time as one dimension, a video may be viewed as a 3d scalar field. In this case, customized methods are needed for capturing both the evolution of moving contours along the time axis and the geometrical distortions of the resulting sweep surfaces. Moreover, it is desirable to work with sparse representations. Indeed, already for basic motions (e.g. rotations, translations), customized methods for the construction of well-adapted sparse video data representations are required. To this end, we propose a novel adaptive approximation algorithm for video data. The utilized nonlinear approximation scheme is based on anisotropic tetrahedralizations of the 3d video domain, whose tetrahedra are adapted locally in space (for contour-like singularities) and locally in time (for anisotropic motions). The key ingredients of our approximation method, **3AT**, are adaptive thinning, a recursive pixel removal scheme, and least squares approximation by linear splines over anisotropic tetrahedralizations. The approximation algorithm **3AT** yields a new concept for the compression of video data. We apply the proposed approximation method first to prototypical geometrical motions, before numerical simulations concerning one natural video are presented.

L. Demaret (✉)
HelmholtzZentrum München, Institut für Biomathematik und Biometrie (IBB), 85764 Neuherberg, Germany
e-mail: laurent.demaret@helmholtz-muenchen.de

A. Iske · W. Khachabi
Department of Mathematics, University of Hamburg, 20146 Hamburg, Germany

A. Iske
e-mail: iske@math.uni-hamburg.de

W. Khachabi
e-mail: khachabi@gmail.com

L. Florack et al. (eds.), *Mathematical Methods for Signal and Image Analysis and Representation*, Computational Imaging and Vision 41,
DOI 10.1007/978-1-4471-2353-8_6,

6.1 Introduction

The production of digital video sequences is usually resulting from different technical processes involving data acquisition, projections of 3d moving objects on a camera matrix, and spatiotemporal digitalization. Natural videos typically display a superposition of anisotropic motions of different geometrical objects into different spatial directions. Simultaneous and independent motions lead to rapidly developing occlusions or disocclusions of background objects by foreground objects [7]. Therefore, the efficient representation of digital video data is a very challenging task, which in particular requires customized computational methods to cope with the very large data complexity.

When discretizing the 3d data with respect to time, a video may be regarded as a sequence of consecutive natural still images, *frames*. In the case of still images, relevant information is given by sharp contours between neighbouring objects. Consequently, in natural videos the evolution of such contours between their adjacent objects plays an important role for the human visual perception. Therefore, it is a crucial task in video processing to perform accurate representations for the moving contours.

When considering time as one (spatial) dimension, a video may also be viewed as a 3d scalar field. In this case, sweep surfaces are resulting from moving contour lines along the time axis. Despite the intrinsic three-dimensional structure of videos, commonly used video-codecs, including the popular standard method MPEG4-H264 [301], work with a still image compression method (for selected *intra-frames*) coupled with a block-oriented motion compensation (for the prediction of *inter-frames*). This strategy, however, often leads to difficulties in the following relevant situations with videos containing

- rigid motions which are not pure translations;
- object deformations (e.g. non-rigid motions by non-uniform scalings);
- object occlusions/disocclusions.

In such cases, much of the coding energy is spent on information corresponding to the compensation of the prediction error. The required information is coded by using a block-based DCT (Discrete Cosine Transform) for the inter-frames as well as for the predicted error compensation. This leads (especially at low bit rates) to strong block artifacts, partly due to discontinuities at the block boundaries, and partly due to typical Gibbs effects of the Fourier analysis method DCT. More recently, alternative wavelet-based methods were proposed to avoid strong block artifacts [308].

Alternative methods for video coding are using triangular meshes [7, 281] to handle the underlying motion field. In these more flexible coding methods, an initial triangulation is designed for the first frame. The initial triangulation is then dynamically updated according to the changes between two consecutive frames. In [7], Altunbasak and Tekalp proposed an occlusion-adaptive and content-based mesh design in combination with a forward tracking procedure. This aims at handling occlusions/disocclusions by the insertion of new triangles or by the removal of old triangles according to the motion estimation. However, coding the corresponding motion vectors is a non-trivial problem. Moreover, in this case additional information is required to indicate covered areas (for occlusions) and uncovered areas (for disocclusions).

In this contribution, we propose a novel concept for adaptive approximation and sparse representation of video data. To this end, we regard the video as a 3d scalar field, where time is taken as one (spatial) dimension. Therefore, we essentially refrain from splitting the video data into separate consecutive image frames, unlike the above mentioned methods. We remark that our interpretation of the 3d video data does not require any sophisticated methods for explicit motion compensation, forward tracking or detection of occlusions. This gives us more flexibility in the data analysis, and moreover it reduces the computational overhead required for the maintenance of various updates between consecutive image frames.

In our approach, the video is viewed as a trivariate function, given by its discrete (grey-scale) values taken at the 3d locations of the video pixels. We approximate the video by a linear spline over an adaptive tetrahedralization of the video domain. To this end, a sparse set of *significant* pixels is first adaptively chosen according to a recursive point removal scheme, adaptive thinning, such that the significant pixels capture the local motion of geometrical components of the video data. The significant pixels define a unique Delaunay tetrahedralization of the video domain, whose tetrahedra are well-adapted *locally* in space (to capture contour-like singularities) and locally in time (to capture anisotropic and irregular motions).

The Delaunay tetrahedralization yields a unique linear spline space for approximation, containing all continuous functions which are piecewise linear over the Delaunay tetrahedralization. From this spline space, we select the unique best approximation in the sense of least squares. The approximating linear spline is a *continuous* function, which can be evaluated at any point in the video domain, in particular at the *discrete* set of pixels. This allows us to reconstruct the entire video data by evaluation of the approximating linear spline at the video pixels. Note that our specific representation of the video (by a continuous function) allows us to display the reconstructed video at any subset of the (continuous) video domain.

The outline of this article is as follows. In Sect. 6.2, we discuss video data representations and Delaunay tetrahedralizations. Moreover, we show some selected prototypical motions (generated by rotations) for videos. Then, in Sect. 6.3 our video approximation scheme is explained in detail, before we discuss important computational aspects concerning its efficient implementation in Sect. 6.4. Numerical simulations are finally presented in Sect. 6.5.

6.2 Videos, Prototypical Motions, and Tetrahedralizations

In video sequences, variations in time are due to displacements of 3d objects projected on the video spatial domain. Many displacements can be described by elements of the Euclidean motion group. The action of Euclidean motions on the object contours produces moving surfaces. Diffusion on the 3d Euclidean motion group for the enhancement of crossing elongated structures is studied in [168].

In this section, Delaunay tetrahedralizations are introduced. Moreover, the utility of tetrahedralizations for relevant tasks of *object occlusions* and *disocclusions*, *rotations*, and *zoomings* are explained. To this end, selected prototypical motions, generated by rotations, of different objects are taken as model problems to demonstrate the enhanced flexibility of tetrahedralizations in video processing.

6.2.1 Representation of Video Data

To explain the representation of video data, let us first fix some notations. A grey-scale video is a sequence of T rectangular planar image frames, each of size $W \times H$ pixels, so that the total number of video pixels is $N = W \times H \times T$, where each video pixel bears a grey-scale (luminance) value.

Therefore, a grey-scale video may be regarded as a mapping

$$V : X \longrightarrow \{0, 1, \ldots, 2^r - 1\}$$

from the 3d video domain

$$X = [0, \ldots, W - 1] \times [0, \ldots, H - 1] \times [0, \ldots, T - 1]$$

of pixel positions to the luminance values of the grey-scale video, where usually $2^r = 256$, i.e., $r = 8$. In other words, a video can be viewed as an element $V \in \{0, 1, \ldots, 2^r - 1\}^X$, where X is the set of pixels and r is the number of bits in the representation of the luminance values.

We regard a video as a trivariate function over the convex hull $[X] \subset \mathbb{R}^3$ of the pixel positions, so that $[X]$ constitutes the (continuous) parallelepipedic video domain. In this setting, each pixel in X is corresponding to a spatial grid point in $[X]$ with integer coordinates. In contrast to standard methods for video processing, our proposed approximation method is based on a pure three-dimensional interpretation of the given video data.

6.2.2 Delaunay Tetrahedralizations

This section introduces Delaunay tetrahedralizations, being one important ingredient of our approximation scheme. Throughout the discussion in this section, let $Y \subset \mathbb{R}^3$ denote a fixed finite point set. Recall that a tetrahedralization $\mathcal{T}_Y$ of Y is a collection of tetrahedra, whose vertex set is Y and whose union is the convex hull $[Y]$. Moreover, we assume that any pair of two distinct tetrahedra in $\mathcal{T}_Y$ intersect at most at one common vertex or along one common edge or across one common triangular face.

Delaunay tetrahedralizations are popular data structures for the efficient implementation of important 3d geometrical queries, such as nearest neighbour search or localization of closest point pairs. In this subsection, we recall some relevant properties of Delaunay tetrahedralizations.

It is convenient to introduce Delaunay tetrahedralizations through their dual *Voronoi diagrams*. To explain the duality between Voronoi diagrams and Delaunay tetrahedralizations, denote by

$$V_Y(y) = \left\{ z \in \mathbb{R}^3 : \|y - z\| = \min_{x \in Y} \|x - z\| \right\} \subset \mathbb{R}^3 \quad \text{for } y \in Y$$

the *Voronoi tile* of $y \in Y$. Note that the Voronoi tile $V_Y(y)$ contains all points which are—w.r.t. the Euclidean norm $\|\cdot\|$—at least as close to y as to any other point in Y. The set $\{V_Y(y)\}_{y \in Y}$ of all Voronoi tiles is called the *Voronoi diagram* of Y, yielding a partitioning of the Euclidean space, i.e.,

$$\mathbb{R}^3 = \bigcup_{y \in X} V_Y(y).$$

Note that each Voronoi tile $V_Y(y)$ is a non-empty, closed and convex polyhedron. Two different Voronoi tiles $V_Y(x)$ and $V_Y(y)$ are either disjoint or they share a vertex, an edge or a triangular face. In the latter case, the points $x \in Y$ and $y \in Y$ are said to be *Voronoi neighbours*.

By connecting all possible Voronoi neighbours, we obtain a graph whose vertex set is Y. This graph defines a *tetrahedral decomposition* $\mathscr{D}_Y$ of the convex hull $[Y]$, provided that no 5 points in X are *co-spherical*. The latter means that no 5 points in Y lie on the 2-dimensional surface of a sphere. For simplicity, we assume this property, the *Delaunay property*, until further notice.

The tetrahedral decomposition $\mathscr{D}_Y$ is said to be the *Delaunay tetrahedralization* of the point set Y. The Delaunay tetrahedralization $\mathscr{D}_X$ of X is unique. For any tetrahedron in $\mathscr{D}_Y$, its circumsphere does not contain any point from Y in its interior, according to the Delaunay property.

Finally, for any $y \in Y$, the Delaunay tetrahedralization $\mathrm{d}(Y \setminus y)$ can be computed from $\mathscr{D}_Y$ by a *local* update. This immediately follows from the Delaunay property, which implies that only the *cell* $\mathscr{C}(y)$ of y in $\mathscr{D}_Y$ needs to be retetrahedralized. Recall that the cell $\mathscr{C}(y)$ of y is the domain consisting of all tetrahedra in $\mathscr{D}_Y$ which contain y as a vertex. For further details on Delaunay tetrahedralizations, see the textbook [348].

6.2.3 *Sparse Representations of Prototypical Motions*

This section concerns the sparse representation of prototypical motions in videos by anisotropic tetrahedralizations. For the sake of presentational simplicity and brevity, we decided to restrict ourselves to rotations of basic planar geometrical objects, ellipses and triangles, with uniform grey-scale values.

We can describe the situation as follows. In the first frame, at time $t = 0$, the projection of the geometrical object is the characteristic function of a domain Ω_0 with

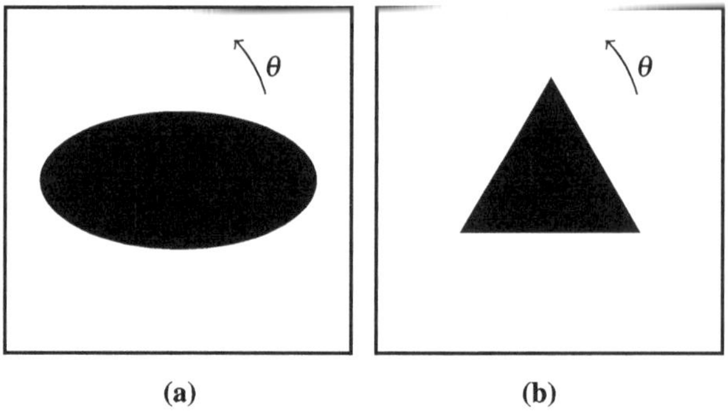

Fig. 6.1 Two prototypical motions. (**a**) Rotating ellipse; (**b**) rotating triangle. In either case, the rotation axis is aligned with the vertical (time) axis, and the origin is the center of the square domain

piecewise smooth boundary Γ_0. In this particular setting, a (general) rigid motion can be viewed as a continuous mapping

$$M : [0, T] \to E^+(2)$$

from time interval $[0, T]$ to the group $E^+(2)$ of rigid motions, i.e., the group generated by translations and rotations.

Now for the special case of rotations, we assume constant angular speed, i.e., no accelerations or decelerations for further simplicity, in which case the motion

$$M(t) = M^t \in E^+(2) \quad \text{for } t \in [0, \dots, T]$$

can be rewritten as a planar transformation $M(u, \theta) \equiv R(\theta)$, where $R(\theta)$ is a (counter-clockwise) rotation about angle θ. In this simple case, the corresponding video sequence is fully described by the parametric surface Σ,

$$\Sigma = \bigcup_{t \in [0,T]} \Gamma_t = \bigcup_{t \in [0,T]} M(t)\Gamma_0. \tag{6.1}$$

With assuming sufficiently small rotation angle θ, the surface Σ is piecewise smooth. In this case, it is reasonable to work with triangular surface elements to approximate Σ. Now the quality of the surface approximation heavily depends on the alignment of the triangular surface elements. Indeed, their long triangular edges should be aligned with directions of smaller surface curvature of Σ, whereas their short edges should point into directions with higher curvature.

Now we consider the approximation of rotational motions of two different geometrical objects, one ellipse and one equilateral triangle, see Figs. 6.1(a), (b). In either case, we let $\theta = \pi/6$ per time unit (the time between two consecutive frames). Figs. 6.2(a), (b) shows the resulting surfaces Σ, respectively, as defined by (6.1).

For the purpose of approximating the rotational triangle E, a suitable Delaunay tetrahedralization may be constructed as follows. First select a time discretization step $t_s = T/n_t$, $n_t \in \mathbb{N}$, then sample the set of vertices $Y \equiv Y(E)$ defined by

$$(E, R(t_s\theta)E, R(2t_s\theta)E, \dots, R(n_t t_s\theta)E).$$

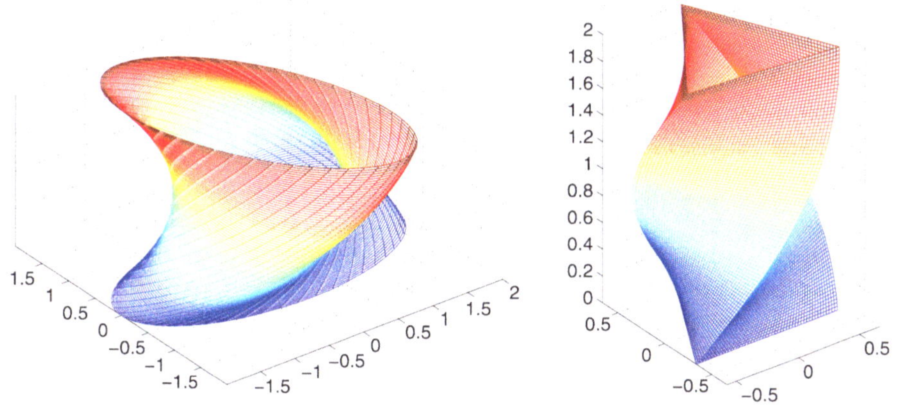

(a) (b)

Fig. 6.2 Two prototypical motions. 3d visualization of the moving contours for the (**a**) rotating ellipse; (**b**) rotating triangle. In either case, the rotation axis is aligned with the vertical (time) axis, and the origin is the center of the square domain

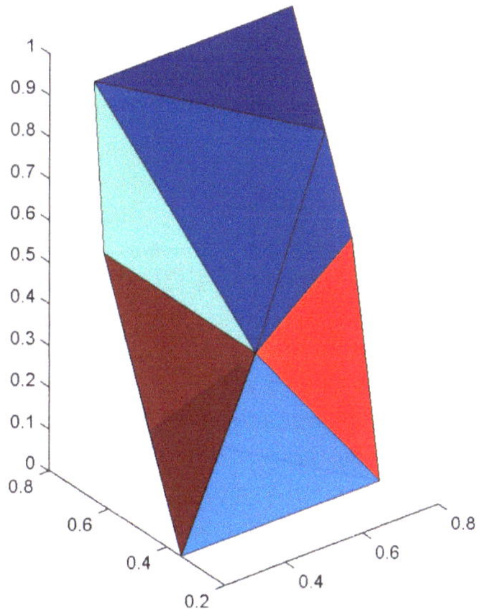

Fig. 6.3 *Rotating triangles*. Triangular surface approximation of the moving contour. Only 12 triangular faces are utilized to represent the entire video sequence, where the surface triangulation is obtained from the Delaunay tetrahedralization of 36 video pixels. In this example, we let $\theta = \pi/6$ per time step

By their construction, the resulting $3 \times (n_t + 1)$ vertices represent the triangular shape for each of the corresponding $n_t + 1$ frames exactly. Therefore, if t_s is sufficiently small, then the corresponding Delaunay tetrahedralization of the 3d vertex set Y recovers the triangular shape throughout all frames. Although this particular choice of vertices is not necessarily optimal, it leads to small approximation error of the sweep surface (moving contour).

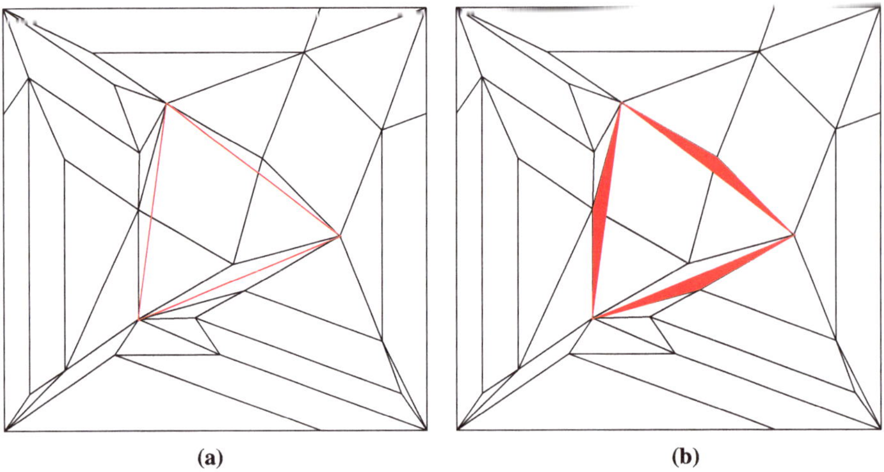

Fig. 6.4 *Rotating triangles* at time $t = 0.75$. (**a**) Projection of tetrahedralization (*black*) and exact solution triangle (*red*); (**b**) the corresponding error area is filled *in red*

To make one special case, let $[0, 1]$ denote the time interval, $t_s = 0.5$ and $n_t = 2$. The corresponding piecewise linear representation of the resulting sweep surface is shown in Fig. 6.3. Moreover, Fig. 6.4(a) displays the projection of the tetrahedralization in the frame plane at intermediate time $t = 0.75$, where (for this particular example) the resulting error is maximal. The resulting error is visualized in Fig. 6.4(b).

6.3 Nonlinear Approximation of Video Data

This section explains our video approximation method, **3AT**, in detail. The method **3AT** combines a recursive pixel removal scheme, adaptive thinning, with least squares approximation by trivariate linear splines. In this approximation method, adaptive thinning is first applied to obtain a sparse set of significant video pixels. The set of significant pixels yields a (unique) anisotropic Delaunay tetrahedralization, where the vertices of the tetrahedralization are given by the significant pixels. This in turn defines a linear approximation space of trivariate linear spline functions over the so obtained anisotropic tetrahedralization. The approximation to the video data is then given by the best approximating linear spline, in the sense of least squares approximation. The overall aim of the resulting adaptive approximation algorithm is to construct a suitable linear spline space of small dimension (being represented by a sparse set of significant pixels), such that the distance between the best approximating linear spline and the given video data is small, thus yielding a small reconstruction error. The construction of the spline space can be viewed as a highly nonlinear approximation process, which generalizes the algorithm proposed in [97, 98] from 2d image data to 3d video data.

In the remainder of this section, we will first discuss trivariate linear splines over tetrahedralizations, before adaptive thinning—the key ingredient of the proposed nonlinear approximation method—is introduced. This includes a discussion on the utilized significance measure, as this is required for the adaptive extraction of the significant pixels.

6.3.1 Linear Splines over Tetrahedralizations

Following our previous paper [98], we work with continuous and piecewise affine approximations over anisotropic tetrahedralizations. Let Π_1 denote the linear space of all trivariate polynomials of degree at most one, and let $\mathscr{D}_Y$ be a fixed Delaunay tetrahedralization. For any subset $Y \subset X$, we denote by $\mathscr{S}_Y$ the linear space of all trivariate continuous functions whose restriction to any tetrahedron $\theta \in \mathscr{D}_Y$ is affine, i.e.,

$$\mathscr{S}_Y = \{f \equiv f(x_1, x_2, x_3) \in \mathscr{C}^0([Y]) : f|_\theta \in \Pi_1 \text{ for all } \theta \in \mathscr{D}_Y\}.$$

Any element in $\mathscr{S}_Y$ is referred to as a *linear spline* over $\mathscr{D}_Y$. For given luminance values at the pixels of Y, $V|_Y = \{V(y) : y \in Y\}$, there is a unique linear spline interpolant $L(Y, V) \in \mathscr{S}_Y$ satisfying

$$L(Y, V)(y) = V(y) \quad \text{for all } y \in Y.$$

The interpolant $L(Y, V)$ can be represented as a linear combination

$$L(Y, V) = \sum_{y \in Y} V(y)\varphi_y$$

of the *Courant elements* $\varphi_y \in \mathscr{S}_Y$, for $y \in Y$, being the unique Lagrangian basis functions in $\mathscr{S}_Y$ satisfying

$$\varphi_y(x) = \begin{cases} 1 & \text{for } y = x; \\ 0 & \text{for } y \neq x; \end{cases} \quad \text{for any } x \in Y.$$

Now, for fixed $Y \subset X$ we can take the spline space $\mathscr{S}_Y$ as an approximation space for the video data $V|_X$, provided that the eight vertices of X (i.e., the four corners of the first frame and the four corners of the last frame) lie in Y.

6.3.2 Sparse Data Selection by Adaptive Thinning

We remark that the approximation quality of the video reconstruction heavily depends on the selection of the (sparse) pixel set Y. A customized construction of the sparse video data representation essentially requires an *adaptive* selection of the pixels in Y. Therefore, let us first explain how the subset $Y \subset X$ is constructed.

To obtain a suitable sparse set $Y = X_n$ of n significant pixels, for some $n \ll N$, adaptive thinning constructs a sequence of nested subsets of pixels

$$X_n \subset X_{n+1} \subset \cdots \subset X_{N-1} \subset X_N = X, \tag{6.2}$$

where the size $|X_p|$ of any subset X_p in (6.2) is p, and so $N = |X|$ is the number of pixels in X.

The utilized adaptive thinning algorithm recursively removes pixels from X, one after the other, where any removal of one pixel depends on the entire video data $V|_X$, as given by the luminance values attached to the video pixels. The pixel removal is done in a greedy way, where at each removal step the removed pixel is a *least significant* pixel. The generic formulation of our recursive pixel removal scheme is as follows.

Algorithm 6.1 (Adaptive Thinning)

(1) *Let* $X_N = X$;
(2) *For* $k = 1, \ldots, N - n$
  (2a) *Find a least significant pixel* $x \in X_{N-k+1}$;
  (2b) *Let* $X_{N-k} = X_{N-k+1} \setminus x$.

To describe a specific thinning strategy, it remains to determine a *significance measure* in order to select a *least significant* pixel in step (2a). Details on this important point are discussed in the following subsection.

6.3.3 Significance Measures for Video Approximation

The quality of video compression schemes is measured in dB (decibel) by the *peak signal to noise ratio*,

$$\mathrm{PSNR} = 10 * \log_{10}\left(\frac{2^r \times 2^r}{\eta^2(Y, X)}\right),$$

where the *mean square error* (MSE) is given by

$$\eta^2(Y, X) = \frac{1}{|X|} \sum_{x \in X} |L(Y, V)(x) - V(x)|^2. \tag{6.3}$$

Therefore, to approximate the video, we wish to construct a subset $Y \subset X$, such that the resulting mean square error $\eta^2(Y, X)$ is small. The construction of a suitable subset $Y \subset X$ is accomplished by Algorithm 6.1, where a natural criterion for a least significant pixel is given by the following definition, already used in our previous papers [97, 98] for pixel removal from images.

Definition 6.1 For $Y \subset X$, a pixel $y^* \in Y$ is said to be `least significant` in Y, iff

$$\eta(y^*) = \min_{y \in Y} \eta(y),$$

where for any $y \in Y$,

$$\eta(y) = \eta(Y \setminus y, X)$$

is the `significance` of the pixel y in Y.

In [122] we have also considered least significant pixel pairs (in images). This leads to a somewhat more sophisticated removal criterion that allows a simultaneous removal of edges (i.e., two-point removals of connected vertices). As was shown in [98], this additional option has improved the resulting image approximation quite significantly. This observation gives rise to lift the two-point removal criterion in [98] from 2d to 3d. We recall the definition of least significant pixel pairs.

Definition 6.2 For $Y \subset X$, a pair $\{y_1^*, y_2^*\} \subset Y$ of two pixels in Y is said to be `least significant` in Y, iff

$$\eta(y_1^*, y_2^*) = \min_{\{y_1, y_2\} \subset Y} \eta(y_1, y_2),$$

where for any pixel pair $\{y_1, y_2\} \subset Y$, we denote its `significance` in Y by

$$\eta(y_1, y_2) = \eta(Y \setminus \{y_1, y_2\}, X).$$

A pixel $y^* \in Y$ is said to be `least significant` in Y, iff it belongs to a least significant pixel pair in Y, $\{y^*, y\} \subset Y$, and satisfies $\eta(y^*) \le \eta(y)$.

As supported by our numerical comparisons in Sect. 6.5, the significance measure of Definition 6.2 (in comparison with that of Definition 6.1) improves the resulting video reconstruction considerably. Indeed, the more sophisticated significance measure of Definition 6.2 allows removals of edges, whose two vertices may have high individual significances (according to Definition 6.1), although their connecting edge may not contribute very much to the approximation to the video data (cf. [98] for more detailed explanations).

6.3.4 Local Optimization by Exchange

In order to further improve the quality of the significant pixels' distribution, we apply the post-processing local optimization procedure proposed in [96]. This local optimization relies on an iterative exchange of pixel pairs. At each exchange step, one current significant pixel is being swapped with one current non-significant pixel. Let us recall the definition of exchangeable pixels.

Definition 6.3 For any $Y \subset X$, let $Z = X \setminus Y$. A pixel pair $(y, z) \in Y \times Z$ satisfying $\eta((Y \cup z) \setminus y; X) < \eta(Y; X)$ is said to be `exchangeable`. A subset $Y \subset X$ is said to be `locally optimal` in X, iff there is no exchangeable pixel pair $(y, z) \in Y \times Z$.

By an exchange of any exchangeable pixel pair $(y, z) \in Y \times Z$, the approximation error $\eta(Y; X)$ is strictly reduced. This leads to the following local optimization algorithm which computes a locally optimal subset in X from any input $Y \subset X$.

Algorithm 6.2 (Exchange)

INPUT: $Y \subset X$;

(1) *Let* $Z = X \setminus Y$;
(2) *WHILE* (Y *not locally optimal in* X)
  (2a) *Locate an exchangeable pair* $(y, z) \in Y \times Z$;
  (2b) *Let* $Y = (Y \setminus y) \cup z$ *and* $Z = (Z \setminus z) \cup y$;

OUTPUT: $Y \subset X$, *locally optimal in* X.

In our numerical experiments, we observed that the local optimization procedure helps improve the shape of the tetrahedra, which are better adapted to the local regularity of the function. Indeed, in areas where the underlying function is smooth and convex, we obtain nearby equilateral tetrahedra, whereas in areas of smaller regularity, long and thin tetrahedra are aligned with the preference directions of the target function.

6.3.5 Minimization of the Mean Square Error

In a post-processing step, we further reduce the mean square error (6.3) by *least squares approximation* [39]. More precisely, we compute from the set $Y \subset X$ of significant pixels, output by Algorithms 6.1 and 6.2, and from the luminance values at the pixels in X the unique *best approximation* $L^*(Y, V) \in \mathscr{S}_Y$ satisfying

$$\sum_{x \in X} |L^*(Y, V)(x) - V(x)|^2 = \min_{s \in \mathscr{S}_Y} \sum_{x \in X} |s(x) - V(x)|^2.$$

Such a best approximation exists and is unique, since $\mathscr{S}_Y$ is a finite dimensional linear space. For numerical aspects of least squares approximation we refer to the textbook [39].

6.4 Computational Aspects

6.4.1 Data Structures for Efficient Implementation

An efficient implementation of our proposed approximation method **3AT** requires suitable data structures. In the implementation of adaptive thinning, Algorithm 6.1,

Fig. 6.5 Removal of a vertex from a Delaunay tetrahedralization. The update of the Delaunay tetrahedralization requires only a local retetrahedralization of the vertex cell

the most critical aspect concerning computational complexity is the removal of one pixel. Moreover, for the exchange of pixel pairs, by the local optimization Algorithm 6.2, also efficient insertions of pixels are needed. In either case, this requires efficient updates of the utilized Delaunay tetrahedralization.

Several methods have been proposed to reduce the complexity for the insertion and removal of vertices in Delaunay tetrahedralization [278, 385]. In [77], for instance, an adaptive divide-and-conquer method is suggested. Unlike the popular *Guibas-Stolfi algorithm* for two dimensions [195], the algorithm in [77] uses an incremental method in the merging step. Our implementation is based on the robust incremental *flip algorithm* of [315].

The flip algorithm in [315] performs an incremental insertion of vertices to compute the (global) Delaunay tetrahedralization. In this method, each insertion of a new vertex requires localizing a tetrahedron from the current Delaunay tetrahedralization which contains the position of that vertex. Therefore, the ordering of the points to be inserted plays an important role for the performance of the incremental insertion algorithm.

Many tetrahedralization methods merely use a random ordering of the point set. To accelerate the point insertion, a preprocessing sorting of the points is proposed in [44]. The concept in [44] relies on space-filling Hilbert curves. In our implementation vertex of removals and insertions, we follow along the lines of the ideas in [44].

Finally, we remark that any insertion or removal of a vertex from a Delaunay tetrahedralization requires only a local update. This important property of Delaunay tetrahedralizations reduces the computational complexity of adaptive thinning, and exchange significantly. For the removal of a vertex y, for instance, only a (local) tetrahedralization of the vertex cell $C(y)$ is required (see Fig. 6.5), where $C(y)$ is given by the union the surrounding tetrahedra of vertex y. The computational complexity of Algorithms 6.1, 6.2 is discussed in Sect. 6.4.3.

6.4.2 Simulation of Simplicity

When five close vertices are co-spherical, the Delaunay tetrahedralization is not unique. Note that this is highly relevant in our particular situation, where the vertex positions are lying on a Cartesian grid. Moreover, uniqueness of the Delaunay tetrahedralization is an important property, which we wish to ensure for many reasons concerning the computational efficiency. In order to solve this problem, we use a generic method called *Simulation of Simplicity*, as proposed in [123]. This method serves to enforce uniqueness, especially in degenerate cases and for more general situations of co-spherical point distributions.

Unlike in other perturbation methods, the simulation of simplicity method allows us to work with integer arithmetic rather than with floating point arithmetic. We have adapted the ideas in [123] to Delaunay tetrahedralizations: by using an lexicographical order of vertices, we always obtain a unique Delaunay tetrahedralization for any given set of 3d points.

6.4.3 Computational Complexity

In this section, we discuss the computational complexity of our nonlinear approximation method. To this end, we show that the complexity of adaptive thinning, Algorithm 6.1, is $\mathcal{O}(N\log(N))$, whereas the performance of one exchange step, Algorithm 6.2, requires only $\mathcal{O}(\log(N))$ operations.

6.4.3.1 Computational Complexity of Adaptive Thinning

Let us start with the discussion on adaptive thinning. For the efficient implementation of Algorithm 6.1, we use two different priority queues, one for the significances of pixels and one for the significances of edges in $\mathcal{D}_Y$. Each priority queue is efficiently implemented by using the data structure *heap*.

For the significances of pixels, we use the significance measure

$$e_\delta(y) = \eta^2(y) - \eta^2(Y,X) \quad \text{for } y \in Y.$$

We remark that the significance measure e_δ is equivalent to the pixel significance measure η of Definition 6.1, i.e. minimizing $\eta(y)$ among all pixels in Y is equivalent to minimizing $e_\delta(y)$ among all pixels in Y [98]. But the significance $e_\delta(y)$ is *local*, since it measures the *anticipated error*, being incurred by the removal of pixel y, on the (local) cell $C(y)$ of y.

As for the significance of pixel pairs, we work with the local significance measures

$$e_\delta(y_1,y_2) = e_\delta(y_1) + e_\delta(y_2) \quad \text{for } [y_1,y_2] \notin \mathcal{D}_Y$$

and

$$e_\delta(y_1,y_2) = \eta^2(y_1,y_2) - \eta^2(Y,X) \quad \text{for } [y_1,y_2] \in \mathcal{D}_Y$$

which are equivalent to the significance measure $\eta(y_1,y_2)$ in Definition 6.2.

Note that $e_\delta(y_1, y_2)$ is a *local* significance measure, since its computation is restricted to the union of the two cells $C(y_1)$ and $C(y_2)$. Due to the simple representation of e_δ, the maintenance of the significances $\{e_\delta(y_1, y_2) : \{y_1, y_2\} \subset Y\}$ can be reduced to the maintenance of the significances $\{e_\delta(y_1, y_2) : [y_1, y_2] \in \mathscr{D}_Y\}$ and $\{e_\delta(y) : y \in Y\}$. To this end, we employ two separate heaps.

By using the *local* significance measures e_δ (rather than η), each pixel removal (according to Algorithm 6.1) costs only $\mathscr{O}(1)$ operations, for the retriangulation of the cell $C(y)$ to obtain $\mathscr{D}_{Y\setminus y}$ and for the required update of the neighbouring pixels' and edges' significances. The required update for each of the two heaps costs $\mathscr{O}(\log(N))$. This makes up $\mathscr{O}(N\log(N))$ operations in total for the removal of at most N pixels.

Theorem 6.1 *The performance of the adaptive thinning algorithm, Algorithm* 6.1, *costs* $\mathscr{O}(N\log(N))$ *operations.*

6.4.3.2 Computational Complexity of Exchange

Now let us turn to the complexity for one pixel exchange. In the efficient implementation of exchange we work with the *local* swapping criterion

$$e_\delta(z; Y \cup z) > e_\delta(y; Y \cup z) \quad \text{for } (y, z) \in Y \times Z, \tag{6.4}$$

which, for unconnected pixels in $\mathscr{D}_{Y\cup z}$, further simplifies to

$$e_\delta(z; Y \cup z) > e_\delta(y; Y) \quad \text{for } [y; z] \notin \mathscr{D}_{Y\cup z}. \tag{6.5}$$

The above swapping criterion is equivalent to the criterion in Definition 6.3 for exchangeable pixel pairs [96].

Moreover, we work with three different heaps: one heap for the pixels in Y, `heapY`, with significances e_δ, one heap for the pixels in $Z = X \setminus Y$, `heapZ`, with significances $e_\delta(z; Y \cup z)$, for $z \in Z$, and one heap for connected pixel pairs, with significances e_δ, `heapE`. For details concerning the different significances e_δ (for pixels and edges), we refer to our previous paper [96].

The local swapping criteria (6.4)–(6.5) enables us to localize an exchangeable pixel pair in only $\mathscr{O}(1)$ operations. The subsequent update of the utilized data structures, i.e., for the three heaps `heapY`, `heapZ`, `heapE`, and the Delaunay tetrahedralization $\mathscr{D}_Y$ costs $\mathscr{O}(\log(N))$ operations. For details, we refer to [96].

Theorem 6.2 *The performance of one pixel exchange, Algorithm* 6.2, *costs* $\mathscr{O}(\log(N))$ *operations.*

6.5 Numerical Simulations

We have applied our video approximation method **3AT** to one popular test example, called `Suzie`. The test video comprises 30 frames. The original video data is shown

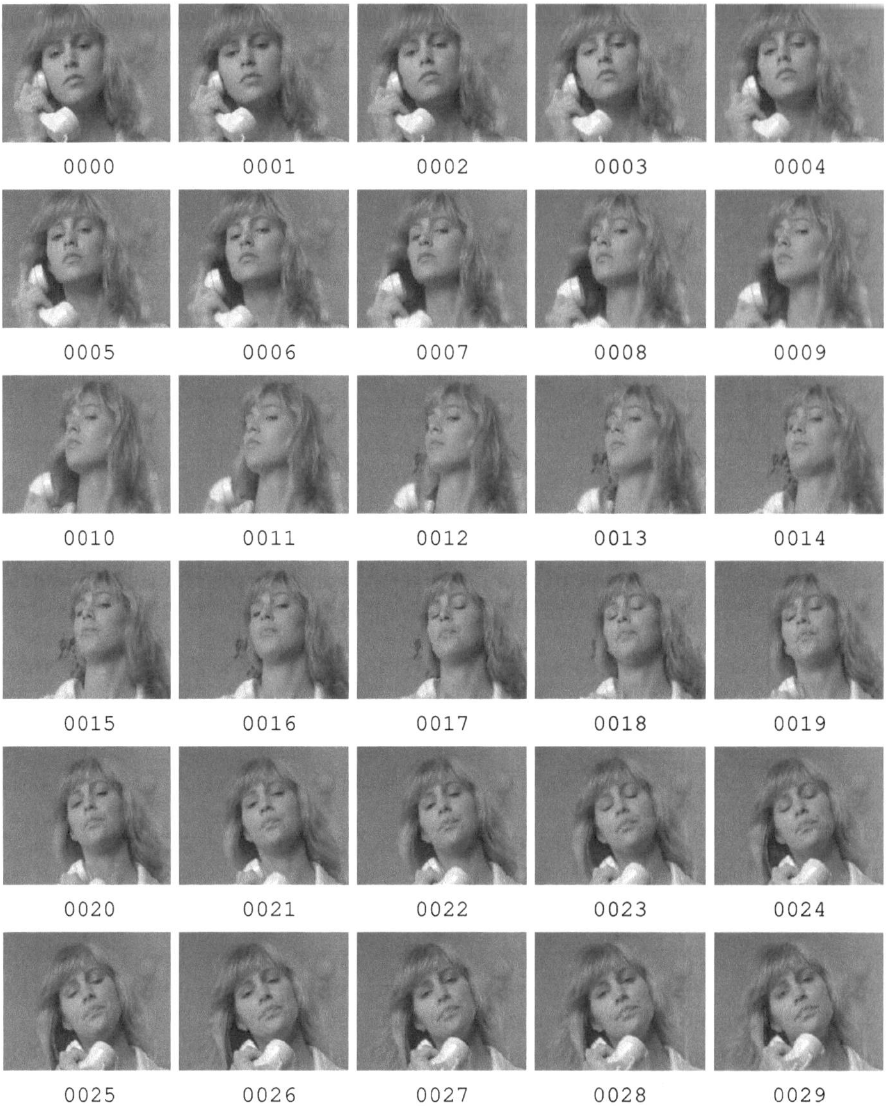

Fig. 6.6 `Suzie`. Video comprising 30 consecutive image frames, *from top left* (frame `0000`) *to bottom right* (frame `0029`)

in Fig. 6.6, whose 30 image frames are displayed in chronological order, from top left to bottom right.

The corresponding reconstruction of the video data by our approximation method **3AT** is shown in Fig. 6.7. Note that **3AT** achieves to reconstruct the test data very well, especially the geometric features of the video. This is due to a well-adapted distribution of the significant pixels, as displayed in Fig. 6.8. Their corresponding tetrahedra are shown in Fig. 6.9, where for each frame plane only their intersecting

Fig. 6.7 `Suzie`. Reconstruction of video data by our approximation method **3AT**. The reconstruction is represented by 30 consecutive image frames, displayed *from top left to down right*. For each image frame, the corresponding PSNR value is shown

tetrahedra are displayed. The tetrahedra are represented by their edges, where those edges lying in the frame plane are displayed by bold solid lines. The other (intersecting) edges are represented by their projections onto the corresponding frame plane, and displayed by thin solid lines.

Note that the representation of the video data by the significant pixels is very sparse. In fact, in comparison with the intermediate frames, only the first and the

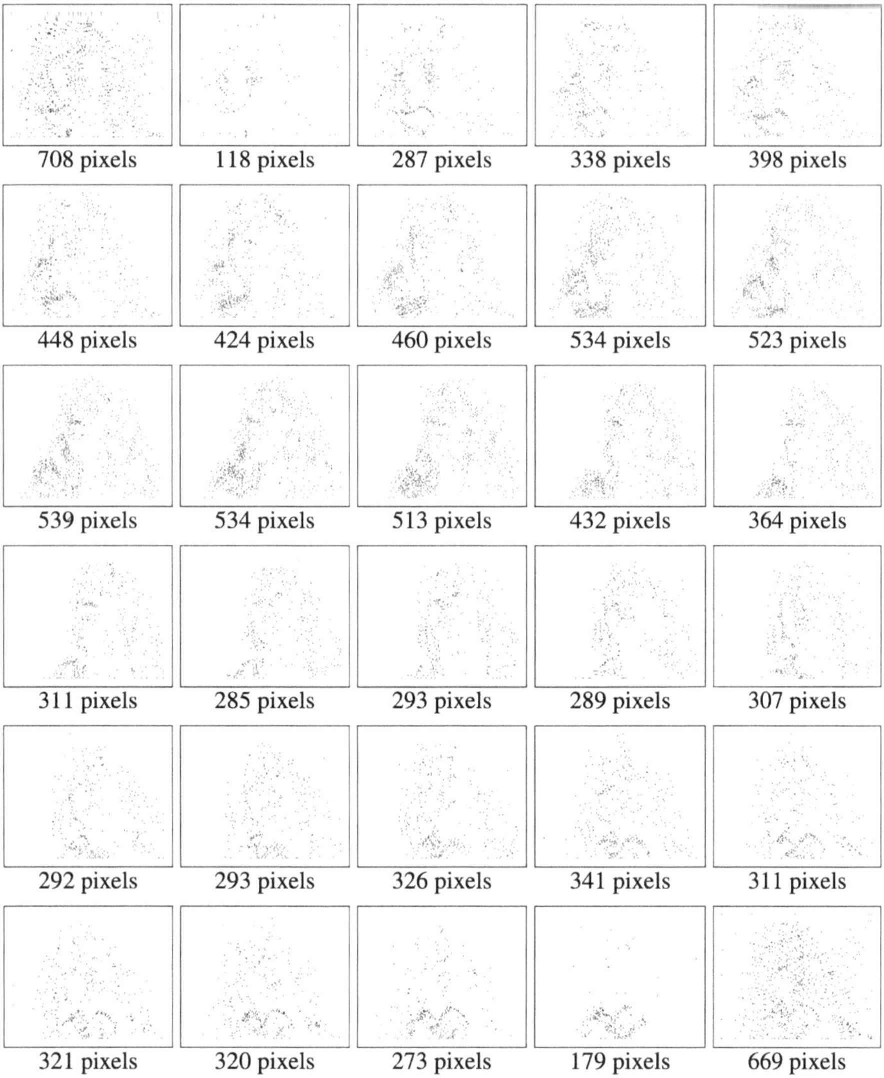

Fig. 6.8 `Suzie`. The significant pixels output by our approximation method **3AT**, and the number of significant pixels per frame

last frame are containing a larger number of significant pixels, see Fig. 6.10 (top). This is because their corresponding Delaunay triangulations (in the image frame plane) are covering the two opposite faces of the video domain, at time $t = 0$ (first frame) and at $t = 29$ (last frame). But the well-adapted distribution of the significant pixels in the intermediate frames is very efficient. Moreover, the different motions of the video's geometrical features are captured very well by the geometry of the significant pixels. The good visual quality of the video reconstruction can be evaluated through Fig. 6.7.

Fig. 6.9 `Suzie`. Tetrahedralization of the significant pixels. For each frame, only their intersecting tetrahedra are displayed. The individual tetrahedra are represented by their edges. Edges lying in the image frame are displayed by *bold solid lines*; edges passing through the image frame are represented by their projections onto the image plane, displayed by *thin solid lines*

Now we measure the quality of the reconstruction by the resulting PSNR value. We obtain a sparse representation of the video data by 11,430 significant pixels, (i.e., 381 significant pixels in average), yielding a PSNR value of the video reconstruction by **3AT** of 35.45 dB.

Fig. 6.10 Suzie. From the video reconstruction by **3AT**, 30 image frames were generated, each corresponding to one image frame of the given video data Suzie. For each of these image frames, as generated by **3AT**, the number of significant pixels and the PSNR value was recorded. *The two graphs* show the number of significant pixels (*top*); the PSNR value (*bottom*), each regarded as a function of the frame indices

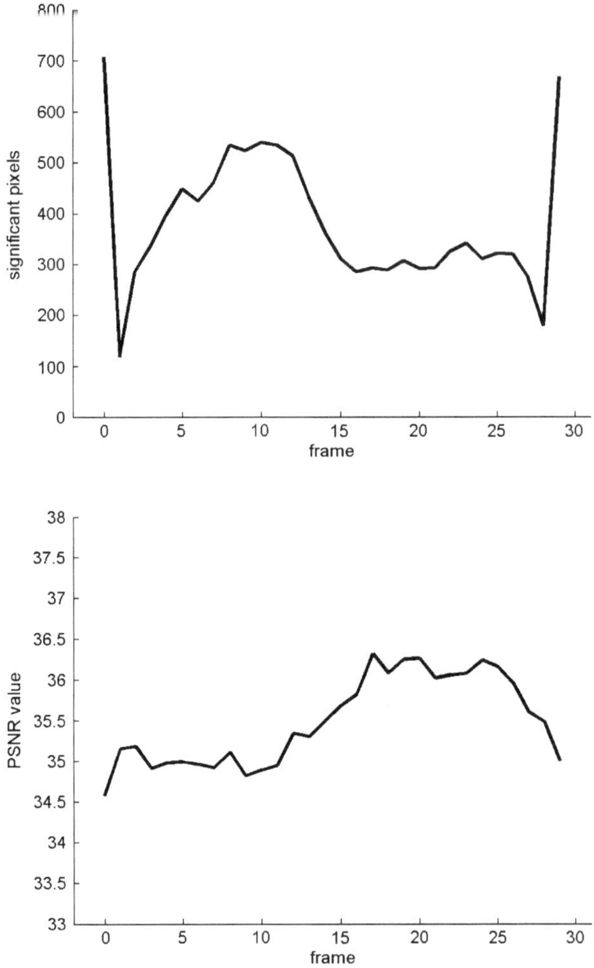

In addition, from the video reconstruction by **3AT**, 30 image frames were generated, each corresponding to one image frame of the given video data Suzie. For each of these image reconstructions, as generated by **3AT**, the corresponding number of significant pixels and the PSNR value was recorded. The two graphs in Fig. 6.10 show the number of significant pixels (top) and the PSNR value (bottom), each regarded as a function of the frame indices.

As regards the evolution of significant pixels, note that the number of pixels is large between times $t = 5$ and $t = 13$. This is due to rapid motions in the video sequence, which requires more energy (in terms of number of significant pixels) to capture the relevant geometrical details.

As regards the PSNR values of the individual frames, the *minimal* PSNR value of 34.58 dB is attained at frame 0000, whereas the *maximal* PSNR value of 36.32 dB is attained at frame 0017. The average PSNR value is 35.49 dB. The resulting distribution of PSNR values (over the 30 frames) reflect a very well-balanced approximation to the video data, at remarkably good reconstruction quality and small number of significant pixels. We finally remark that our proposed approximation method features a surface-preserving smoothing process, with surfaces being understood as moving contours. This helps to avoid misaligned aliasing artifacts between consecutive frames.

6.6 Conclusion and Final Remarks

We have proposed a novel adaptive approximation algorithm for sparse representations of video data. Our nonlinear approximation method works with linear splines over anisotropic tetrahedralizations of the 3d video domain, whose tetrahedra are adapted locally in space and in time. The sparse representation of the video data relies on the construction of a small set of least significant pixels, which are selected by a recursive pixel removal scheme, termed adaptive thinning. We have applied our resulting video approximation method, **3AT**, to prototypical geometrical motions and to a natural video. The good performance of the proposed video compression method is supported by our numerical simulations.

We finally remark that alternative approaches to multiscale representations with local geometry adaptation can be found in Chap. 9. Future research should include investigations concerning the usage of adaptive smoothing and denoising methods in combination with anisotropic diffusion and adaptive filtering strategies, cf. Chaps. 1, 2, 3, 4, 7, 12, and 13 for related topics.

Acknowledgements The second author was supported by the priority program DFG-SPP 1324 of the *Deutsche Forschungsgemeinschaft* (DFG) within the project "Adaptive Approximation Algorithms for Sparse Data Representation". We wish to thank Nira Dyn for several fruitful discussions concerning the design of the presented video approximation method.

Chapter 7
Continuous Diffusion Wavelet Transforms and Scale Space over Euclidean Spaces and Noncommutative Lie Groups

Hartmut Führ

Abstract Scale space representations over Euclidean space are closely related to continuous wavelet transforms. E.g., taking the time derivative of a Gaussian scale space representation yields a continuous wavelet transform with respect to the Mexican Hat wavelet. Departing from this observation, we introduce a continuous diffusion wavelet transform associated to any diffusion semigroup acting on an abstract Hilbert space. Under quite general assumptions this wavelet transform is norm-preserving, thus giving rise to wavelet-type inversion formulae. For the diffusion associated to the heat equation on the Heisenberg group $\mathbb{H}$, we recover the so-called Mexican Hat wavelet previously obtained by Mayeli. Our approach provides an alternative short proof of admissibility for the Mexican Hat wavelet on $\mathbb{H}$, as well as a new family of wavelets for $\mathbb{H}$ and more general homogeneous Lie groups.

7.1 Introduction

For a given image $f \in \mathrm{L}^2(\mathbb{R}^2)$, let

$$u_f(x,t) = f * g_t(x) \quad (x \in \mathbb{R}^2, t > 0)$$

denote the Gaussian scale space representation of f, with

$$g_t(x) = \frac{\exp(-\|x\|^2/2t)}{2\pi t}.$$

The standard interpretation of $u_f(\cdot,t)$ is that it contains all details of f of size at least $\sqrt{t}$. By contrast, given a continuous wavelet transform

$$\mathscr{W}_\psi f(x,t) = f * \psi_t(x)$$

with $\psi_t(x) = t^{-1}\psi(t^{-1}x)$, the standard interpretation of $\mathscr{W}_\psi f(\cdot,t)$ is that of containing all details of size *precisely* t.

H. Führ (✉)
Lehrstuhl A für Mathematik, RWTH Aachen, 52056 Aachen, Germany
e-mail: fuehr@matha.rwth-aachen.de

L. Florack et al. (eds.), *Mathematical Methods for Signal and Image Analysis and Representation*, Computational Imaging and Vision 41,
DOI 10.1007/978-1-4471-2353-8_7, © Springer-Verlag London Limited 2012

Hence, it is tempting to think of wavelet transform as a scale derivative of the scale space transform u_f. And indeed: Up to a suitable power of t and a rescaling of the time axis, $\partial_t u_f$ coincides with $\mathscr{W}_\psi f$ with respect to the Mexican Hat wavelet $\psi = \Delta g_1$, where Δ denotes the Laplacian on $\mathbb{R}^d$. A discussion of this phenomenon for α-scale spaces, which contains this as a special case, can be found in Sect. 7.2.

As I will explain shortly, there is a general principle at work here, and this chapter is devoted to explaining this principle in detail. To this end, I introduce a rather general definition of a diffusion wavelet transform associated to an operator semigroup acting on a Hilbert space $\mathscr{H}$. Given a suitable infinitesimal generator Q (assumed self-adjoint and negative definite), the associated continuous diffusion wavelet transform of $f \in \mathscr{H}$ is the $\mathscr{H}$-valued function $W_Q f$ defined as

$$W_Q f(t) = \partial_s(s \mapsto e^{sQ} f)(t) = Qe^{tQ} f \ (t > 0).$$

Whenever the evolution equation allows a meaningful interpretation of the scale parameter, the associated diffusion wavelet transform should resemble the usual continuous wavelet transform. Besides this rather vague parallel to continuous wavelet transform, there is also a concrete and useful one, and that has to do with reconstruction: Assuming that the generator Q is negative definite, with 0 not contained in its point spectrum, it follows that W_Q is norm-preserving in the sense that, for all $f \in \mathscr{H}$,

$$\|f\|^2 = 4 \int_{\mathbb{R}^+} \|W_Q f(t)\|^2 t \, dt; \qquad (7.1)$$

see Theorem 7.2 below. The formulation given here has the advantage that it provides a unified and simple view of both wavelet-like transforms and diffusion semigroups.

Equation (7.1) also points out an alternative means of recovering f from u_f: Instead of employing

$$f = \lim_{t \to 0} u_f(\cdot, t),$$

using that u_f solves the initial value problem, we can also invert W_Q by its (renormalized) adjoint, which yields

$$f = 4 \int_{\mathbb{R}^+} Qe^{tQ} W_Q(f) t \, dt. \qquad (7.2)$$

It should be stressed that the main mathematical ideas presented in the following have been developed elsewhere: In particular, the paper by Geller and Mayeli [182] contains our main abstract result, Theorem 7.2, though in somewhat different formulation and with slightly different assumptions. The idea to introduce scales and associated wavelet systems via the spectral theory of a suitable operator semigroup also underlies the construction of Coifman and Maggioni in [80]. In fact the continuous diffusion wavelet transform can be viewed as continuous analogue of their approach. Nonetheless, it seems worthwhile to elucidate the connection between continuous wavelet transforms and diffusion explicitly, and specifically with reference to recent work in signal and image processing such as [109, 111].

This chapter is structured as follows: In Sect. 7.2 I investigate the connections between α-scale space over $\mathbb{R}^d$, as studied in [109, 170], and continuous wavelet transform. This section is intended to provide an illustration and a review of notions from wavelet and scale space theory. In particular, it comments on different versions of wavelet inversion. In Sect. 7.3, I prove (7.1), and show how the results of Sect. 7.2 follow directly from this. The final Sect. 7.4 considers evolution equations arising from left-invariant differential operators on Lie groups. For homogeneous groups, the associated wavelet system can also be interpreted in terms of representation theory. This observation allows to explicitly relate recent developments in the construction of scale spaces on non-commutative Lie groups such as the Heisenberg group [111], and wavelet systems on those groups [302, 303].

7.2 Scale Space and Wavelet Transforms on Euclidean Space

In this section we focus on the family of α-scale space representations, as investigated in [109], see also Chap. 1 for background on linear diffusion. In order to describe these representations, let us introduce some notation. We let Δ denote the Laplacian on $\mathbb{R}^d$, i.e. $\Delta = \sum_{i=1}^{d} \frac{d^2}{dx_i^2}$, first defined on the test function, then extended uniquely to its maximal domain as self-adjoint operator on $\mathrm{L}^2(\mathbb{R}^d)$. Then Δ is a negative-definite self-adjoint (unbounded) operator, and thus spectral calculus allows to define arbitrary powers of $-\Delta$. By slight abuse of notation, we let $\partial_t u$ denote the partial derivative of $u : \mathbb{R}^d \times \mathbb{R}^+ \to \mathbb{C}$ with respect to the last (time) variable, and let $(-\Delta)^\alpha$ stand for the action of the power of $-\Delta$ acting on the first d (spatial) variables.

Then, given an arbitrary function $f \in \mathrm{L}^2(\mathbb{R}^d)$, the scale space representation of f is the solution $u : \mathbb{R}^d \times \mathbb{R}^+ \to \mathbb{C}$ to the initial value problem

$$\begin{aligned} \partial_t u &= -(-\Delta)^\alpha u, \\ \lim_{t \to 0} u(\cdot, t) &= f. \end{aligned} \tag{7.3}$$

In this chapter we focus on L^2-theory, hence the limit is understood in the L^2-norm. The solutions of (7.3) are easily described in terms of Green's calculus: One has that

$$u(\cdot, t) = f * K_t, \tag{7.4}$$

where the K_t are suitable kernel functions. Outside of exceptional values, such as $\alpha = 1/2, 1$ corresponding to the Poisson and the Gaussian kernels, respectively, these functions are not known explicitly. However, they are easy to write down on the Fourier transform side, using the fact that the Fourier transform acts as the spectral decomposition of $(-\Delta)^\alpha$. In this chapter, the Fourier transform of a function $f \in \mathrm{L}^1(\mathbb{R}^d)$ is defined as

$$\widehat{f}(\xi) = \int_{\mathbb{R}^d} f(x) e^{i\langle x, \xi\rangle}\, dx.$$

Then, at least formally, the equations $u(\cdot,t) = e^{-t(-\Delta)^\alpha} f$ as well as

$$((-\Delta)^\alpha f)^\wedge(\xi) = |\xi|^{2\alpha} \widehat{f}(\xi)$$

lead to the equation

$$\mathscr{F}_1(u)(\xi,t) = \widehat{f}(\xi) e^{-t|\xi|^{2\alpha}},$$

where $\mathscr{F}_1$ denotes partial Fourier transform with respect to the first d coordinates. Thus

$$\widehat{K_t}(\xi) = e^{-t|\xi|^{2\alpha}}. \tag{7.5}$$

This allows in particular the observation that the K_t are obtained from K_1 via dilation. We let $D_t : \mathrm{L}^2(\mathbb{R}^d) \to \mathrm{L}^2(\mathbb{R}^d)$ denote the (L^1-normalized) dilation operator $D_t f(x) = |t|^{-d} f(t^{-1}x)$, for $t \neq 0$. Then $\widehat{K_t}(\xi) = \widehat{K_1}(t^{1/2\alpha}\xi)$ translates to

$$K_t(x) = D_{t^{1/2\alpha}} K_1. \tag{7.6}$$

For the following, it will also be useful to note that Δ is 2-homogeneous with respect to dilation:

$$\forall t \neq 0 : D_t \circ \Delta = |t|^2 \Delta \circ D_t. \tag{7.7}$$

We next describe the relationship to continuous wavelet transforms. For this purpose, we introduce the translation operators T_x via $T_x f(y) = f(y-x)$. Now fix a vector $\eta \in \mathrm{L}^2(\mathbb{R}^d)$. We associate to η a wavelet system $(\eta_{x,t})_{x\in\mathbb{R}^d, t>0}$ given by $\eta_{x,t} = T_x D_t \eta$. Then the wavelet transform of $f \in \mathrm{L}^2(\mathbb{R}^d)$ is given by the family of scalar products with the wavelet system, i.e. $\mathscr{W}_\eta f : \mathbb{R}^d \times \mathbb{R}^+ \to \mathbb{C}$, with

$$\mathscr{W}_\eta f(x,t) = \langle f, T_x D_t \eta \rangle = f * D_t \eta^*(x), \tag{7.8}$$

where we used the notation $\eta^*(x) = \overline{\eta(-x)}$. $\mathscr{W}_\eta f$ is called the *continuous wavelet transform of f with respect to the analyzing wavelet η*. The precise choice of η is motivated by various requirements, among them the possibility of stable recovery of f from $\mathscr{W}_\eta f$. In the following, we consider two related, but usually different notions:

Definition 7.1

(a) η is called admissible if $\mathscr{W}_\eta : \mathrm{L}^2(\mathbb{R}^d) \to \mathrm{L}^2(\mathbb{R}^d \times \mathbb{R}^+, dx\frac{dt}{t})$ is isometric up to a scalar, i.e., if

$$\|f\|_2^2 = C_\eta \int_{\mathbb{R}^d} \int_{\mathbb{R}^+} |\mathscr{W}_\eta f(x,t)|^2 \, dx \, \frac{dt}{t}$$

holds, for all $f \in \mathrm{L}^2(\mathbb{R}^d)$, with some fixed $C_\eta > 0$. In this case, f can be recovered via the (weak-sense) inversion formula

$$f = \frac{1}{C_\eta} \int_{\mathbb{R}^d} \int_{\mathbb{R}^+} \mathscr{W}_\eta f(x,t) \eta_{x,t} \, dx \, \frac{dt}{t}. \tag{7.9}$$

(b) η is called Morlet-admissible if for all $f \in \mathrm{L}^2(\mathbb{R}^d)$

$$f = \lim_{\varepsilon \to 0, A \to \infty} \frac{1}{\tilde{C}_\eta} \int_\varepsilon^A W_\eta f(\cdot, t) \frac{dt}{t} \tag{7.10}$$

holds in the weak sense, with a fixed constant $\tilde{C}_\eta \neq 0$.

The notion of admissibility (at least for $d = 1$) is at the roots of wavelet analysis [193]. The notion of Morlet-admissibility was defined in [121], and further investigated in [222, 223].

The wavelet inversion formula (7.9) can be read as an expansion of the function f with respect to the wavelet system $\eta_{x,t}$. By comparison, Morlet-admissibility allows reconstruction by integration over scale alone, and thus has been used as a means of effective wavelet reconstruction [222, 317].

Admissible and Morlet-admissible vectors can be characterized by the Fourier transform, as the following lemma shows.

Lemma 7.1 *Let $\eta \in \mathrm{L}^2(\mathbb{R}^d)$ be given.*

(a) *η is admissible with constant C_η iff*

$$\int_{\mathbb{R}^+} \frac{|\widehat{\eta}(t\xi)|^2}{|t|} dt = C_\eta \quad (a.e.\ \xi \in \mathbb{R}^d). \tag{7.11}$$

(b) *η is Morlet-admissible with constant $\tilde{C}_\eta \neq 0$ iff*

$$\int_{\mathbb{R}^+} \frac{\overline{\widehat{\eta}(t\xi)}}{t} dt = \tilde{C}_\eta \quad (a.e.\ \xi \in \mathbb{R}^d), \tag{7.12}$$

with absolute convergence of the integrals.

For the proof of part (a), we refer to [174, Theorem 5.8]. Part (b) is a straightforward adaptation of the argument for (a).

Remark 7.1 Continuous wavelet transforms are closely related to group theory, in the following way: Let G be any locally compact group, and π a strongly continuous unitary representation of G in the Hilbert space $\mathscr{H}_\pi$. We let $\mathrm{L}^2(G)$ denote the L^2-space with respect to left Haar measure μ_G, and $C_b(G)$ the space of bounded continuous functions. Given a vector $\eta \in \mathscr{H}_\pi$, the coefficient operator $V_\eta : \mathscr{H}_\pi \to C_b(G)$ assigns each element $f \in \mathscr{H}$ its matrix coefficient,

$$V_\eta f(x) = \langle f, \pi(x)\eta \rangle.$$

Now η is called *admissible* whenever V_η is an isometric embedding into $\mathrm{L}^2(G)$. In this case, one has the inversion formula

$$f = \int_G V_\eta f(x)\, d\mu_G(x).$$

Expansions of this type are useful since the group element $x \in G$ often allows an interpretation of the vector $\pi(x)\eta$ as an elementary building block with special properties encoded by x. For instance, the Heisenberg group underlying the Gabor transform used in Chap. 8 allows to associate time and frequency parameters with Gabor atoms, whereas the Euclidean motion group employed in Chap. 10 produces atoms characterized by position and orientation; see Sect. 10.2.

A systematic treatment of admissible vectors, containing characterizations of admissible vectors and those representations for which they exist, can be found in [174]. It should be pointed out that in principle the approach of [174] allows to decide existence of admissible vectors for all type I groups; however, for the explicit construction of admissible vectors these methods are rather heavy-handed. In this respect, the techniques presented here will be seen to provide an attractive alternative.

Let us now rewrite the Euclidean wavelet transform in group theoretic terms. The underlying group is the semidirect product group $G = \mathbb{R}^d \rtimes \mathbb{R}^+$, which is the set $\mathbb{R}^d \times \mathbb{R}^+$ with the group law

$$(x,t)(y,r) = (x + ty, tr).$$

The representation is provided by

$$\pi(x,t) = t^{d/2} T_x D_t.$$

π is a unitary representation of G, and the above wavelet transform as a renormalized version of the group-theoretic one:

$$(\mathscr{W}_\eta f)(x,t) = t^{-d/2}\langle f, \pi(x,t)\eta\rangle = t^{-d/2} V_\eta f(x,t).$$

Left Haar measure of G is given by $d\mu_G(x,t) = dx\,\frac{dt}{t^{d+1}}$. Hence admissibility as defined in Definition 7.1 coincides with group-theoretic admissibility.

The chief reason for preferring V_η over $\mathscr{W}_\eta$ is its better compatibility with group theory: It guarantees that V_η intertwines the action of π with left shifts on G. This observation has some significance for the uses of continuous wavelet transforms in signal and image processing, a fact pointed out and used extensively, e.g., in [109]. E.g., the use of left-invariant processing or enhancement operators on continuous wavelet transforms guarantees covariance with respect to G of the net operator. However, when discussing relations to scale space, $\mathscr{W}_\eta$ is more convenient, and will thus be our preferred object.

Now we can state the precise relationship between scale space and wavelet representations:

Theorem 7.1 *Let u_f denote the scale space representation of $f \in \mathrm{L}^2(\mathbb{R}^d)$, and let $\Phi = \partial_t u_f$. Then*

$$\Phi(x,t) = t^{-1}\mathscr{W}_\eta f(x, t^{1/2\alpha}), \tag{7.13}$$

where $\eta = -(-\Delta)^\alpha K_1$ is both admissible and Morlet-admissible.

Proof We first observe that, since u solves the initial value problem

$$\partial_t u = -(-\Delta)^\alpha u = -(-\Delta)^\alpha (f * K_t) = f * (-(-\Delta)^\alpha K_t).$$

Since Δ is 2-homogeneous with respect to dilations, it follows that $-(-\Delta)^\alpha$ is 2α-homogeneous. Hence, using (7.6), we find

$$-(-\Delta)^\alpha K_t = t^{-1} D_{t^{1/2\alpha}} \eta$$

which shows (7.13). (Note also that $\eta = \eta^*$, since $\widehat{\eta}$ is real-valued.) In order to check Morlet-admissibility, we observe that

$$\widehat{\eta}(\xi) = |\xi|^{2\alpha} e^{|\xi|^{2\alpha}}$$

and compute

$$\int_{\mathbb{R}^+} |t\xi|^{2\alpha} e^{-|t\xi|^{2\alpha}} \frac{dt}{t} = \int_{\mathbb{R}^+} t^{2\alpha} e^{-t^{2\alpha}} \frac{dt}{t} < \infty,$$

where we used that the measure dt/t is invariant under multiplications, and $2\alpha - 1 > -1$. Admissibility is checked by a similar computation. □

7.3 Continuous Diffusion Wavelet Transform

Throughout this section, we consider an abstract Hilbert space $\mathscr{H}$ and a semigroup of operators $(S_t)_{t\in\mathbb{R}^+}$ on $\mathscr{H}$ given by

$$S_t = e^{tQ}.$$

Here Q is self-adjoint and negative definite. Given $f \in \mathscr{H}$, we let $u_f : \mathbb{R}^+ \to \mathscr{H}$ be defined as

$$u_f(t) = S_t(f).$$

Thus, u_f can be understood as the solution of the evolution equation

$$(\partial_t u_f)(t) = Q(u_f(t)).$$

We then define the associated continuous diffusion wavelet transform as

$$W_Q f(t) = (\partial_t u_f)(t) = Qe^{tQ}(f).$$

Note that this is well-defined: The negative definiteness of Q guarantees that Qe^{tQ} is well-defined, self-adjoint and bounded. By definition, the diffusion wavelet transform captures the rate of change in u_f at scale t. An often used interpretation of continuous wavelet transform as a succession of regularization/smoothing (corresponding to e^{tQ}) followed by differentiation (corresponding to Q) can also be transferred to our construction. Alternatively, note that for $f \in \mathrm{dom}(Q)$, the inclusion $e^{tQ}Q \subset Qe^{tQ}$ allows the observation that $W_Q f$ is the result of running the evolution with initial data Qf.

The following theorem is essentially [182, Lemma 2.1]. We present a somewhat different formulation, intended to emphasize the connections between diffusion and wavelet transforms.

Theorem 7.2 *Assume that* 0 *is not in the point spectrum of* Q. *Then, for all* $f \in \mathscr{H}$,

$$f = - \lim_{\varepsilon \to 0, A \to \infty} \int_{\varepsilon}^{A} W_Q f(s)\, ds, \tag{7.14}$$

where the integral on the right hand side is understood in the weak sense, and convergence holds in the norm. Furthermore, for all $f \in \mathscr{H}$, *the diffusion wavelet transform* $W_Q : \mathscr{H} \to \mathrm{L}^2(\mathbb{R}^+, t\,dt; \mathscr{H})$ *is isometric up to a constant*:

$$\|f\|^2 = 4 \int_0^\infty \|W_Q f(t)\|^2 t\, dt. \tag{7.15}$$

The inverse transform is provided by the weak-sense integral

$$f = 4 \int_0^\infty Q e^{tQ} [(W_Q f)(t)] t\, dt. \tag{7.16}$$

Proof Equation (7.14) is just an application of the fundamental theorem of calculus: We find

$$\int_{\varepsilon}^{A} W_Q f(s)\, ds = \int_{\varepsilon}^{A} \partial_t u_f(s)\, ds = u_f(A) - u_f(\varepsilon) = e^{AQ} f - e^{\varepsilon Q} f.$$

As Q is negative definite, $e^{AQ} f \to 0$ as $A \to \infty$, whereas $e^{\varepsilon Q} f \to f$ as $\varepsilon \to 0$, since we assume that 0 is not in the point spectrum of Q.

For the isometry property, let μ denote the spectral measure of Q. Given $f \in \mathscr{H}$, let μ_f denote the induced scalar Borel measure on $\mathbb{R}$, defined by $\mu_f(A) = \langle \mu(A) f, f \rangle$. Then the spectral representations of Q and e^{tQ} allow to compute

$$\begin{aligned}
\int_0^\infty \|W_Q f(t)\|^2 t\, dt &= \int_0^\infty \int_{-\infty}^0 \lambda^2 e^{2t\lambda}\, d\mu_f(\lambda) t\, dt \\
&= \lim_{\varepsilon \to 0, A \to \infty} \int_{\varepsilon}^{A} \int_{-\infty}^{0} \lambda^2 t e^{2t\lambda}\, d\mu_f(\lambda)\, dt \\
&= \lim_{\varepsilon \to 0, A \to \infty} \int_{-\infty}^{0} \lambda \int_{\varepsilon}^{A} \lambda t e^{2\lambda t}\, dt\, d\mu_f(\lambda) \\
&= \lim_{\varepsilon \to 0, A \to \infty} \int_{-\infty}^{0} \int_{\lambda\varepsilon}^{\lambda A} t e^{2t}\, dt\, d\mu_f(\lambda) \\
&= \lim_{\varepsilon \to 0, A \to \infty} \int_{-\infty}^{0} \left(\frac{2t-1}{4} e^{2t} \right) \Big|_{\lambda\varepsilon}^{\lambda A} d\mu_f(\lambda) = \frac{1}{4} \|f\|^2,
\end{aligned}$$

where we used Lebesgue's theorem as well as the assumption $\mu(\{0\}) = 0$.

For the proof of (7.16) note that (7.15) implies that W_Q is inverted by $4W_Q^*$. For $f \in \mathscr{H}$ and $G \in \mathrm{L}^2(\mathbb{R}^+, t\,dt; \mathscr{H})$, self-adjointness of Qe^{tQ} allows to compute that

$$\begin{aligned}\langle W_Q(f), G\rangle &= \int_0^\infty \langle Qe^{tQ} f, G(t)\rangle t\,dt\\ &= \int_0^\infty \langle f, Qe^{tQ}G(t)\rangle t\,dt\\ &= \left\langle f, \int_0^\infty Qe^{tQ}G(t)t\,dt\right\rangle,\end{aligned}$$

showing that, in the weak sense,

$$W_Q^* G = \int_0^\infty Qe^{tQ}G(t)t\,dt. \qquad \square$$

Clearly, (7.14) and (7.16) are analogues of Morlet reconstruction and wavelet inversion. The ease and generality with which these results could be formulated and proved render them somewhat suspicious. However, the next remark shows that the Morlet admissibility and admissibility statements from Theorem 7.1 are special cases of Theorem 7.2. Further interesting examples can be found in the more general context of left-invariant diffusion on Lie groups, as sketched below.

Remark 7.2 In the special case $Q = -(-\Delta)^\alpha$, as considered in Theorem 7.1, Eq. (7.13) yields

$$\mathscr{W}_\eta f(\cdot, t) = t^{2\alpha} W_Q f(\cdot, t^{2\alpha}).$$

Hence,

$$\begin{aligned}\int_0^\infty \mathscr{W}_\eta f(\cdot, t)\frac{dt}{t} &= \int_0^\infty t^{2\alpha} W_Q(\cdot, t^{2\alpha})\frac{dt}{t}\\ &= \frac{1}{2\alpha}\int_0^\infty W_Q f(\cdot, s)\,ds\\ &= -\frac{1}{2\alpha} f,\end{aligned}$$

where the last equation was due to (7.14). This shows Morlet admissibility. By a similar argument, we can prove admissibility:

$$\begin{aligned}\int_0^\infty \|\mathscr{W}_\eta f(\cdot, t)\|^2 \frac{dt}{t} &= \int_0^\infty t^{4\alpha}\|W_Q(\cdot, t^{2\alpha})\|^2 \frac{dt}{t}\\ &= \frac{1}{2\alpha}\int_0^\infty s^2\|W_Q(\cdot, s)\|^2 \frac{ds}{s}\\ &= \frac{1}{2\alpha}\int_0^\infty \|W(\cdot, s)\|^2 s\,ds = \frac{1}{8\alpha}\|f\|^2.\end{aligned}$$

Note also that here we need not appeal to the admissibility criteria formulated in Lemma 7.1. In fact, we now recognize these criteria to be based on the spectral resolution of the generator as well.

Remark 7.3 The framework of Theorem 7.2 can be extended to semigroups with a normal generator Q, as long as the spectrum of Q lies in a suitable cone contained in the half-space $\{z \in \mathbb{C} : \mathrm{Re}(z) < 0\}$. More precisely, by analogous calculations as in the proof of Theorem 7.2, one obtains for the operator W_Q and arbitrary $f \in \mathcal{H}$ that

$$\|W_Q f\|^2 = \frac{1}{4} \int_{\mathbb{C}} \frac{|z|^2}{|\mathrm{Re}(z)|^2} \, d\mu_f(z).$$

Hence, as long as the spectrum of Q is confined to the cone defined by $|\mathrm{Im}(z)| \leq |\mathrm{Re}(z)| C$, for some fixed C, W_Q is bounded, and $W_Q^* W_Q$ is given by $h(Q)$, where $h(z) = \frac{|z|^2}{|\mathrm{Re}(z)|^2}$. In particular, W_Q is inverted by $g(Q) W_Q^*$, with $g = 1/h$.

7.4 Scale Space and Continuous Wavelet Transforms over Nilpotent Lie Groups

From now on, we let N denote a simply connected, connected nilpotent Lie group. As a manifold, $N = \mathbb{R}^d$, and the left Haar measure of N is given by Lebesgue measure. We write $\mathrm{L}^2(N)$ for the associated space of square-integrable functions. Let Q be a negative self-adjoint operator on $\mathrm{L}^2(N)$, assumed to be left-invariant.

Given $f \in \mathrm{L}^2(N)$, the *left-invariant scale space representation of f arising from Q* is the solution $u_f : N \times \mathbb{R}^+ \to \mathbb{C}$ of the evolution equation

$$\partial_t u = Qu, \qquad \lim_{t \to 0} u(\cdot, t) = f. \tag{7.17}$$

Note that we abused notation in the same way as in Sect. 7.2.

The semigroup $t \mapsto e^{tQ}$ consists of left-invariant operators, and under suitable assumptions, the solutions can be obtained by convolution with Green's functions: Since the operator e^{tQ} is bounded on $\mathrm{L}^2(N)$, the Schwartz kernel theorem guarantees a distributional kernel for e^{tQ}, and since e^{tQ} commutes with left shifts, it follows that this kernel acts via convolution. Thus, for all $t > 0$ and all $f \in \mathrm{L}^2(N)$,

$$(e^{tQ} f)(x) = (f * K_t)(x) = \int_G f(y) K_t(y^{-1}x) \, d\mu_G(y), \tag{7.18}$$

with suitable kernel functions K_t, and therefore $u(\cdot, t) = f * K_t$. For the associated diffusion wavelet transform, we need to make the additional assumption that $K_t \in \mathrm{dom}(Q)$, which then allows to observe that

$$W_Q f(x,t) = Q(f * K_t)(x) = f * QK_t(x),$$

which exhibits the wavelet transform at scale t as a convolution product with the "wavelet" QK_t. To further strengthen the analogy to wavelets, observe that we can write

$$f * QK_t = \langle f, L_x \psi_t \rangle, \qquad \psi_t(x) = \overline{(QK_t)(x)},$$

where L_x denotes left translation. This suggests to read $L_x \psi_t$ as a detail of scale t located at position x. Furthermore, the inversion formula (7.16) immediately translates to

$$f = 2 \int_{\mathbb{R}^+} \int_N \langle f, L_x \psi_t \rangle L_x \psi_t \, dx t \, dt,$$

which then becomes an expansion of f into details of varying sizes and positions. Clearly, it remains to be verified whether the wavelets ψ_t actually justify this reading, i.e. whether it is meaningful to regard t as scale parameter; and this will depend on the group, as well as on the choice of the infinitesimal generator. We will now discuss a specific example of scale space representations, introduced in [111], and show how it is related to group-theoretic constructions considered in [302, 303].

7.4.1 The Heisenberg Group

The Heisenberg group is a non-commutative example where the continuous diffusion wavelet transform can also be interpreted as a group-theoretic wavelet transform. The following is entirely parallel to the discussion of Sect. 7.2 in terms of continuous diffusion wavelet transform, as worked out in Remark 7.2.

As a set, the Heisenberg group is given as $\mathbb{H} = \mathbb{R}^3$, with group law given by

$$(p_1, q_1, s_1)(p_2, q_2, s_2) = (p_1 + p_2, q_1 + q_2, s_1 + s_2 + (p_1 q_2 - p_2 q_1)/2).$$

The Lebesgue measure on $\mathbb{R}^3$ turns out to be the Haar measure of $\mathbb{H}$, and we let $\mathrm{L}^2(\mathbb{H})$ denote the associated L^2-space.

We let $\mathfrak{h}$ denote the Lie algebra of $\mathbb{H}$. Let $P, Q \in \mathfrak{h}$ denote the infinitesimal generators associated to the subgroups $\mathbb{R} \times \{0\} \times \{0\}$ and $\{0\} \times \mathbb{R} \times \{0\}$. Then, identified in the usual manner as left-invariant differential operators, P and Q are given, say for $f \in C_c^\infty(\mathbb{H})$, by

$$Pf(p,q,s) = \lim_{h \to 0} \frac{f((p,q,s)(h,0,0)) - f(p,q,s)}{h} = \left(\partial_p f - \frac{q}{2} \partial_s f \right)(p,q,s),$$

$$Qf(p,q,s) = \lim_{h \to 0} \frac{f((p,q,s)(0,h,0)) - f(p,q,s)}{h} = \left(\partial_q f + \frac{p}{2} \partial_s f \right)(p,q,s).$$

We let the *sub-Laplacian* on $\mathbb{H}$ be defined as

$$\mathscr{L} = P^2 + Q^2.$$

It can be shown that $\mathscr{L}$ has a unique operator closure (denoted by the same symbol), which is negative definite and self-adjoint. We can therefore use spectral calculus to define the operator $Q = -(-\mathscr{L})^\alpha$, with $\alpha > 0$ fixed from now on. 0 is not in the point spectrum of $\mathscr{L}$, and therefore also not of Q.

Hence Theorem 7.2 supplies an associated isometric diffusion wavelet transform. In order to see that this transform is in fact also a group-theoretic wavelet transform, we need a dilation action of $\mathbb{R}^+$ on $\mathbb{H}$. This is supplied by

$$r.(p,q,s) = (rp, rq, r^2 s).$$

It is easily verified that this defines a left action of $\mathbb{R}^+$ on $\mathbb{H}$ by automorphisms. The associated L^1-normalized dilation operator on $\mathrm{L}^2(\mathbb{H})$ is given by

$$D_r f(p,q,t) = r^{-4} f(r^{-1}p, r^{-1}q, r^{-2}t).$$

It is easily verified that the operator $\mathscr{L}$ is 2-homogeneous with respect to this dilation action, $D_t \circ \mathscr{L} = t^2 \mathscr{L} \circ D_t$, which entails that Q is 2α-homogeneous. But this in turn implies for the Green's kernels $(K_t)_{t>0}$ that

$$K_t = D_{t^{1/2\alpha}} K_1.$$

We next define the group-theoretic wavelet transform on $\mathrm{L}^2(\mathbb{H})$. The dilation action allows to introduce the semidirect product group $G = \mathbb{H} \rtimes \mathbb{R}^+$. Given a function $\eta \in \mathrm{L}^2(\mathbb{H})$, the group-theoretic wavelet transform associated to η is given as

$$\mathscr{W}_\eta f(x,t) = \langle f, L_x D_t \eta \rangle \quad (x \in \mathbb{H}, t > 0).$$

We note in passing that once again, $\mathscr{W}_\eta$ is a matrix coefficient up to normalization, associated to the unitary representation

$$\pi(x,t) = t^2 L_x D_t$$

of the semidirect product $G = \mathbb{H} \rtimes \mathbb{R}^+$. In particular, an inversion formula for $\mathscr{W}_\eta$ will establish admissibility (in the group-theoretic sense) of $\psi = \eta^*$.

Now a verbatim repetition of the arguments in the proof of Theorem 7.1 and in Remark 7.2 establishes the following result:

Theorem 7.3 *Let $(K_t)_{t>0}$ denote the Green's kernels for the evolution equation associated to $Q = -(-\mathscr{L})^\alpha$. Assume that $K_1 \in \mathrm{dom}(Q)$. Then $\eta = QK_1$ fulfills, for all $f \in \mathrm{L}^2(\mathbb{H})$,*

$$W_Q f(x,t) = t^{-1} \mathscr{W}_\eta f(x,t).$$

Moreover, $\mathscr{W}_\eta$ can be inverted by Morlet reconstruction

$$f = -2\alpha \lim_{\varepsilon \to 0, A \to \infty} \int_\varepsilon^A \mathscr{W}_\eta f(x,t) \frac{dt}{t}$$

as well as by wavelet inversion

$$f = 8\alpha \int_{\mathbb{R}^+} \int_{\mathbb{H}} \mathscr{W}_\eta f(x,t) L_x D_t \eta \, dx \, \frac{dt}{t}.$$

Moreover, $\mathscr{W}_\eta$ *is isometric up to a constant*:

$$\|f\|^2 = 8\alpha \int_{\mathbb{R}^+} \int_{\mathbb{H}} |\mathscr{W}_\eta f(x,t)|^2 \, dx \, \frac{dt}{t}.$$

Remark 7.4 The Heisenberg group, as well as the Euclidean cases discussed in Sect. 7.2 are examples of a whole class of Lie groups to which the scheme presented here can be applied, namely *homogeneous Lie groups*. A simply connected Lie group N is called homogeneous if its Lie algebra $\mathfrak{n}$ admits a one-parameter group of Lie algebra automorphisms $\delta_t = \exp(tA)$, with A diagonalizable with strictly positive entries. This group is called the dilation group; for many purposes, it plays the role of the scalars in the Abelian case. The group then acts on N via Lie group automorphisms. Finally, one can define a sub-Laplacian on N by taking a suitable set of generators $X_1, \ldots, X_d$ of $\mathfrak{n}$ (identified, as usual, with the corresponding left-invariant differential operators) and letting

$$\mathscr{L} = \sum_{i=1}^{d} X_i^2.$$

For a large class of homogeneous groups, called *stratified* Lie groups, this operator can be shown to have a self-adjoint, negative definite extension, whose point spectrum does not contain 0. In addition, a suitable choice of the X_i makes $\mathscr{L}$ 2-homogeneous. Hence, all that was needed to make the Heisenberg group case work in full analogy to the Euclidean case is present here as well, and the same reasoning as before provides a family of admissible wavelets associated to the powers of $-\mathscr{L}$, provided the technical condition that $K_1 \in \mathrm{dom}(Q)$ is fulfilled.

I expect this condition to be fulfilled for all values of α, but current literature (as far as I have been able to check) only allows to comment on $\alpha \in \{1/2, 1\}$. For $\alpha = 1$, one obtains an admissible wavelet $\mathscr{L} K_1$, where K_1 is the heat kernel. This wavelet was first described in [302]; it can be viewed as a direct generalization of the so-called Mexican Hat wavelet to the homogeneous group setting. These wavelets are Schwartz functions with vanishing moments, and thus the analogy to Euclidean wavelets is rather apt. For $\alpha = 1/2$, the kernel K_1 is the Poisson kernel studied in [158]. By the results in that paper, the condition $K_1 \in \mathrm{dom}(Q)$ is fulfilled, and the resulting wavelet is a smooth function with polynomial decay.

The construction of wavelets on homogeneous groups has recently been studied by various authors [88, 174, 181, 233, 302]. The existence of such functions was observed in [174]. However, the purely representation-theoretic techniques of [174] did not allow to establish the existence of "nice" wavelet functions, with suitable smoothness and decay properties, and similar comments apply to the arguments in [88, 233]. The first nice wavelet to be constructed for this context was the Mexican

Hat wavelet on the Heisenberg group, introduced by Mayeli [302, 303], though by a somewhat more intricate argument.

7.5 Conclusion

The chief purpose of this chapter was to exhibit the close connection between continuous wavelet transforms and linear diffusion, both by discussion of concrete examples and by the definition of a general wavelet transform associated to a semigroup of operators, with rather mild conditions on the infinitesimal generator. Informally, the connection may be summarized by the statement that diffusion up to time t retains the details of the input data of size $\geq \sqrt{t}$, whereas wavelet coefficients at scale t describe the details precisely of size t (in a suitably understood sense). Thus wavelet decomposition can be viewed as a close relative of the description provided by the diffusion process. Since most details of size precisely t are not visible at significantly smaller or larger scales, it may be argued that the wavelet transform provides a sparser description of the input data than the diffusion output. Nonetheless, the wavelet description is complete, as witnessed by the inversion formula.

Chapter 8
Left Invariant Evolution Equations on Gabor Transforms

Remco Duits, Hartmut Führ, and Bart Janssen

Abstract By means of the unitary Gabor transform one can relate operators on signals to operators on the space of Gabor transforms. In order to obtain a translation and modulation invariant operator on the space of signals, the corresponding operator on the reproducing kernel space of Gabor transforms must be left invariant, i.e. it should commute with the left regular action of the reduced Heisenberg group H_r. By using the left invariant vector fields on H_r and the corresponding left-invariant vector fields on phase space in the generators of our transport and diffusion equations on Gabor transforms we naturally employ the essential group structure on the domain of a Gabor transform. Here we mainly restrict ourselves to non-linear adaptive left-invariant convection (reassignment), while maintaining the original signal.

8.1 Introduction

The Gabor transform of a signal $f \in \mathbb{L}_2(\mathbb{R}^d)$ is a function $\mathscr{G}_\psi[f] : \mathbb{R}^d \times \mathbb{R}^d \to \mathbb{C}$ that can be roughly understood as a musical score of f, with $\mathscr{G}_\psi[f](p, q)$ describing the contribution of frequency q to the behavior of f near p [177, 224]. This interpretation is necessarily of limited precision, due to the various uncertainty principles, but it has nonetheless turned out to be a very rich source of mathematical theory as well as practical signal processing algorithms.

The use of a window function for the Gabor transform results in a smooth, and to some extent blurred, time-frequency representation; though keep in mind that by

R. Duits (✉)
Department of Mathematics and Computer Science & Department of Biomedical Engineering, Eindhoven University of Technology, Den Dolech 2, 5600 MB Eindhoven, The Netherlands
e-mail: R.Duits@tue.nl

H. Führ
Lehrstuhl A für Mathematik, RWTH Aachen, 52056 Aachen, Germany
e-mail: fuehr@MathA.rwth-aachen.de

B. Janssen
Department of Biomedical Engineering, Eindhoven University of Technology, Den Dolech 2, 5600 MB Eindhoven, The Netherlands
e-mail: B.J.Janssen@tue.nl

L. Florack et al. (eds.), *Mathematical Methods for Signal and Image Analysis and Representation*, Computational Imaging and Vision 41,
DOI 10.1007/978-1-4471-2353-8_8, © Springer-Verlag London Limited 2012

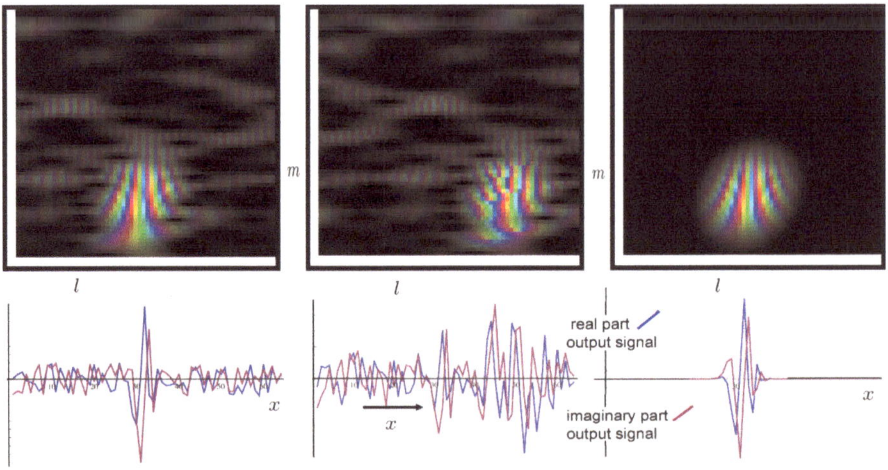

Fig. 8.1 *Top row from left to right*, (1) the Gabor transform of original signal f, (2) processed Gabor transform $\Phi_t(\mathcal{W}_\psi f)$ where Φ_t denotes a phase invariant shift (for more elaborate adaptive convection/reassignment operators see Sect. 8.6 where we operationalize the theory in [93]) using a discrete Heisenberg group, where l represents discrete spatial shift and m denotes discrete local frequency, (3) processed Gabor transform $\Phi_t(\mathcal{W}_\psi f)$ where Φ_t denotes a *phase covariant* diffusion operator on Gabor transforms with stopping time $t > 0$. *Bottom row, from left to right*: (1) Original complex-valued signal f, (2) output signal $\Upsilon_\psi f = \mathcal{W}_\psi^* \Phi_t \mathcal{W}_\psi f$ where Φ_t denotes a *phase-invariant* spatial shift (due to phase invariance the output signal looks bad and clearly phase invariant spatial shifts in the Gabor domain do not correspond to spatial shifts in the signal domain), (3) Output signal $\Upsilon_\psi f = \mathcal{W}_\psi^* \Phi_t \mathcal{W}_\psi f$ where Φ_t denotes *phase-covariant* adaptive diffusion in the Gabor domain with stopping time $t > 0$

the uncertainty principle, there is no such thing as a "true time-frequency representation". For purposes of signal analysis, say for the extraction of instantaneous frequencies, various authors tried to improve the resolution of the Gabor transform, literally in order to sharpen the time-frequency picture of the signal; this type of procedure is often called "reassignment" in the literature. For instance, Kodera et al. [260] studied techniques for the enhancement of the spectrogram, i.e. the squared modulus of the short-time Fourier transform. Since the phase of the Gabor transform is neglected, the original signal is not easily recovered from the reassigned spectrogram. Since then, various authors developed reassignment methods that were intended to allow (approximate) signal recovery [15, 69, 93].

We claim that a proper treatment of phase may be understood as *phase covariance*, rather than *phase invariance*, as advocated previously. An illustration of this claim is contained in Fig. 8.1, where phase-covariance is preferable over phase invariance. For example restoration of the old phase in the phase invariant shift (the same holds for the adaptive phase-invariant convection) creates noisy artificial patterns (Fig. 8.1, middle image) in the phase of the transported strong responses in the Gabor domain.

We adapt the group theoretical approach developed for the Euclidean motion groups in the recent works [109, 112, 113, 115, 117, 166], thus illustrating the scope of the methods devised for general Lie groups in [111] in signal and image processing. Regarding the 2D-Euclidean motion group case $SE(2)$, see also Chap. 10 of this book. Furthermore, regarding the $SE(2)$ case where we consider invertible orientation scores in stead of Gabor transforms, we note that there exists an interesting connection with orientation channel representations (that are explained in Chap. 2) and invertible orientation scores, for details see [117]. Finally, the efficient X-junction preserving flows in Chap. 1 of this book corresponds to crossing preserving flow via invertible orientation scores [112, 113, 117, 167] using only the $m = 0$, and $m = 2$ angular frequencies in orientation [166, Chaps. 3, 4.3, 6]. Without this restriction the crossing preserving flow via invertible orientation scores generically deals with all possible crossing situations.

In this chapter we restrict ourselves to Gabor transforms defined on the Heisenberg group, where reassignment will be seen to be a special case of left-invariant convection. A useful source of ideas specific to Gabor analysis and reassignment was the paper [93]. For details on phase covariant diffusions on Gabor transforms (see the right column in Fig. 8.1), we refer to [118, Chap. 7] and [236, Chap. 6]. For the construction of wavelets on the Heisenberg group, see Chap. 7. For an extension of the theory in this chapter with applications in cardiac imaging we refer to [119].

The chapter is structured as follows: Sect. 8.2 collects basic facts concerning the Gabor transform and its relation to the Heisenberg group. Section 8.3 contains the formulation of the convection-diffusion schemes. We explain the rationale behind these schemes, and comment on their interpretation in differential-geometric terms. Section 8.4 is concerned with a transfer of the schemes from the full Heisenberg group to phase space, resulting in a dimension reduction that is beneficial for implementation. The resulting scheme on phase space is described in Sect. 8.5. For a suitable choice of Gaussian window, it is possible to exploit Cauchy-Riemann equations for the analysis of the algorithms, and the design of more efficient alternatives. Section 8.6 describes a discrete implementation, and presents some experiments.

8.2 Gabor Transforms and the Reduced Heisenberg Group

Throughout the paper, we fix integers $d \in \mathbb{N}$ and $n \in \mathbb{Z} \setminus \{0\}$. The continuous Gabor-transform $\mathcal{G}_\psi[f] : \mathbb{R}^d \times \mathbb{R}^d \to \mathbb{C}$ of a square integrable signal $f : \mathbb{R}^d \to \mathbb{C}$ is commonly defined as

$$\mathcal{G}_\psi[f](p,q) = \int_{\mathbb{R}^d} f(\xi)\overline{\psi(\xi - p)} e^{-2\pi n i(\xi - p)\cdot q}\, d\xi, \tag{8.1}$$

where $\psi \in \mathbb{L}_2(\mathbb{R}^d)$ is a suitable window function. For window functions centered around zero both in space and frequency, the Gabor coefficient $\mathcal{G}_\psi[f](p,q)$ expresses the contribution of the frequency nq to the behaviour of f near p.

This interpretation is suggested by the Parseval formula associated to the Gabor transform, which reads

$$\int_{\mathbb{R}^d}\int_{\mathbb{R}^d} |\mathscr{G}_\psi[f](p,q)|^2\,\mathrm{d}p\,\mathrm{d}q = C_\psi \int_{\mathbb{R}^d} |f(p)|^2\,\mathrm{d}p,$$

$$\text{where } C_\psi = \frac{1}{n}\|\psi\|^2_{\mathbb{L}_2(\mathbb{R}^d)} \tag{8.2}$$

for all $f, \psi \in \mathbb{L}_2(\mathbb{R}^d)$. This property can be rephrased as an inversion formula:

$$f(\xi) = \frac{1}{C_\psi}\int_{\mathbb{R}^d}\int_{\mathbb{R}^d} \mathscr{G}_\psi[f](p,q)\, e^{i2\pi n(\xi-p)\cdot q}\,\psi(\xi-p)\,\mathrm{d}p\,\mathrm{d}q, \tag{8.3}$$

to be read in the weak sense. The inversion formula is commonly understood as the decomposition of f into building blocks, indexed by a time and a frequency parameter; most applications of Gabor analysis are based on this heuristic interpretation. For many such applications, the phase of the Gabor transform is of secondary importance (see, e.g., the characterization of function spaces via Gabor coefficient decay [192]). However, since the Gabor transform uses highly oscillatory complex-valued functions, its phase information is often crucial, a fact that has been specifically acknowledged in the context of reassignment for Gabor transforms [93].

For this aspect of Gabor transform, as for many others, the group-theoretic viewpoint becomes particularly beneficial. The underlying group is the *reduced Heisenberg group* H_r. As a set, $H_r = \mathbb{R}^{2d} \times \mathbb{R}/\mathbb{Z}$, with the group product

$$(p,q,s+\mathbb{Z})(p',q',s'+\mathbb{Z}) = \left(p+p', q+q', s+s' + \frac{1}{2}(q\cdot p' - p\cdot q') + \mathbb{Z}\right).$$

This makes H_r a connected (non-Abelian) nilpotent Lie group. The Lie algebra is spanned by vectors $A_1, \ldots, A_{2d+1}$ with Lie brackets $[A_i, A_{i+d}] = -A_{2d+1}$, and all other brackets vanishing.

H_r acts on $\mathbb{L}_2(\mathbb{R}^d)$ via the *Schrödinger representations* $\mathscr{U}^n : H_r \to \mathscr{B}(\mathbb{L}_2(\mathbb{R}))$,

$$\mathscr{U}^n_{g=(p,q,s+\mathbb{Z})}\psi(\xi) = e^{2\pi i n(s+q\xi-\frac{pq}{2})}\psi(\xi-p), \quad \psi \in \mathbb{L}_2(\mathbb{R}). \tag{8.4}$$

The associated matrix coefficients are defined as

$$\mathscr{W}^n_\psi f(p,q,s+\mathbb{Z}) = (\mathscr{U}^n_{(p,q,s+\mathbb{Z})}\psi, f)_{\mathbb{L}_2(\mathbb{R}^d)}. \tag{8.5}$$

This reveals that, with some change of notation, the Gabor transform is a group-theoretical wavelet transform of the type sketched in Remark 7.1. In the following, we will often omit the superscript n from U and $\mathscr{W}_\psi$, implicitly assuming that we use the same choice of n as in the definition of $\mathbb{G}_\psi$. Then a simple comparison of (8.5) with (8.1) reveals that

$$\mathscr{G}_\psi[f](p,q) = \mathscr{W}_\psi f\left(p,q,s=-\frac{pq}{2}\right). \tag{8.6}$$

Since $\mathscr{W}_\psi f(p,q,s+\mathbb{Z}) = e^{2\pi i n s}\mathscr{W}_\psi f(p,q,0+\mathbb{Z})$, the phase variable s does not affect the modulus, and (8.2) can be rephrased as

$$\int_0^1 \int_{\mathbb{R}^d} \int_{\mathbb{R}^d} |\mathscr{W}_\psi[f](p,q,s+\mathbb{Z})|^2 \,\mathrm{d}p\,\mathrm{d}q\,\mathrm{d}s = C_\psi \int_{\mathbb{R}^d} |f(p)|^2 \,\mathrm{d}p. \tag{8.7}$$

Just as before, this induces a weak-sense inversion formula, which reads

$$f = \frac{1}{C_\psi} \int_0^1 \int_{\mathbb{R}^d} \int_{\mathbb{R}^d} \mathscr{W}_\psi[f](p,q,s+\mathbb{Z}) \mathscr{U}^n_{(p,q,s+\mathbb{Z})} \psi \,\mathrm{d}p\,\mathrm{d}q\,\mathrm{d}s.$$

As a byproduct of (8.7), we note that the Schrödinger representation is irreducible. Furthermore, the orthogonal projection $\mathbb{P}_\psi$ of $\mathbb{L}_2(H_r)$ onto the range $\mathscr{R}(\mathscr{W}_\psi)$ turns out to be right convolution with a suitable (reproducing) kernel function,

$$(\mathbb{P}_\psi U)(h) = U * K(h) = \int_{H_r} U(g) K(g^{-1}h)\,\mathrm{d}g,$$

with $\mathrm{d}g$ denoting the left Haar measure (which is just the Lebesgue measure on $\mathbb{R}^{2d} \times \mathbb{R}/\mathbb{Z}$) and $K(p,q,s) = \frac{1}{C_\psi}\mathscr{W}_\psi \psi(p,q,s) = \frac{1}{C_\psi}(U_{(p,q,s)}\psi, \psi)$.

The chief reason for choosing the somewhat more redundant function $\mathscr{W}_\psi f$ over $\mathscr{G}_\psi[f]$ is that $\mathscr{W}_\psi$ translates time-frequency shifts acting on the signal f to shifts in the argument. If $\mathscr{L}$ and $\mathscr{R}$ denote the left and right regular representation, i.e., for all $g, h \in H_r$ and $F \in \mathbb{L}_2(H_r)$,

$$(\mathscr{L}_g F)(h) = F(g^{-1}h), \qquad (\mathscr{R}_g F)(h) = F(hg),$$

then $\mathscr{W}_\psi$ intertwines $\mathscr{U}$ and $\mathscr{L}$,

$$\mathscr{W}_\psi \circ \mathscr{U}^n_g = \mathscr{L}_g \circ \mathscr{W}_\psi. \tag{8.8}$$

Thus the additional group parameter s in H_r keeps track of the phase shifts induced by the non-commutativity of time-frequency shifts. By contrast, *right shifts* on the Gabor transform corresponds to changing the window:

$$\mathscr{R}_g(\mathrm{W}^n_\psi(h)) = (\mathscr{U}_{hg}\psi, f) = \mathscr{W}_{\mathscr{U}_g \psi} f(h). \tag{8.9}$$

8.3 Left Invariant Evolutions on Gabor Transforms

We relate operators $\Phi : \mathscr{R}(\mathscr{W}_\psi) \to \mathbb{L}_2(H_r)$ on Gabor transforms, which actually use and change the relevant phase information of a Gabor transform, in a well-posed manner to operators $\Upsilon_\psi : \mathbb{L}_2(\mathbb{R}^d) \to \mathbb{L}_2(\mathbb{R}^d)$ on signals via

$$\begin{aligned}(\Upsilon_\psi f)(\xi) &= (\mathcal{W}_\psi^{-1} \circ \Phi \circ \mathcal{W}_\psi f)(\xi) \\ &= \frac{1}{C_\psi} \int_{[0,1]} \int_{\mathbb{R}^d} \int_{\mathbb{R}^d} (\Phi(\mathcal{W}_\psi f))(p,q,s) \\ &\quad \times e^{i2\pi n[(\xi,q)+(s)-(1/2)(p,q)]} \psi(\xi - p)\, \mathrm{d}p\, \mathrm{d}q\, \mathrm{d}s. \qquad (8.10)\end{aligned}$$

Our aim is to design operators Υ_ψ that address signal processing problems such as denoising or detection.

8.3.1 Design Principles

We now formulate a few desirable properties of Υ_ψ, and sufficient conditions for Φ to guarantee that Υ_ψ meets these requirements.

1. *Covariance with respect to time-frequency-shifts*: The operator Υ_ψ should commute with time-frequency shifts. This requires a proper treatment of the phase.
   One easy way of guaranteeing covariance of Υ_ψ is to ensure left invariance of Φ: If Φ commutes with $\mathcal{L}_g$, for all $g \in H_r$, it follows from (8.8) that
   $$\Upsilon_\psi \circ \mathcal{U}_g^n = \mathcal{W}_\psi^* \circ \Phi \circ \mathcal{W}_\psi \circ \mathcal{U}_g^n = \mathcal{W}_\psi^* \circ \Phi \circ \mathcal{L}_g \circ \mathcal{W}_\psi = \mathcal{U}_g^n \circ \Upsilon_\psi.$$
   Generally speaking, left invariance of Φ is not a *necessary* condition for invariance of Υ_ψ: Note that $\mathcal{W}_\psi^* = \mathcal{W}_\psi^* \circ \mathbb{P}_\psi$. Thus if Φ is left-invariant, and $A : \mathbb{L}_2(H_r) \to \mathcal{R}(\mathcal{W}_\psi^n)^\perp$ an arbitrary operator, then $\Phi + A$ cannot be expected to be left-invariant, but the resulting operator on the signal side will be the same as for Φ, thus covariant with respect to time-frequency shifts.
   The authors [93] studied reassignment procedures that leave the phase invariant, whereas we shall put emphasis on phase covariance. Note however that the two properties are not mutually exclusive; convection along equiphase lines fulfills both. (See also the discussion in Sect. 8.3.4.)
2. *Nonlinearity:* The requirement that Υ_ψ commute with $\mathcal{U}^n$ immediately rules out linear operators Φ. Recall that $\mathcal{U}^n$ is irreducible, and by Schur's lemma [105], any linear intertwining operator is a scalar multiple of the identity operator.
3. By contrast to left invariance, right invariance of Φ is undesirable. By a similar argument as for left-invariance, it would provide that $\Upsilon_\psi = \Upsilon_{\mathcal{U}_g^n \psi}$.

We stress that one cannot expect that the processed Gabor transform $\Phi(\mathcal{W}_\psi f)$ is again the Gabor transform of some function constructed by the same kernel ψ, i.e. we do not expect that $\Phi(\mathcal{R}(\mathcal{W}_\psi^n)) \subset \mathcal{R}(\mathcal{W}_\psi^n)$.

8.3.2 Invariant Differential Operators on H_r

The basic building blocks for the evolution equations are the left-invariant differential operators on H_r of degree one. These operators are conveniently obtained by differentiating the *right* regular representation, restricted to one-parameter subgroups through the generators $\{A_1, \ldots, A_{2d+1}\} = \{\partial_{p_1}, \ldots, \partial_{p_d}, \partial_{q_1}, \ldots, \partial_{q_d}, \partial_s\} \subset T_e(H_r)$,

$$\mathrm{d}\mathcal{R}(A_i)U(g) = \lim_{\epsilon \to 0} \frac{U(ge^{\epsilon A_i}) - U(g)}{\epsilon}, \tag{8.11}$$

for all $g \in H_r$ and smooth $U \in \mathcal{C}^\infty(H_r)$. The resulting differential operators

$$\{\mathrm{d}\mathcal{R}(A_1), \ldots, \mathrm{d}\mathcal{R}(A_{2d+1})\} =: \{\mathcal{A}_1, \ldots, \mathcal{A}_{2d+1}\}$$

denote the left-invariant vector fields on H_r, and brief computation of (8.11) yields:

$$\mathcal{A}_i = \partial_{p_i} + \frac{q_i}{2}\partial_s, \qquad \mathcal{A}_{d+i} = \partial_{q_i} - \frac{p_i}{2}\partial_s, \qquad \mathcal{A}_{2d+1} = \partial_s, \quad \text{for } i = 1, \ldots, d.$$

The differential operators obey the same commutation relations as their Lie algebra counterparts $A_1, \ldots, A_{2d+1}$

$$[\mathcal{A}_i, \mathcal{A}_{d+i}] := \mathcal{A}_i\mathcal{A}_{d+i} - \mathcal{A}_{d+i}\mathcal{A}_i = -\mathcal{A}_{2d+1}, \tag{8.12}$$

and all other commutators are zero. I.e. $\mathrm{d}\mathcal{R}$ is a Lie algebra isomorphism.

8.3.3 Setting up the Equations

For the effective operator Φ, we will choose left-invariant evolution operators with stopping time $t > 0$. To stress the dependence on the stopping time we shall write Φ_t rather than Φ. Typically, such operators are defined by $W(p, q, s, t) = \Phi_t(\mathcal{W}_\psi f)(p, q, s)$ where W is the solution of

$$\begin{cases} \partial_t W(p, q, s, t) = Q(|\mathcal{W}_\psi f|, \mathcal{A}_1, \ldots, \mathcal{A}_{2d})W(p, q, s, t), \\ W(p, q, s, 0) = \mathcal{W}_\psi f(p, q, s), \end{cases} \tag{8.13}$$

where we note that the left-invariant vector fields $\{\mathcal{A}_i\}_{i=1}^{2d+1}$ on H_r are given by

$$\mathcal{A}_i = \partial_{p_i} + \frac{q_i}{2}\partial_s, \qquad \mathcal{A}_{d+i} = \partial_{q_i} - \frac{p_i}{2}\partial_s, \qquad \mathcal{A}_{2d+1} = \partial_s, \quad \text{for } i = 1, \ldots, d,$$

with left-invariant quadratic differential form

$$\begin{aligned} &Q(|\mathcal{W}_\psi f|, \mathcal{A}_1, \ldots, \mathcal{A}_{2d}) \\ &\quad = -\sum_{i=1}^{2d} a_i(|\mathcal{W}_\psi f|)(p, q)\mathcal{A}_i + \sum_{i=1}^{2d}\sum_{j=1}^{2d} \mathcal{A}_i D_{ij}(|\mathcal{W}_\psi f|)(p, q)\mathcal{A}_j. \end{aligned} \tag{8.14}$$

Here $a_i(|\mathcal{W}_\psi f|)$ and $D_{ij}(|\mathcal{W}_\psi f|)$ are functions such that $(p, q) \mapsto a_i(|\mathcal{W}_\psi f|)(p, q) \in \mathbb{R}$ and $(p, q) \mapsto a_i(|\mathcal{W}_\psi f|)(p, q) \in \mathbb{R}$ are smooth and either $D = 0$ (pure convection) or $D^T = D > 0$ holds pointwise (with $D = [D_{ij}]$) for all $i = 1, \ldots, 2d$,

$j = 1, \dots, 2d$. Moreover, in order to guarantee left-invariance, the mappings $a_i : \mathcal{W}_\psi f \mapsto a_i(|\mathcal{W}_\psi f|)$ need to fulfill the covariance relation

$$a_i(|\mathcal{L}_h \mathcal{W}_\psi f|)(g) = a_i(|\mathcal{W}_\psi f|)(p - p', q - q'), \tag{8.15}$$

for all $f \in \mathbb{L}_2(\mathbb{R})$, and all $g = (p, q, s + \mathbb{Z}), h = (p', q', s' + \mathbb{Z}) \in H_r$.

For $a_1 = \cdots = a_{2d+1} = 0$, the equation is a diffusion equation, whereas if $D = 0$, the equation describes a convection. We note that existence, uniqueness and square-integrability of the solutions (and thus well-definedness of Υ) are issues that will have to be decided separately for each particular choice of a_i and D. In general existence and uniqueness are guaranteed, see Sect. 8.7.

This definition of Φ_t satisfies the criteria we set up above:

1. Since the evolution equation is left-invariant (and provided uniqueness of the solutions), it follows that Φ_t is left-invariant. Thus the associated Υ_ψ is invariant under time-frequency shifts.
2. In order to ensure non-linearity, not all of the functions a_i, D_{ij} should be constant, i.e. the schemes should be *adaptive convection* and/or *adaptive diffusion*, via *adaptive* choices of convection vectors $(a_1, \dots, a_{2d})^T$ and/or conductivity matrix D. We will use ideas similar to our previous work on adaptive diffusions on invertible orientation scores [112–114, 117, 169], and to those presented in Chap. 10, where we employed evolution equations for the Euclidean motion group. We use the absolute value to adapt the diffusion and convection to avoid oscillations.
3. The two-sided invariant differential operators of degree one correspond to the center of the Lie algebra, which is precisely the span of A_{2d+1}. Both in the cases of diffusion and convection, we consistently removed the $\mathcal{A}_{2d+1} = \partial_s$-direction, and we removed the s-dependence in the coefficients $a_i(|\mathcal{W}_\psi f|)(p, q)$, $D_{ij}(|\mathcal{W}_\psi f|)(p, q)$ of the generator $Q(|\mathcal{W}_\psi f|, \mathcal{A}_1, \dots, \mathcal{A}_{2d})$ by taking the absolute value $|\mathcal{W}_\psi f|$, which is independent of s. A more complete discussion of the role of the s-variable is contained in the following subsection.

8.3.4 Convection and Diffusion Along Horizontal Curves

So far our motivation for (8.13) has been group theoretical. There is one issue we did not address yet, namely the omission of $\partial_s = \mathcal{A}_{2d+1}$ in (8.13). Here we first motivate this omission and then consider the differential geometrical consequence that (adaptive) convection and diffusion takes place along so-called horizontal curves.

The reason for the removal of the $\mathcal{A}_{2d+1}$ direction in our diffusions and convections is simply that this direction leads to a scalar multiplication operator mapping the space of Gabor transform to itself, since $\partial_s \mathcal{W}_\psi f = -2\pi i n \mathcal{W}_\psi f$. Moreover, we adaptively steer the convections and diffusions by the modulus of a Gabor transform $|\mathcal{W}_\psi f(p, q, s)| = |\mathcal{G}_\psi f(p, q)|$, which is independent of s, and clearly a vector field $(p, q, s) \mapsto F(p, q)\partial_s$ is left-invariant iff F is constant. Consequently it does *not* make sense to include the separate ∂_s in our convection-diffusion equations, as it can only yield a scalar multiplication, as for all constant $\alpha > 0$, $\beta \in \mathbb{R}$ we have

$$
\begin{aligned}
&[\partial_s, Q(|\mathcal{W}_\psi f|, \mathcal{A}_1, \ldots, \mathcal{A}_{2d})] = 0 \quad \text{and} \quad \partial_s \mathcal{W}_\psi f = -2\pi i n \mathcal{W}_\psi f \\
&\Rightarrow \quad e^{t((\alpha\partial_s^2 + \beta\partial_s) + Q(|\mathcal{W}_\psi f|, \mathcal{A}_1, \ldots, \mathcal{A}_{2d}))} \\
&\qquad = e^{-t\alpha(2\pi n)^2 - t\beta 2\pi i n}\, e^{tQ(|\mathcal{W}_\psi f|, \mathcal{A}_1, \ldots, \mathcal{A}_{2d})}.
\end{aligned}
$$

In other words ∂_s is a redundant direction in each tangent space $T_g(H_r)$, $g \in H_r$. This however does *not* imply that it is a redundant direction in the group manifold H_r itself, since clearly the s-axis represents the relevant phase and stores the non-commutative nature between position and frequency, [118, Chap. 1].

The omission of the redundant direction ∂_s in $T(H_r)$ has an important geometrical consequence. Akin to our framework of linear evolutions on orientation scores, cf. [112, 169], this means that we enforce horizontal diffusion and convection, i.e. transport and diffusion only takes place along so-called *horizontal* curves in H_r which are curves $t \mapsto (p(t), q(t), s(t)) \in H_r$, with $s(t) \in (0, 1)$, along which

$$
s(t) = \frac{1}{2}\int_0^t \sum_{i=1}^{d} q_i(\tau)p_i'(\tau) - p_i(\tau)q_i'(\tau)\, \mathrm{d}\tau,
$$

see Theorem 8.1. This gives a nice geometric interpretation to the phase variable $s(t)$, since by the Stokes theorem it represents the net surface area between a straight line connection between $(p(0), q(0), s(0))$ and $(p(t), q(t), s(t))$ and the actual horizontal curve connection $[0, t] \ni \tau \mapsto (p(\tau), q(\tau), s(\tau))$. For details, see [118].

In order to explain why the omission of the redundant direction ∂_s from the tangent bundle $T(H_r)$ implies a restriction to horizontal curves, we consider the dual frame associated to our frame of reference $\{\mathcal{A}_1, \ldots, \mathcal{A}_{2d+1}\}$. We will denote this dual frame by $\{\mathrm{d}\mathcal{A}^1, \ldots, \mathrm{d}\mathcal{A}^{2d+1}\}$ and it is uniquely determined by $\langle \mathrm{d}\mathcal{A}^i, \mathcal{A}_j\rangle = \delta^i_j$, $i, j = 1, 2, 3$ where δ^i_j denotes the Kronecker delta. A brief computation yields

$$
\begin{aligned}
\mathrm{d}\mathcal{A}^i|_{g=(p,q,s)} &= \mathrm{d}p^i, \qquad \mathrm{d}\mathcal{A}^{d+i}|_{g=(p,q,s)} = \mathrm{d}q^i, \quad i = 1, \ldots, d, \\
\mathrm{d}\mathcal{A}^{2d+1}|_{g=(p,q,s)} &= \mathrm{d}s + \frac{1}{2}(p \cdot \mathrm{d}q - q \cdot \mathrm{d}p).
\end{aligned}
\tag{8.16}
$$

Consequently a smooth curve $t \mapsto \gamma(t) = (p(t), q(t), s(t))$ is horizontal iff

$$
\langle \mathrm{d}\mathcal{A}^{2d+1}|_{\gamma(s)}, \gamma'(s)\rangle = 0 \quad \Leftrightarrow \quad s'(t) = \frac{1}{2}(q(t)\cdot p'(t) - p(t)\cdot q'(t)).
$$

Theorem 8.1 *Let $f \in \mathbb{L}_2(\mathbb{R})$ be a signal and $\mathcal{W}_\psi f$ be its Gabor transform associated to the Schwartz function ψ. If we just consider convection and no diffusion* (*i.e.* $D = 0$) *then the solution of* (8.13) *is given by*

$$
W(g, t) = \mathcal{W}_\psi f(\gamma_f^g(t)), \quad g = (p, q, s) \in H_r,
$$

where the characteristic horizontal curve $t \mapsto \gamma_f^{g_0}(t) = (p(t), q(t), s(t))$ *for each* $g_0 = (p_0, q_0, s_0) \in H_r$ *is given by the unique solution of the following ODE*:

$$\begin{cases} \dot{p}(t) = -a^1(|\mathscr{W}_\psi f|)(p(t), q(t)), & p(0) = p_0, \\ \dot{q}(t) = -a^2(|\mathscr{W}_\psi f|)(p(t), q(t)), & q(0) = q_0, \\ \dot{s}(t) = \frac{q(t)}{2}\dot{p}(t) - \frac{p(t)}{2}\dot{q}(t), & s(0) = s_0. \end{cases}$$

Consequently, the operator $\mathscr{W}_\psi f \mapsto W(\cdot, t)$ *is phase covariant* (*the phase moves along with the characteristic curves of transport*):

$$\arg\{W(g,t)\} = \arg\{\mathscr{W}_\psi f(\gamma_f^g)\} \quad \textit{for all } t > 0.$$

Proof For proof see [118, pp. 30, 31]. □

Also for the (degenerate) diffusion case with $D = D^T = [D_{ij}]_{i,j=1,\dots,d} > 0$, the omission of the $(2d+1)$th direction $\partial_s = \mathscr{A}_{2d+1}$ implies that diffusion takes place along horizontal curves. Moreover, the omission does not affect the smoothness and uniqueness of the solutions of (8.13), since the initial condition is infinitely differentiable (if ψ is a Schwarz function) and the Hörmander condition [111, 228] is by (8.12) still satisfied.

The removal of the ∂_s direction from the tangent space does *not* imply that one can entirely ignore the ∂_s-axis in the domain of a (processed) Gabor transform. The domain of a (processed) Gabor transform $\Phi_t(\mathscr{W}_\psi f)$ should *not*[1] be considered as $\mathbb{R}^{2d} \equiv H_r/\Theta$. Simply, because $[\partial_p, \partial_q] = 0$ whereas we should have (8.12). For further differential geometrical details see the appendices of [118], analogous to the differential geometry on orientation scores, [112], [118, Appendices D, C.1].

8.4 Towards Phase Space and Back

As pointed out in the introduction it is very important to keep track of the phase variable $s > 0$. The first concern that arises here is whether this results in slower algorithms. In this section we will show that this is not the case. As we will explain next, one can use an *invertible* mapping $\mathscr{S}$ from the space $\mathscr{H}_n$ of Gabor transforms to phase space (the space of Gabor transforms restricted to the plane $s = \frac{pq}{2}$). As a result by means of conjugation with $\mathscr{S}$ we can map our diffusions on $\mathscr{H}_n \subset \mathbb{L}_2(\mathbb{R}^2 \times [0,1])$ uniquely to diffusions on $\mathbb{L}_2(\mathbb{R}^2)$ simply by conjugation with $\mathscr{S}$. From a geometrical point of view it is better/easier to consider the diffusions on $\mathscr{H}_n \subset \mathbb{L}_2(\mathbb{R}^{2d} \times [0,1])$ than on $\mathbb{L}_2(\mathbb{R}^{2d})$, even though all our numerical PDE-Algorithms take place in phase space in order to gain speed.

---

[1] As we explain in [118, Appendices B and C] the Gabor domain is a principal fiber bundle $P_T = (H_r, \mathbb{T}, \pi, \mathscr{R})$ equipped with the Cartan connection form $\omega_g(X_g) = \langle \mathrm{d}s + \frac{1}{2}(p\,\mathrm{d}q - q\,\mathrm{d}p), X_g\rangle$, or equivalently, it is a contact manifold, cf. [54, p. 6], [118, Appendix B, Definition B.14], $(H_r, \mathrm{d}\mathscr{A}^{2d+1})$).

Definition 8.1 Let $\mathscr{H}_n$ denote the space of all complex-valued functions F on H_r such that $F(p,q,s+\mathbb{Z}) = e^{-2\pi i n s} F(p,q,1)$ and $F(\cdot,\cdot,s+\mathbb{Z}) \in \mathbb{L}_2(\mathbb{R}^{2d})$ for all $s \in \mathbb{R}$, then clearly $\mathscr{W}_\psi f \in \mathscr{H}_n$ for all $f, \psi \in \mathscr{H}_n$.

In fact $\mathscr{H}_n$ is the closure of the space $\{\mathscr{W}_\psi^n f \,|\, \psi, f \in \mathbb{L}_2(\mathbb{R})\}$ in $\mathbb{L}_2(H_r)$. The space $\mathscr{H}_n$ is bi-invariant, since:

$$\mathscr{W}_\psi^n \circ \mathscr{U}_g^n = \mathscr{L}_g \circ \mathscr{W}_\psi^n \quad \text{and} \quad \mathscr{W}_{\mathscr{U}_g^n \psi}^n = \mathscr{R}_g \circ \mathscr{W}_\psi^n, \tag{8.17}$$

where again $\mathscr{R}$ denotes the right regular representation on $\mathbb{L}_2(H_r)$ and $\mathscr{L}$ denotes the left regular representation of H_r on $\mathbb{L}_2(H_r)$. We can identify $\mathscr{H}_n$ with $\mathbb{L}_2(\mathbb{R}^{2d})$ by means of the following operator $\mathscr{S} : \mathscr{H}_n \to \mathbb{L}_2(\mathbb{R}^{2d})$ given by

$$(\mathscr{S}F)(p,q) = F\left(p,q,\frac{pq}{2}+\mathbb{Z}\right) = e^{i\pi n p q} F(p,q,0+\mathbb{Z}).$$

Clearly, this operator is invertible and its inverse is given by

$$(\mathscr{S}^{-1}F)(p,q,s+\mathbb{Z}) = e^{-2\pi i s n} e^{-i\pi n p q} F(p,q).$$

The operator $\mathscr{S}$ simply corresponds to taking the section $s(p,q) = -\frac{pq}{2}$ in the left cosets H_r/Θ where $\Theta = \{(0,0,s+\mathbb{Z}) \mid s \in \mathbb{R}\}$ of H_r. Furthermore we recall the common Gabor transform $\mathscr{G}_\psi^n$ given by (8.1) and its relation (8.6) to the full Gabor transform. This relation is simply $\mathscr{G}_\psi^n = \mathscr{S} \circ \mathscr{W}_\psi^n$.

Theorem 8.2 *Let the operator Φ map the closure $\mathscr{H}_n$, $n \in \mathbb{Z}$, of the space of Gabor transforms into itself, i.e. $\Phi : \mathscr{H}_n \to \mathscr{H}_n$. Define the left and right-regular rep's of H_r on $\mathscr{H}_n$ by restriction*

$$\mathscr{R}_g^{(n)} = \mathscr{R}_g|_{\mathscr{H}_n} \quad \text{and} \quad \mathscr{L}_g^{(n)} = \mathscr{L}_g|_{\mathscr{H}_n} \quad \text{for all } g \in H_r. \tag{8.18}$$

Define the corresponding left and right-regular rep's of H_r on phase space by

$$\tilde{\mathscr{R}}_g^{(n)} := \mathscr{S} \circ \mathscr{R}_g^{(n)} \circ \mathscr{S}^{-1}, \qquad \tilde{\mathscr{L}}_g^{(n)} := \mathscr{S} \circ \mathscr{L}_g^{(n)} \circ \mathscr{S}^{-1}.$$

For explicit formulas see [118, p. 9]. *Let $\tilde{\Phi} := \mathscr{S} \circ \Phi \circ \mathscr{S}^{-1}$ be the corresponding operator on $\mathbb{L}_2(\mathbb{R}^{2d})$ and*

$$\Upsilon_\psi = (\mathscr{W}_\psi^n)^* \circ \Phi \circ \mathscr{W}_\psi^n = (\mathscr{S}\mathscr{W}_\psi^n)^{-1} \circ \tilde{\Phi} \circ \mathscr{S}\mathscr{W}_\psi^n = (\mathscr{G}_\psi^n)^* \circ \tilde{\Phi} \circ \mathscr{G}_\psi^n.$$

Then one has the following correspondence:

$$\Upsilon_\psi \circ \mathscr{U}^n = \mathscr{U}^n \circ \Upsilon_\psi \quad \Leftarrow \quad \Phi \circ \mathscr{L}^n = \mathscr{L}^n \circ \Phi \quad \Leftrightarrow \quad \tilde{\Phi} \circ \tilde{\mathscr{L}}^n = \tilde{\mathscr{L}}^n \circ \tilde{\Phi}. \tag{8.19}$$

If moreover $\Phi(\mathscr{R}(\mathscr{W}_\psi)) \subset \mathscr{R}(\mathscr{W}_\psi)$ then the left implication may be replaced by an equivalence. If Φ does not satisfy this property then one may replace $\Phi \to \mathscr{W}_\psi \mathscr{W}_\psi^* \Phi$ in (8.19) *to obtain full equivalence. Note that $\Upsilon_\psi = \mathscr{W}_\psi^* \Phi \mathscr{W}_\psi = \mathscr{W}_\psi^* (\mathscr{W}_\psi \mathscr{W}_\psi^* \Phi) \mathscr{W}_\psi$.*

Proof For details see our technical report [118, Theorem 2.2]. □

8.5 Left-Invariant Evolutions on Phase Space

For the remainder of the paper, for the sake of simplicity, we fix $d = 1$.

Now we would like to apply Theorem 8.2 to our left invariant evolutions (8.13) to obtain the left-invariant diffusions on phase space (where we reduce 1 dimension in the domain). To this end we first compute the left-invariant vector fields $\{\tilde{\mathscr{A}_i}\} := \{\mathscr{S}\mathscr{A}_i\mathscr{S}^{-1}\}_{i=1}^3$ on phase space. The left-invariant vector fields on phase space are

$$\begin{aligned}\tilde{\mathscr{A}_1}U(p',q') &= \mathscr{S}\mathscr{A}_1\mathscr{S}^{-1}U(p',q') = ((\partial_{p'} - 2n\pi i q')U)(p',q'),\\ \tilde{\mathscr{A}_2}U(p',q') &= \mathscr{S}\mathscr{A}_2\mathscr{S}^{-1}U(p',q') = (\partial_{q'}U)(p',q'),\\ \tilde{\mathscr{A}_3}U(p',q') &= \mathscr{S}\mathscr{A}_3\mathscr{S}^{-1}U(p',q') = -2in\pi U(p',q'),\end{aligned} \tag{8.20}$$

for all $(p,q) \in \mathbb{R}$ and all locally defined smooth functions $U : \Omega_{(p,q)} \subset \mathbb{R}^2 \to \mathbb{C}$.

Now that we have computed the left-invariant vector fields on phase space, we can express our left-invariant evolution equations (8.13) on phase space

$$\begin{cases}\partial_t \tilde{W}(p,q,t) = \tilde{Q}(|\mathscr{G}_\psi f|, \tilde{\mathscr{A}_1}, \tilde{\mathscr{A}_2})\tilde{W}(p,q,t),\\ \tilde{W}(p,q,0) = \mathscr{G}_\psi f(p,q),\end{cases} \tag{8.21}$$

with left-invariant quadratic differential form

$$\tilde{Q}(|\mathscr{G}_\psi f|, \tilde{\mathscr{A}_1}, \tilde{\mathscr{A}_2}) = -\sum_{i=1}^{2} a_i(|\mathscr{G}_\psi f|)(p,q)\tilde{\mathscr{A}_i} + \sum_{i=1}^{2}\sum_{j=1}^{2} \tilde{\mathscr{A}_i} D_{ij}(|\mathscr{G}_\psi f|)(p,q)\tilde{\mathscr{A}_j}. \tag{8.22}$$

Similar to the group case, the a_i and D_{ij} are functions such that

$$(p,q) \mapsto a_i(|\mathscr{G}_\psi f|)(p,q) \in \mathbb{R}$$

and

$$(p,q) \mapsto a_i(|\mathscr{G}_\psi f|)(p,q) \in \mathbb{R}$$

are smooth and either $D = 0$ (pure convection) or $D^T = D > 0$ (with $D = [D_{ij}]$ $i, j = 1, \ldots, 2d$), so Hörmander's condition [228] (which guarantees smooth solutions $\tilde{W}$, provided the initial condition $\tilde{W}(\cdot,\cdot,0)$ is smooth) is satisfied because of (8.12).

Theorem 8.3 *The unique solution $\tilde{W}$ of* (8.21) *is obtained from the unique solution W of* (8.13) *by means of*

$$\tilde{W}(p,q,t) = (\mathscr{S}W(\cdot,\cdot,\cdot,t))(p,q), \quad \textit{for all } t \geq 0 \textit{ and for all } (p,q) \in \mathbb{R}^2,$$

with in particular $\tilde{W}(p,q,0) = \mathscr{G}_\psi(p,q) = (\mathscr{S}\mathscr{W}_\psi)(p,q) = (\mathscr{S}W(\cdot,\cdot,\cdot,0))(p,q)$.

Proof This follows by the fact that the evolutions (8.13) leave the function space $\mathscr{H}_n$ invariant and the fact that the evolutions (8.21) leave the space invariant $\mathbb{L}_2(\mathbb{R}^2)$ invariant, so that we can apply direct conjugation with the invertible operator $\mathscr{S}$ to relate the unique solutions, where we have

$$\begin{aligned}\tilde{W}(p,q,t) &= (e^{t\tilde{Q}(|\mathscr{G}_\psi f|,\tilde{\mathscr{A}_1},\tilde{\mathscr{A}_2})}\mathscr{G}_\psi f)(p,q)\\ &= (e^{t\tilde{Q}(|\mathscr{G}_\psi f|,\mathscr{S}\mathscr{A}_1\mathscr{S}^{-1},\mathscr{S}\mathscr{A}_2\mathscr{S}^{-1})}\mathscr{S}\mathscr{W}_\psi f)(p,q)\\ &= (e^{\mathscr{S}\circ tQ(|\mathscr{W}_\psi f|,\mathscr{A}_1,\mathscr{A}_2)\circ\mathscr{S}^{-1}}\mathscr{S}\mathscr{W}_\psi f)(p,q)\\ &= (\mathscr{S}\circ e^{tQ(|\mathscr{W}_\psi f|,\mathscr{A}_1,\mathscr{A}_2)}\circ\mathscr{S}^{-1}\mathscr{S}\circ\mathscr{W}_\psi f)(p,q)\\ &= (\mathscr{S}W(\cdot,\cdot,\cdot,t))(p,q)\end{aligned} \tag{8.23}$$

for all $t>0$ on densely defined domains. For every $\psi\in\mathbb{L}_2(\mathbb{R})\cap S(\mathbb{R})$, the space of Gabor transforms is a reproducing kernel space with a bounded and smooth reproducing kernel, so that $\mathscr{W}_\psi f$ (and thereby $|\mathscr{W}_\psi f|=|\mathscr{G}_\psi f|=\sqrt{(\Re\mathscr{G}_\psi f)^2+(\Im\mathscr{G}_\psi f)^2}$) is uniformly bounded and continuous and equality (8.23) holds for all $p,q\in\mathbb{R}^2$. □

8.5.1 The Cauchy Riemann Equations on Gabor Transforms

As previously observed in [93], the Gabor transforms associated to Gaussian windows obey Cauchy-Riemann equations which are particularly useful for the analysis of convection schemes, as well as for the design of more efficient algorithms.

More precisely, if $\psi(\xi)=\psi_a(\xi):=e^{-\pi n\frac{(\xi-c)^2}{a^2}}$ and f is some arbitrary signal in $\mathbb{L}_2(\mathbb{R})$ then we have

$$\begin{aligned}&(a^{-1}\mathscr{A}_2+ia\mathscr{A}_1)\mathscr{W}_\psi(f)=0 \quad\Leftrightarrow\quad (a^{-1}\tilde{\mathscr{A}_2}+ia\tilde{\mathscr{A}_1})\mathscr{G}_\psi(f)=0,\\ &(a^{-1}\mathscr{A}_2+ia\mathscr{A}_1)\log\mathscr{W}_\psi(f)=0 \quad\Leftrightarrow\quad (a^{-1}\tilde{\mathscr{A}_2}+ia\tilde{\mathscr{A}_1})\log\mathscr{G}_\psi(f)=0,\end{aligned} \tag{8.24}$$

where we recall that $\mathscr{G}_\psi(f)=\mathscr{S}\mathscr{W}_\psi(f)$ and $\mathscr{A}_i=\mathscr{S}^{-1}\tilde{\mathscr{A}_i}\mathscr{S}$ for $i=1,2,3$. For details see [118], [236, Chap. 5], where the essential observation is that we can write

$$\mathscr{G}_{\psi_a}f(p,q)=\sqrt{a}\mathscr{G}_{\mathscr{D}_{\frac{1}{a}}\psi}(f)(p,q)=\sqrt{a}\mathscr{G}_\psi\mathscr{D}_{\frac{1}{a}}f\left(\frac{p}{a},aq\right)$$

with $\psi=\psi_{a=1}$ and where the unitary dilation operator $\mathscr{D}_a:\mathbb{L}_2(\mathbb{R})\to\mathbb{L}_2(\mathbb{R})$ is given by $\mathscr{D}_a(\psi)(x)=a^{-\frac{1}{2}}f(x/a)$, $a>0$. For the case $a=1$, Eq. (8.24) was noted in [93]. As a direct consequence of (8.24) we have

$$\begin{aligned}&|\tilde{U}^a|\partial_q\tilde{\Omega}^a=-a^2\partial_p|\tilde{U}^a| \quad\text{and}\quad |\tilde{U}^a|\partial_p\tilde{\Omega}^a=a^{-2}\partial_q|\tilde{U}^a|+2\pi q,\\ &\mathscr{A}_2\Omega^a=a^2\mathscr{A}_1|U^a| \quad\text{and}\quad \mathscr{A}_1\Omega^a=a^{-2}\mathscr{A}_1|U^a|,\end{aligned} \tag{8.25}$$

where $\tilde{U}^a$ resp. U^a is short notation for $\tilde{U}^a = \mathscr{G}_{\psi_a}(f)$, $U^a = \mathscr{W}_{\psi_a}(f)$, $\tilde{\Omega}^a = \arg\{\mathscr{G}_{\psi_a}(f)\}$ and $\Omega^a = \arg\{\mathscr{W}_{\psi_a}(f)\}$.

If one equips the *contact-manifold* (for general definition see cf. [54, p. 6] or [118, Appendix B, Definition B.14]), given by the pair $(H_r, \mathrm{d}\mathscr{A}^3)$, recall (8.16) with the following *non-degenerate*[2] left-invariant metric tensor

$$\mathscr{G}_\beta = g_{ij}\mathrm{d}\mathscr{A}^i \otimes \mathrm{d}\mathscr{A}^j = \beta^4 \mathrm{d}\mathscr{A}^1 \otimes \mathrm{d}\mathscr{A}^1 + \mathrm{d}\mathscr{A}^2 \otimes \mathrm{d}\mathscr{A}^2, \tag{8.26}$$

which is bijectively related to the linear operator $G : \mathfrak{H} \to \mathfrak{H}'$, where $\mathfrak{H} = \mathrm{span}\{\mathscr{A}_1, \mathscr{A}_2\}$ denotes the horizontal part of the tangent space, that maps $\mathscr{A}_1$ to $\beta^4 \mathrm{d}\mathscr{A}^1$ and $\mathscr{A}_2$ to $\mathrm{d}\mathscr{A}^2$. The inverse operator of G is bijectively related to

$$\mathscr{G}_\beta^{-1} = g^{ij}\mathscr{A}_i \otimes \mathscr{A}_j = \beta^{-4}\mathscr{A}_1 \otimes \mathscr{A}_1 + \mathscr{A}_2 \otimes \mathscr{A}_2.$$

Here the fundamental positive parameter β^{-1} has physical dimension length, so that this first fundamental form is consistent with respect to physical dimensions. Intuitively, the parameter β sets a global balance between changes in frequency space and changes in position space. The Cauchy-Riemann relations (8.25) that hold between local phase and local amplitude can be written in geometrical form:

$$\mathscr{G}^{-1}_{\beta=\frac{1}{a}}(\mathrm{d}\log|U|, \mathbb{P}_{\mathfrak{H}^*}\mathrm{d}\Omega) = 0, \tag{8.27}$$

where $U = \mathscr{W}_{\psi_a} f = |U|e^{i\Omega}$ and where the left-invariant gradient equals $\mathrm{d}\Omega = \sum_{i=1}^{3} \mathscr{A}_i \Omega \, \mathrm{d}\mathscr{A}^i$ whose horizontal part equals $\mathbb{P}_{\mathfrak{H}^*}\mathrm{d}\Omega = \sum_{i=1}^{2} \mathscr{A}_i \Omega \, \mathrm{d}\mathscr{A}^i$. This gives us a geometric understanding. The horizontal part $\mathbb{P}_{\mathfrak{H}^*}\mathrm{d}\Omega|_{g_0}$ of the normal co-vector $\mathrm{d}\Omega|_{g_0}$ to the surface $\{(p,q,s) \in H_r \mid \Omega(p,q,s) = \Omega(g_0)\}$ is $\mathscr{G}_\beta$-orthogonal to the normal co-vector $\mathrm{d}|U||_{g_0}$ to the surfaces $\{(p,q,s) \in H_r \mid |U|(p,q,s) = |U|(g_0)\}$.

8.6 Phase Invariant Convection on Gabor Transforms

First we derive left-invariant and phase-invariant differential operators on Gabor transforms $U := \mathscr{W}_\psi(f)$, which will serve as generators of left-invariant phase-invariant convection (i.e. set $D = 0$ in (8.13) and (8.21)) equations on Gabor transforms. This type of convection is also known as differential reassignment, cf. [69, 93], where the practical goal is to sharpen Gabor distributions towards lines (close to minimal energy curves [118, Appendix D]) in H_r, while maintaining the signal as much as possible.

On the group H_r it directly follows by the product rule for differentiation that the following differential operators $\mathscr{C} : \mathscr{H}_n \to \mathscr{H}_n$ given by

$$\mathscr{C}(U) = \mathscr{M}(|U|)(-\mathscr{A}_2\Omega\mathscr{A}_1 U + \mathscr{A}_1\Omega\mathscr{A}_2 U), \quad \text{where } \Omega = \arg\{U\}$$

[2]The metric tensor is degenerate on H_r, but we consider a contact manifold $(H_3, \mathrm{d}\mathscr{A}^3)$ where tangent vectors along horizontal curves do not have an $\mathscr{A}_3$-component.

are phase invariant, where $\mathcal{M}(|U|)$ denotes a multiplication operator on $\mathcal{H}_n$ with the modulus of U naturally associated to a bounded monotonically increasing differentiable function $\mu : [0, \max(U)] \to [0, \mu(\max(U))] \subset \mathbb{R}$ with $\mu(0) = 0$, i.e. $(\mathcal{M}(|U|)V)(p,q) = \mu(|U|(p,q))V(p,q)$ for all $V \in \mathcal{H}_n$, $(p,q) \in \mathbb{R}^2$.

The absolute value of Gabor transform is almost everywhere smooth (if ψ is a Schwarz function) bounded and $\mathcal{C}$ can be considered as an unbounded operator from $\mathcal{H}_n$ *into* $\mathcal{H}_n$, as the bi-invariant space $\mathcal{H}_n$ is invariant under bounded multiplication operators which do not depend on $z = e^{2\pi i s}$. Concerning phase invariance, direct computation yields: $\mathcal{C}(e^{i\Omega}|U|) = \mathcal{M}(|U|)\, e^{i\Omega}(-\mathcal{A}_2\Omega\mathcal{A}_1|U| + \mathcal{A}_1\Omega\mathcal{A}_2|U|)$. For Gaussian kernels $\psi_a(\xi) = e^{-a^{-2}\xi^2 n\pi}$ we may apply the Cauchy Riemann relations (8.24) which simplifies for the special case $\mathcal{M}(|U|) = |U|$ to

$$\mathcal{C}(e^{i\Omega}|U|) = (a^2(\partial_p|U|)^2 + a^{-2}(\partial_q|U|)^2)\, e^{i\Omega}. \tag{8.28}$$

Now consider the following phase-invariant adaptive convection equation on H_r,

$$\begin{cases} \partial_t W(g,t) = -\mathcal{C}(W(\cdot,t))(g), \\ W(g,0) = U(g) \end{cases} \tag{8.29}$$

with either

$$\begin{aligned} &1.\quad \mathcal{C}(W(\cdot,t)) = \mathcal{M}(|U|)(-\mathcal{A}_2\Omega, \mathcal{A}_1\Omega)\cdot(\mathcal{A}_1 W(\cdot,t), \mathcal{A}_2 W(\cdot,t)) \quad \text{or} \\ &2.\quad \mathcal{C}(W(\cdot,t)) = e^{i\Omega}\left(a^2\frac{(\partial_p|W(\cdot,t)|)^2}{|W(\cdot,t)|} + a^{-2}\frac{(\partial_q|W(\cdot,t)|)^2}{|W(\cdot,t)|}\right). \end{aligned} \tag{8.30}$$

In the first choice we stress that $\arg(W(\cdot,t)) = \arg(W(\cdot,0)) = \Omega$, since transport only takes place along iso-phase surfaces. Initially, in case $\mathcal{M}(|U|) = 1$ the two approaches are the same since at $t = 0$ the Cauchy Riemann relations (8.25) hold, but as time increases the Cauchy-Riemann equations are violated (this directly follows by the preservation of phase and non-preservation of amplitude), which has been more or less overlooked in the single step convection schemes in [69, 93].

The second choice in (8.30) in (8.29) is just a phase-invariant inverse Hamilton Jakobi equation on H_r, with a Gabor transform as initial solution. Rather than computing the viscosity solution of this non-linear PDE, we may as well store the phase and apply an inverse Hamilton Jakobi system on $\mathbb{R}^2$ with the amplitude $|U|$ as initial condition and multiply with the stored phase factor afterwards.

With respect to the first choice in (8.30) in (8.29), which is much more cumbersome to implement, the authors in [93] considered the equivalent equation on phase space:

$$\begin{cases} \partial_t \tilde{W}(p,q,t) = -\tilde{\mathcal{C}}(\tilde{W}(\cdot,t))(p,q), \\ \tilde{W}(p,q,0) = \mathcal{G}_\psi f(p,q) =: \tilde{U}(p,q) = e^{i\tilde{\Omega}(p,q)}|\tilde{U}(p,q)| \\ \qquad\qquad\quad = e^{i\tilde{\Omega}(p,q)}|U|(p,q) \end{cases} \tag{8.31}$$

with $\tilde{C}(\tilde{W}(\cdot,t)) = \mathcal{M}(|U|)(-\tilde{\mathcal{A}_2}\tilde{\Omega}\tilde{\mathcal{A}_1}\tilde{W}(\cdot,t) + (\partial_q\tilde{\Omega} - 2\pi q)\tilde{\mathcal{A}_2}\tilde{W}(\cdot,t))$, where we recall $\mathcal{G}_\psi = \mathcal{S}\mathcal{W}_\psi$ and $\mathcal{A}_i = \mathcal{S}^{-1}\tilde{\mathcal{A}_i}\mathcal{S}$ for $i = 1,2,3$. Note that the authors in [93] consider the case $\mathscr{M}' = 1$. However the case $\mathscr{M} = 1$ and the earlier mentioned case $\mathscr{M}(|U|) = |U|$ are equivalent:

$$\frac{\partial}{\partial t}|U| = a^2 \frac{(\partial_p |U|)^2}{|U|} + a^{-2} \frac{(\partial_q |U|)^2}{|U|}$$
$$\Leftrightarrow \quad \frac{\partial}{\partial t} \log |U| = a^2 (\partial_p \log |U|)^2 + a^{-2} (\partial_q \log |U|)^2.$$

Although the approach in [93] is highly plausible, the authors did not provide an explicit computational scheme like we provide in the next section.

On the other hand with the second approach in (8.30) one does not need the technicalities of the previous section, since here the viscosity solution of the system (8.31), [128, Chap. 10], is given by a basic inverse convolution over the (max, +) algebra, [57], (also known as *erosion operator* in image analysis)

$$\tilde{W}(p,q,t) = (K_t \ominus |U|)(p,q) e^{i\Omega(p,q,t)}, \tag{8.32}$$

with the kernel $K_t(p,q) = -\frac{a^{-2}p^2 + a^2 q^2}{4t}$ and where the erosion operator is given by

$$(f \ominus g)(p,q) = \inf_{(p',q') \in \mathbb{R}^2} [g(p',q') - f(p-p', q-q')].$$

Here the homomorphism between dilation/erosion and diffusion/inverse diffusion is given by the Cramer transform $C = \mathfrak{F} \circ \log \circ \mathscr{L}$, [6, 57], which is a concatenation of the multi-variate Laplace transform, logarithm and Fenchel transform. The Fenchel transform maps a convex function $c : \mathbb{R}^2 \to \overline{\mathbb{R}}$ onto $\mathbf{x} \mapsto [\mathfrak{F}c](\mathbf{x}) = \sup\{\mathbf{y} \cdot \mathbf{x} - c(\mathbf{y}) \mid \mathbf{y} \in \mathbb{R}^2\}$. The isomorphic property of the Cramer transform is

$$\mathscr{C}(f * g) = \mathfrak{F} \log \mathscr{L}(f * g) = \mathfrak{F}(\log \mathscr{L} f + \log \mathscr{L} g) = \mathscr{C} f \oplus \mathscr{C} g,$$

with convolution on the (max, +)-algebra given by

$$f \oplus g(\mathbf{x}) = \sup_{\mathbf{y} \in \mathbb{R}^d} [f(\mathbf{x} - \mathbf{y}) + g(\mathbf{y})].$$

8.7 Existence and Uniqueness of the Evolution Solutions

The convection diffusion systems (8.13) have unique solutions, since the coefficients a_i and D_{ij} depend smoothly on the modulus of the initial condition $|\mathscr{W}_\psi f| = |\mathscr{G}_\psi f|$. So for a *given* initial condition $\mathscr{W}_\psi f$ the left-invariant convection diffusion generator $Q(|\mathscr{W}_\psi f|, \mathscr{A}_1, \ldots, \mathscr{A}_{2d})$ is of the type

$$Q(|\mathscr{W}_\psi f|, \mathscr{A}_1, \ldots, \mathscr{A}_{2d}) = \sum_{i=1}^{d} \alpha_i \mathscr{A}_i + \sum_{i,j=1}^{d} \mathscr{A}_i \beta_{ij} \mathscr{A}_j.$$

Such hypo-elliptic operators with almost everywhere smooth coefficients given by $\alpha_i(p,q) = a_i(|\mathscr{G}_\psi f|)(p,q)$ and $\beta_{ij}(p,q) = D_{ij}(|\mathscr{G}_\psi f|)(p,q)$ generate strongly continuous, semigroups on $\mathbb{L}_2(\mathbb{R}^2)$, as long as we keep the functions α_i and β_{ij} fixed, [401], yielding unique solutions where at least formally we may write

$$W(p,q,s,t) = \Phi_t(\mathscr{W}_\psi f)(p,q,s) = e^{t(\sum_{i=1}^d \alpha_i \mathscr{A}_i + \sum_{i,j=1}^d \mathscr{A}_i \beta_{ij} \mathscr{A}_j)} \mathscr{W}_\psi f(p,q,s)$$

with $\lim_{t\downarrow 0} W(\cdot,t) = \mathscr{W}_\psi f$ in $\mathbb{L}_2$-sense. Note that if ψ is a Gaussian kernel and $f \neq 0$ the Gabor transform $\mathscr{G}_\psi f \neq 0$ is real analytic on $\mathbb{R}^{2d}$, so it can not vanish on a set with positive measure, so that $\alpha_i : \mathbb{R}^2 \to \mathbb{R}$ are almost everywhere smooth.

This applies in particular to the first reassignment approach in (8.30) (mapping everything consistently into phase space using Theorem 8.3), where we have set

$$a_1(|\mathscr{G}_\psi f|) = \mathscr{M}(|\mathscr{G}_\psi f|)|\mathscr{G}_\psi f|^{-1}\partial_p|\mathscr{G}_\psi f|,$$

$$a_2(|\mathscr{G}_\psi f|) = \mathscr{M}(|\mathscr{G}_\psi f|)|\mathscr{G}_\psi f|^{-1}\partial_q|\mathscr{G}_\psi f|,$$

and $D = 0$. Now we have to be careful with the second approach in (8.30), as here the operator $\tilde{U} \mapsto \tilde{\mathscr{C}}(\tilde{U})$ is *non-linear* and we are not allowed to apply the general theory. Nevertheless the operator $\tilde{U} \mapsto \tilde{\mathscr{C}}(\tilde{U})$ is left-invariant and maps the space $\mathbb{L}_2^+(\mathbb{R}^2) = \{f \in \mathbb{L}_2(\mathbb{R}^2) \mid f \geq 0\}$ into itself again. In these cases the erosion solutions (8.32) are the *unique viscosity solutions*, of (8.29), see [85].

Remark 8.1 For the diffusion case, [118, Chap. 7], [236, Chap. 6], we have $D = [D_{ij}]_{i,j=1,\dots,2d} > 0$, in which case the (horizontal) diffusion generator

$$Q(|\mathscr{W}_\psi f|, \mathscr{A}_1, \dots, \mathscr{A}_{2d})$$

on the group is *hypo-elliptic*, whereas the corresponding generator

$$\tilde{Q}(|\mathscr{G}_\psi f|, \tilde{\mathscr{A}}_1, \dots, \tilde{\mathscr{A}}_{2d})$$

on phase space is *elliptic*. By the results [128, Chap. 7.1.1] and [291] we conclude that there exists a unique weak solution $\tilde{W} = \mathscr{S} W \in \mathbb{L}_2(\mathbb{R}^+, \mathbb{H}_1(\mathbb{R}^2)) \cap \mathbb{H}_1(\mathbb{R}^+, \mathbb{L}_2(\mathbb{R}^2))$ and thereby we can apply continuous point evaluation in time and operator $\mathbb{L}_2(\mathbb{R}^2) \ni \mathscr{G}_\psi f \mapsto \tilde{\Phi}_t(\mathscr{G}_\psi f) := \tilde{W}(\cdot,\cdot,t) \in \mathbb{L}_2(\mathbb{R}^2)$ is well-defined, for all $t > 0$. By means of the combination of Theorems 8.2 and 8.3 we can transfer the existence and uniqueness result for the elliptic diffusions on phase space to the existence and uniqueness result for the hypo-elliptic diffusions on the group H_r (that is via conjugation with $\mathscr{S}$).

8.7.1 Algorithm for the PDE Approach to Differential Reassignment

Here we provide an explicit algorithm on the discrete Gabor transform $G_\psi^D \mathbf{f}$ of the discrete signal $\mathbf{f}$, that consistently corresponds to the theoretical PDE's on the continuous case as proposed in [93], i.e. convection equation (8.29) where we apply the first choice (8.30). Although that the PDE by [93] is not as simple as the second approach in (8.30) (which corresponds to a standard erosion step on the absolute value $|\mathcal{G}_\psi f|$ followed by a restoration of the phase afterwards) we do provide an explicit numerical scheme of this PDE, where we stay entirely in the *discrete phase space*.

It should be stressed that taking straightforward central differences of the continuous differential operators of Sect. 8.6 does not work. For details and non-trivial motivation of left-invariant differences on discrete Heisenberg groups see [118].

Explicit upwind scheme with left-invariant finite differences in pseudo-code for $\mathcal{M} = 1$

For $l = 1, \ldots, K-1$, $m = 1, \ldots, M-1$ set $\tilde{W}[l,m,0] := G^D_{\boldsymbol{\psi}}\mathbf{f}[l,m]$.
For $t = 1, \ldots, T$.
For $l = 0, \ldots, K-1$, for $m = 1, \ldots, M-1$ set
$\tilde{v}^1[l,m] := -\frac{aK}{2}(\log|\tilde{W}|[l+1,m,t=0] - \log|\tilde{W}|[l-1,m,t=0])$
$\tilde{v}^2[l,m] := -\frac{aM}{2}(\log|\tilde{W}|[l,m+1,t=0]| - \log|\tilde{W}|[l,m-1,t=0])$
$\tilde{W}[l,m,t] := \tilde{W}[l,m,t-1] + K\Delta t(z^+(\tilde{v}^1)[l,m][\tilde{\mathcal{A}}_1^{D^-}\tilde{W}][l,m,t] + z^-(\tilde{v}^1) \times [l,m][\tilde{\mathcal{A}}_1^{D^+}\tilde{W}][l,m,t]) + M\Delta t(z^+(\tilde{v}^2)[l,m][\tilde{\mathcal{A}}_2^{D^-}\tilde{W}][l,m,t] + z^-(\tilde{v}^2)[l,m] \times [\tilde{\mathcal{A}}_2^{D^+}\tilde{W}][l,m,t])$.

Subsequently, we provide a brief explanation of the involved variables:

l	discrete position variable $l = 0, \ldots, K-1$.
m	discrete frequency variable $m = 1, \ldots, M-1$.
t	discrete time $t = 1, \ldots T$, where T is the stopping time.
$\boldsymbol{\psi}$	discrete kernel $\boldsymbol{\psi} = \boldsymbol{\psi}_a^C = \{\psi_a(nN^{-1})\}_{n=-(N-1)}^{N-1}$ or $\boldsymbol{\psi} = \{\boldsymbol{\psi}_a^D[n]\}_{n=-(N-1)}^{N-1}$ see below.
$G^D_{\boldsymbol{\psi}}\mathbf{f}[l,m]$	discrete Gabor transform computed by diagonalization via Zak transform [235].
$\tilde{W}[l,m,t]$	discrete evolving Gabor transform evaluated at position l, frequency m and time t.
$\tilde{\mathcal{A}}_i^{D^\pm}$	forward (+), backward (−) left-invariant position ($i = 1$) and frequency ($i = 2$) shifts.
$z^\pm$	$z^+(\phi)[l,m,t] = \max\{\phi(l,m,t), 0\}$, $z^-(\phi)[l,m,t] = \min\{\phi(l,m,t), 0\}$ for upwind.

The discrete left-invariant shifts on discrete phase space are given by

$$
\begin{aligned}
(\tilde{\mathcal{A}}_1^{D^+}\tilde{\Phi})[l,m] &= K(e^{-\frac{2\pi i L m}{M}}\tilde{\Phi}[l+1,m] - \tilde{\Phi}[l,m]),\\
(\tilde{\mathcal{A}}_1^{D^-}\tilde{\Phi})[l,m] &= K(\tilde{\Phi}[l,m] - e^{\frac{2\pi i L m}{M}}\tilde{\Phi}[l,m]),\\
(\tilde{\mathcal{A}}_2^{D^+}\tilde{\Phi})[l,m] &= MN^{-1}(\tilde{\Phi}[l,m+1] - \tilde{\Phi}[l,m]),\\
(\tilde{\mathcal{A}}_2^{D^-}\tilde{\Phi})[l,m] &= MN^{-1}(\tilde{\Phi}[l,m] - \tilde{\Phi}[l,m-1]).
\end{aligned}
\tag{8.33}
$$

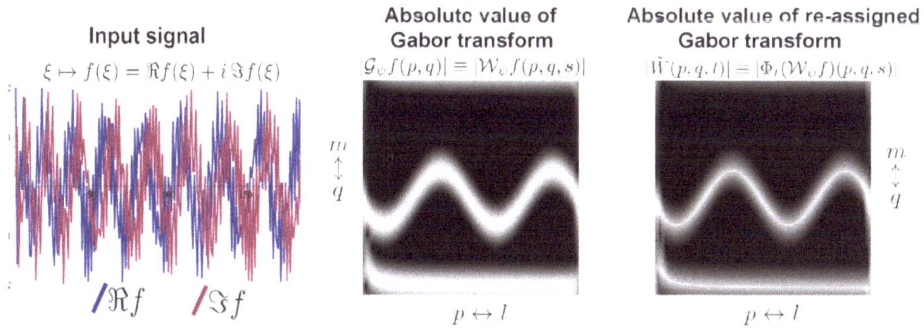

Fig. 8.2 Illustration of reassignment by adaptive phase-invariant convection explained in Sect. 8.6, using the upwind scheme of Sect. 8.7.1 applied on a Gabor transform

The discrete Gabor transform equals

$$G_{\psi}^{D}\mathbf{f}[l,m] = \frac{1}{N}\sum_{n=0}^{N-1} \overline{\psi[n-lL]} f[n] e^{-\frac{2\pi i n(n-lL)}{M}},$$

where M/L denotes the (integer) oversampling factor and $N = KL$. The discrete Cauchy Riemann kernel ψ_a^D is derived in [118] and satisfies the system

$$\forall_{l=0,\dots,K-1}\forall_{m=0,\dots,M-1}\forall_{\mathbf{f}\in\ell_2(I)}:$$
$$\frac{1}{a}(\tilde{\mathscr{A}}_2^{D^+} + \tilde{\mathscr{A}}_2^{D^-}) + i\,a(\tilde{\mathscr{A}}_1^{D^+} + \tilde{\mathscr{A}}_1^{D^-})(G_{\psi_a^D}^{D}\mathbf{f})[l,m] = 0, \qquad (8.34)$$

which has a unique solution in case of extreme over-sampling $K = M = N$, $L = 1$.

For an illustration of how our left-invariant finite difference (upwind)-scheme for differential reassignment sharpens the Gabor coefficients, see Fig. 8.2.

8.7.2 Evaluation of Reassignment

We distinguished between two approaches to apply left-invariant adaptive convection on discrete Gabor-transforms.[3] Either we apply the numerical upwind PDE-scheme described in Sect. 8.7.1 using the discrete left-invariant vector fields (8.33), or we apply erosion (8.32) on the modulus and restore the phase afterwards. Within each of the two approaches, we can use the discrete Cauchy-Riemann kernel ψ_a^D or the sampled continuous Cauchy-Riemann kernel ψ_a^C.

To evaluate these 4 methods we apply the reassignment scheme to the reassignment of a linear chirp that is multiplied by a modulated Gaussian and is sampled

[3] The induced frame operator can be efficiently diagonalized by Zak-transform, [235], boiling down to diagonalization of inverse Fourier transform on H_r, [118, Chap. 2.3]. We used this in our algorithms.

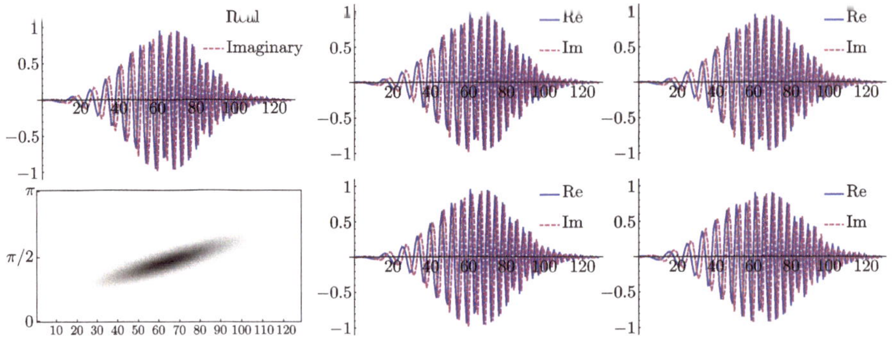

Fig. 8.3 Reconstructions of the reassigned Gabor transforms of the original signal that is depicted *on the top left* whose absolute value of the Gabor transform is depicted *on bottom left*. *In the right*: *1st row* corresponds to reassignment by the upwind scheme ($\mathscr{M} = 1$) of Sect. 8.7.1, where again *left* we used ψ_a^C and *right* we used ψ_a^D. Parameters involved are grid constants $K = M = N = 128$, window scale $a = 1/6$, time step $\Delta t = 10^{-3}$ and time $t = 0.1$. *2nd row* to reassignment by morphological erosion where *in the left* we used kernel ψ_a^C and *in the right* we used ψ_a^D. The goal of reassignment is achieved; all reconstructed signals are close to the original signal, whereas their corresponding Gabor transforms depicted in Fig. 8.4 are much sharper than the absolute value of the Gabor transform of the original depicted on *the bottom left* of this figure

using $N = 128$ samples. The input signal is an analytic signal so it suffices to show its Gabor transform from 0 to π. A visualization of this complex valued signal can be found Fig. 8.3 (top). The other signals in this figure are the reconstructions from the reassigned Gabor transforms that are given in Fig. 8.4. Here the topmost image shows the Gabor transform of the original signal. One can also find the reconstructions and reassigned Gabor transforms respectively using the four methods of reassignment. The parameters involved in generating these figures are $N = 128$, $K = 128$, $M = 128$, $L = 1$. Furthermore $a = 1/6$ and the time step for the PDE based method is set to $\Delta t = 10^{-3}$. All images show a snapshot of the reassignment method stopped at $t = 0.1$. The signals are scaled such that their energy equals the energy of the input signal. This is needed to correct for the numerical diffusion the discretization scheme suffers from. Clearly the reassigned signals resemble the input signal quite well. The PDE scheme that uses the sampled continuous window shows some defects. In contrast, the PDE scheme that uses ψ_a^D resembles the modulus of the original signal the most. Table 8.1 shows the relative ℓ_2-errors for all 4 experiments.

Advantages of the erosion scheme (8.32) over the PDE-scheme of Sect. 8.7.1 are:

1. The erosion scheme does not produce numerical approximation-errors in the phase, which is evident since the phase is not used in the computations.
2. The erosion scheme does not involve numerical diffusion as it does not suffer from finite step-sizes.
3. The separable erosion scheme is much faster.

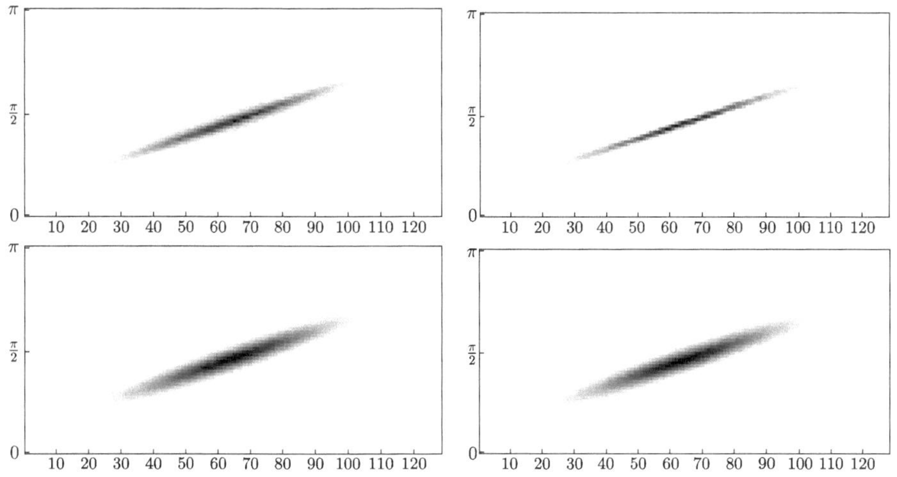

Fig. 8.4 Absolute value of the reassigned Gabor transforms of the signals depicted in the right two columns of Fig. 8.3

Table 8.1 The first column shows $\epsilon_1 = (\|\mathbf{f} - \tilde{\mathbf{f}}\|_{\ell_2(I)})\|\mathbf{f}\|_{\ell_2}^{-1}$, the relative error of the complex valued reconstructed signal compared to the input signal. In the second column $\epsilon_2 = (\||\mathbf{f}| - |\tilde{\mathbf{f}}|\|_{\ell_2(I)})\|\mathbf{f}\|_{\ell_2}^{-1}$ can be found which represents the relative error of the modulus of the signals. Parameters involved are $K = M = N = 128$, window scale $a = \frac{1}{8}$ and convection time $t = 0.1$, with times step $\Delta t = 10^{-3}$ if applicable. PDE stand for the upwind scheme presented in Sect. 8.7.1 and erosion means the morphological erosion method given by (8.32)

	ϵ_1	ϵ_2	t
Erosion continuous window	2.41×10^{-2}	8.38×10^{-3}	0.1
Erosion discrete window	8.25×10^{-2}	7.89×10^{-2}	0.1
PDE continuous window	2.16×10^{-2}	2.21×10^{-3}	0.1
PDE discrete window	1.47×10^{-2}	3.32×10^{-4}	0.1
PDE discrete window	2.43×10^{-2}	6.43×10^{-3}	0.16

The convection time in the erosion scheme is different than the convection time in the upwind-scheme, due to violation of the Cauchy-Riemann equations. Typically, to get similar visual sharpening of the re-assigned Gabor transforms, the convection time of the PDE-scheme should be taken larger than the convection time of the erosion scheme (due to numerical blur in the PDE-scheme). For example $t = 1.6$ for the PDE-scheme roughly corresponds to $t = 1$ in the sense that the ℓ_2-errors nearly coincide, see Table 8.1. The method that uses a sampled version of the continuous window shows large errors in Fig. 8.4 the defects are clearly visible. This shows the importance of the window selection, i.e. *in the PDE-schemes* it is better to use window ψ_a^D rather than window ψ_a^C. However, Fig. 8.5 and Table 8.1 clearly indicate that *in the erosion schemes* it is better to choose window ψ_a^C than ψ_a^D.

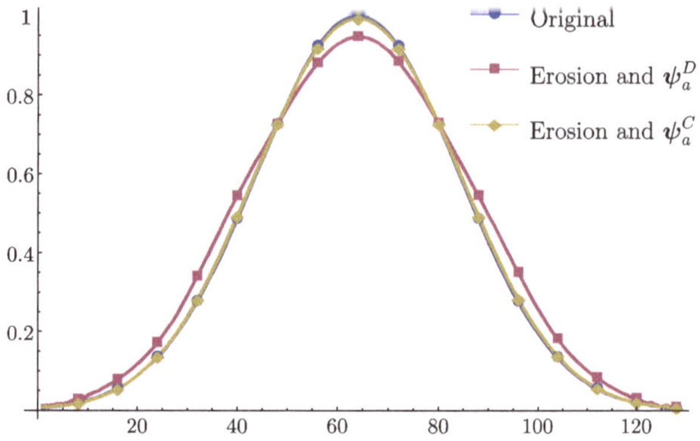

Fig. 8.5 The modulus of the signals *in the bottom row* of Fig. 8.3. For erosion (8.32) ψ_a^C performs better than erosion applied on a Gabor transform constructed by ψ_a^D

8.8 Conclusion

We have presented a framework for signal processing via evolutions on Gabor transforms, which are to be considered as functions on the (full) Heisenberg group. These evolutions must be left-invariant for appropriate treatment of the phase in the Gabor domain. Moreover, (non-linear) left-invariant operators in the Gabor domain correspond to (nonlinear) operators in the signal domain that commute with translations and modulations. We have considered (and compared) several approaches and implementations for differential reassignment (the convection case). For extensions to Gabor transform of 2D-images and for signal enhancement via left-invariant diffusion we refer to [118, 119].

Acknowledgements The Netherlands Organisation for Scientific Research (NWO) is gratefully acknowledged for financial support.

Chapter 9
Scale Space Representations Locally Adapted to the Geometry of Base and Target Manifold

Luc Florack

Abstract We generalize the Gaussian multi-resolution image paradigm for a Euclidean domain to general Riemannian base manifolds and also account for the codomain by considering the extension into a fibre bundle structure. We elaborate on aspects of parametrization and gauge, as these are important in practical applications. We subsequently scrutinize two examples that are of interest in biomathematical modeling, viz. scale space on the unit sphere, used among others for codomain regularization in the context of high angular resolution diffusion imaging (HARDI), and retino-cortical scale space, proposed as a biologically plausible model of the human visual pathway from retina to striate cortex.

9.1 Introduction

Physical field observables (referred to as signals or images henceforth) have an intrinsically limited resolution. Resolution invariance can be realized (for levels below measurement resolution) by generating a one-parameter family of gracefully degraded copies of the fiducial image in a causal fashion, subject to certain invariance axioms, of which scale invariance is the one most relevant in this context. The canonical way to operationalize this is by convolution with a normalized Gaussian, yielding a so-called Gaussian scale space [151, 261, 290, 391, 404]. Traditionally this is done against a Euclidean background geometry, and in a homogeneous and isotropic fashion. This also holds for certain generalizations proposed in the literature, such as the so-called α-scale spaces by Duits et al. [116], the relativistic scale spaces by Burgeth et al. [58], etc. Moreover, scale space images are viewed as smooth functions with a globally defined codomain, cf. the various PDE formulations elsewhere in this book (Chaps. 1, 3, 5, 7, 8, 11). That is to say, image values are expressed relative to a universal standard (unit of intensity), which implies that they can be compared irrespective of the base point to which they are attached.

L. Florack (✉)
Department of Mathematics and Computer Science & Department of Biomedical Engineering, Eindhoven University of Technology, Den Dolech 2, 5600 MB Eindhoven, The Netherlands
e-mail: L.M.J.Florack@tue.nl

L. Florack et al. (eds.), *Mathematical Methods for Signal and Image Analysis and Representation*, Computational Imaging and Vision 41,
DOI 10.1007/978-1-4471-2353-8_9,

Local adaptation mechanisms have been proposed to control the resolution degradation process *locally* so as to account for the differential structure of the image, e.g. for the purpose of "edge enhancement" [154, 334, 403, 433]. Such mechanisms rely on and make use of the specific local structure of the image function, and always yield nonlinear filtering paradigms.

Here we consider local adaptation mechanisms of a different kind, viz. based on the geometry of the domain and codomain of the image. We depart from the (typically implicit) assumption that space is Euclidean, and even from the assumption that the image can be modeled as a function from a spatial, temporal, or spatiotemporal domain into a globally defined codomain. Instead we adapt the canonical paradigm to a general (Riemannian) geometry of its domain, and regard the codomain as a *point-wise* entity, leading to a so-called fibre bundle, recall Chap. 5. Linearity is manifest throughout.

9.2 Theory

9.2.1 Preliminaries on Riemannian Geometry

Let $S : \Omega^m \to \mathbb{R}$ denote a given raw signal defined on an m-dimensional Riemannian surface Ω^m, say. This could be a surface embedded in an n-dimensional Euclidean space, in which case it inherits its (generally non-flat) Riemannian metric from that of the embedding space. In this case we may parametrize Ω^m using coordinates ξ^μ, $\mu = 1, \ldots, m \leq n$, and derive the components of the metric tensor of Ω^m from that of $\mathbb{R}^n$ as follows:

$$g_{\mu\nu} = \frac{\partial x^i}{\partial \xi^\mu} \eta_{ij} \frac{\partial x^j}{\partial \xi^\nu}, \tag{9.1}$$

in which η_{ij}, $1 \leq i, j \leq n$, denote the components of the Euclidean metric tensor of the embedding space, and $g_{\mu\nu}$, $1 \leq \mu, \nu \leq m$ the ones induced on Ω^m, recall the definition of the pullback metric in Chap. 1, Sect. 1.2.3, and Chap. 5, Sect. 5.4.1.1, notably Eq. (5.8). For notational convenience we set $g = \det g_{\mu\nu}$ and $\eta = \det \eta_{ij}$. These are relative scalars. More precisely, $\sqrt{g}\, d\xi^1 \cdots d\xi^m$ and $\sqrt{\eta}\, dx^1 \cdots dx^n$ are the basic volume elements on Ω^m, respectively $\mathbb{R}^n$. Henceforth we will concentrate on the intrinsic properties of Ω^m without reference to its embedding in $\mathbb{R}^n$. In fact, no such embedding will be required.

By D_μ we shall denote the covariant derivative with respect to x^μ induced by the metric tensor $g_{\mu\nu}$. By construction the metric is "covariantly constant": $D_\rho g_{\mu\nu} = 0$, whence also $D_\mu g = 0$. (Under the additional assumption that space is free of so-called torsion, one can take this as the definition of the covariant derivative [390].) The covariant derivative defines a connection that allows us to compare neighbouring quantities despite locally varying (an)holonomic bases by "correcting" for the rate of change of the basis vectors in any direction along the base manifold. If $\{\mathbf{e}_\mu\}$ is a basis of the local tangent space at some implicit point of Ω^m, then by definition

$$D_\nu \mathbf{e}_\mu = \Gamma^\rho_{\mu\nu} \mathbf{e}_\rho. \tag{9.2}$$

The coefficients can be constructed so as to be "compatible with the metric", i.e. so as to respect the requirement of covariant constancy. "Metric compatibility" is tantamount to the applicability of the product rule despite a spatially varying metric. That is, if $\mathbf{v}, \mathbf{w}$ are differentiable vector fields with components v^μ, w^μ relative to the local basis $\{\mathbf{e}_\mu\}$, then

$$\nabla(\mathbf{v}, \mathbf{w}) = (\nabla \mathbf{v}, \mathbf{w}) + (\mathbf{v}, \nabla \mathbf{w}), \tag{9.3}$$

or, in terms of components,

$$D_\rho(g_{\mu\nu} v^\mu w^\nu) = D_\rho v^\mu g_{\mu\nu} w^\nu + v^\mu g_{\mu\nu} D_\rho w^\nu. \tag{9.4}$$

This uniquely establishes the (symmetric part of the) so-called Christoffel symbols in Eq. (9.2):

$$\Gamma^\rho_{\mu\nu} = \frac{1}{2} g^{\rho\lambda}(\partial_\mu g_{\lambda\nu} + \partial_\nu g_{\lambda\mu} - \partial_\lambda g_{\mu\nu}). \tag{9.5}$$

The components of $\nabla \mathbf{v}$ are given by

$$D_\mu v^\nu = \partial_\mu v^\nu + \Gamma^\nu_{\rho\mu} v^\rho. \tag{9.6}$$

(It can be shown that the antisymmetric part of the Christoffel symbols captures the torsion of the underlying manifold [390], which, in our case, will be assumed to vanish.)

Formulae for the component representation of $\nabla \mathbf{T}$ for any type of tensor field $\mathbf{T}$ are now uniquely determined by application of the product rule using covariant constancy of the metric in a similar fashion, together with the observation that ∇f coincides with the familiar gradient if f is a scalar function (no "Γ-corrections" need to be carried out since function values are absolute, i.e. do not rely on a local basis). Just fill in the (co-)vector slots of $\mathbf{T}$ so as to produce a scalar field, and differentiate. In particular, if ω is a covector with components ω_μ relative to a local basis $\{\mathbf{e}^\mu\}$ of the dual tangent space, then the components of $\nabla\omega$ are obtained by working out the identity $\nabla(\omega(\mathbf{v})) = \nabla\omega(\mathbf{v}) + \omega(\nabla \mathbf{v})$ in terms of components. As a result one finds

$$D_\mu \omega_\nu = \partial_\mu \omega_\nu - \Gamma^\rho_{\nu\mu} \omega_\rho. \tag{9.7}$$

For further details on Riemannian geometry and tensor calculus the reader is referred to the literature [390].

9.2.2 Scale Space

The above geometric elaboration allows us to extend the scale space paradigm beyond the usual, Euclidean context [151, 261, 290, 391, 404]. To this end, consider the following functional, in which (the minimizer) $S_t : \Omega^m \to \mathbb{R}$ will be seen to act as the scale space extension of the raw signal S (cf. Sect. 5.4.1, Chap. 5):

$$E(S_t) = \int_{\Omega^m} \left[(S(\xi) - S_t(\xi))^2 + \sum_{k\geq 1} \frac{t^k}{k!} D_{\mu_1} \cdots D_{\mu_k} S_t(\xi) D^{\mu_1} \cdots D^{\mu_k} S_t(\xi) \right] D\xi. \tag{9.8}$$

Here, $D^\mu S_t = g^{\mu\nu} D_\nu S_t$, and $D\xi = \sqrt{g}\, d\xi^1 \cdots d\xi^m$ is a parametrization invariant measure. Equivalently, one may write (notice the appearance of minus signs)

$$E(S_t) = \int_{\Omega^m} \left[(S(\xi) - S_t(\xi))^2 + \sum_{k\geq 1} \frac{(-t)^k}{k!} S_t(\xi) \Delta_g^k S_t(\xi) \right] D\xi, \tag{9.9}$$

in which $\Delta_g = g^{\mu\nu} D_\mu D_\nu$ denotes the self-adjoint Laplace-Beltrami operator on Ω^m, i.e. if $f : \Omega^m \to \mathbb{R}$ is a scalar function on Ω^m, then we have

$$\Delta_g f = \frac{1}{\sqrt{g}} \partial_\mu (g^{\mu\nu} \sqrt{g} \partial_\nu f). \tag{9.10}$$

Using covariant constancy of the metric, and the fact that Δ_g^k is self-adjoint for any $k = \mathbb{Z}_0^+$, it is straightforward to derive the corresponding Euler-Lagrange equations. For Eq. (9.8) this yields an "infinite order" PDE for the family S_t, $t \geq 0$, given the data function S:

$$S = \sum_{k\geq 0} \frac{(-t)^k}{k!} \Delta_g^k S_t = \exp(-t\Delta_g) S_t. \tag{9.11}$$

Inversion yields the solution[1]

$$S_t = \sum_{k\geq 0} \frac{t^k}{k!} \Delta_g^k S = \exp(t\Delta_g) S. \tag{9.12}$$

Notice that it satisfies the heat equation on Ω^m:

$$\partial_t S_t - \Delta_g S_t = 0, \tag{9.13}$$

with initial condition $S_0 = S$. This is analogous to the standard scale space representation in the Euclidean plane, with the Euclidean Laplace operator Δ on $\mathbb{R}^n$ formally replaced by the Laplace-Beltrami operator, Δ_g, cf. similar constructs in Chaps. 1 and 5. The parameter $t \geq 0$ controls the (inverse) resolution.

The asymptotic cases are as follows:

$$\lim_{t\to 0^+} S_t = S, \tag{9.14}$$

$$\lim_{t\to\infty} S_t = \frac{\int_{\Omega^m} S(\xi)\, d\xi}{\int_{\Omega^m} d\xi}. \tag{9.15}$$

[1]The operator $\exp(t\Delta_g)$ is bounded and defines a strongly continuous semigroup for $t \in \mathbb{R}^+ \cup \{0\}$.

The latter expression holds for compact Ω^m, but can be generalized with some care. The physical significance of S_t is that it represents the raw data S resolved at finite scale $t \in (0, \infty)$. For this reason S_t is referred to as the scale space representation of S.

9.2.3 Parametrization Aspects

In practice one must account for parametrization aspects. For instance, the assumption of linearity typically fails to hold due to nonlinear signal transfer in real physical or biological systems. Typically such systems are characterized by a compact or semi-infinite range, which is obviously inconsistent with linearity. Nevertheless, linearity implies no loss of generality in cases that can be understood in terms of codomain mappings of the type $S_t = \gamma(U_t)$ for some monotonic function $\gamma : \mathbb{R} \to \mathbb{R}$, with $\gamma' > 0$, say. Recall that the incorporation of a general Riemannian metric instead of a Euclidean one may likewise account for a physically or biologically relevant reparametrization of the base manifold, or, more generally, of a Riemannian metric transform (which may or may not be induced by a domain reparametrization). As a result one should study a modification of the basic paradigm, Eq. (9.13), by instead considering

$$\partial_t U_t - \Delta_g U_t - \mu \|\nabla U_t\|_g^2 = 0, \tag{9.16}$$

in which $\mu = (\ln \gamma')'$ and $\|\nabla u\|_g^2 = g^{\mu\nu}\, \partial_\mu u \partial_\nu u$.

9.2.4 Gauge Aspects

A more subtle aspect is that of gauging signal values obtained at different locations on the base manifold. This aspect has received virtually no attention in the image processing literature, with a few notable exceptions [184, 263, 264]. Especially in biological vision it is not self-evident that signals of equal *physical* magnitude at different locations in the visual field are perceived as such, i.e. have the same *perceptual* magnitude. Indeed, there are well-known illusions refuting the validity of this ubiquitous assumption, such as the "bright sun" illusion, Fig. 9.1, and the Craik-O'Brien-Cornsweet illusion, Fig. 9.2.

Illusions such as those of Figs. 9.1–9.2 cannot be understood if we insist on treating the signal as a spatial function. Instead, we must somehow gauge its values by selecting appropriate *perceptual* units of intensity *pointwise*. Georgiev argued that this can be achieved by modeling the signal as a section of a fibred space [184]. Koenderink exploited the same idea in an image processing context [264]. Each fibre represents the local codomain of possible intensity values at one point of the image domain. The "vertical" projection of a fibre onto the base manifold is well-defined, and yields the base point to which it is attached. In this view, however, one

Fig. 9.1 Bright sun illusion. The sun appears brighter than the paper background. Are we deceived because of our bias to recognize the bright object in the middle as the sun, or as a light bulb?

Fig. 9.2 Craik-O'Brien-Cornsweet illusion. The left part of the bar seems brighter than the right part, which is however an illusion. This example questions the position that the bright sun illusion arises (exclusively) at a semantic level due to our predisposition to judge sunlight as brightest. But what could possibly be Nature's "purpose" to trigger such illusions a priori?

declines from a "horizontal" projection of intensity values, i.e. from the possibility to compare signal values at distinct locations. Instead, one furnishes the fibred space with a so-called *connection*, to be defined below, which in turn induces a *gauge field* that allows us to "correct" for differences of neighbouring signal values by accounting for the spatial variation of the perceptual units of intensity.

The heuristic argument above suggests that instead of the raw, physical value $S(x)$ of the signal function we consider its value relative to the local unit of intensity, $\sigma(x)$ say, i.e. we consider perceived intensity $S(x)\sigma(x)$ for some perceptual (locally adaptive) unit $\sigma(x)$. This selection of a local unit, $\sigma(x)$, respectively any local multiple of this, $S(x)\sigma(x)$, for each x, is referred to as a *section* of the fibred space. We introduce a connection via a *covariant derivative* operator $\mathscr{D}$—as a genuine extension of the covariant derivative used in the foregoing—which we subject to the usual requirement posed by the product rule:

$$\mathscr{D}(S(x)\sigma(x)) = \mathscr{D}S(x)\sigma(x) + S(x)\mathscr{D}\sigma(x).$$

We then identify $\mathscr{D}S(x) = \nabla S(x)$, the usual gradient of the scalar function S evaluated at x. The unit $\sigma(x)$ is treated as a local basis vector for the one-dimensional fibre at x, and so the quantity $\mathscr{D}\sigma(x)$ must itself be a section, i.e. a local multiple of the basis vector $\sigma(x)$. In this way we obtain a local gauge field A (a covector field) with amplitude $A(x)$ at x:

$$\mathscr{D}\sigma(x) = A(x)\sigma(x).$$

All in all, the components of the covariant derivative of the signal relative to a local basis can be written as

$$\mathscr{D}_\mu(S(x)\sigma(x))\,\mathbf{e}^\mu(x) = D^A_\mu S(x)\sigma(x)\,\mathbf{e}^\mu(x),$$

with[2]

$$D^A_\mu S(x) = (\partial_\mu + A_\mu(x))S(x).$$

Effectively the whole procedure thus boils down to what is known as the "principle of minimal substitution" in physics, whereby one replaces ordinary differentiation of the signal function (i.e. the coefficient of the local basis section $\sigma(x)$), by covariant differentiation:[3] $\partial_\mu S(x) \to (\partial_\mu + A_\mu(x))S(x)$.

It remains an outstanding problem how the visual system gauges itself after exposure to a given signal. In any case, prolonged exposure to a stationary scene ($S(x) > 0$ independent of time) and deprived from the possibility to make saccadic eye movements "nullifies" one's percept completely, which can be understood as an adaptation mechanism in which the input stimulus has somehow evolved towards a "covariantly constant" steady state, i.e. $D^A_\mu S(x) = 0$, inducing a gauge field

$$A_\mu(x) = -\partial_\mu \ln S(x). \tag{9.17}$$

Intuitively it makes sense from an economic and ecological point of view to consider a stationary background stimulus as "void".

Our paradigmatic equation, Eq. (9.13), can now be modified so as to account for a background gauge field through minimal substitution of the Laplace-Beltrami operator:

$$\partial_t S_t - \Delta^A_g S_t = 0, \tag{9.18}$$

with self-dual gauge adapted Laplace-Beltrami operator[4]

$$\Delta^A_g = g^{\mu\nu}(D_\mu - A_\mu)(D_\nu + A_\nu) = D^\mu_A D^A_\mu.$$

[2]Additional Γ-corrections will be needed for non-scalar signals, recall Sect. 9.2.1, e.g. $D^{A,\Gamma}_\mu v^\nu(x) = (\delta^\nu_\rho(\partial_\mu + A_\mu(x)) + \Gamma^\nu_{\rho\mu}(x))v^\rho(x)$.

[3]In physics, however, one typically considers Hermitean operators $i\partial_\mu$, respectively $i\partial_\mu + A_\mu$.

[4]This definition differs from the one proposed by Georgiev, which fails to be self-dual [184].

In this case ordinary partial derivatives have been replaced by ones that are covariant in *both domain as well as codomain*: $\partial_\mu \to D_\mu \to D^A_\mu$, $\partial^\mu \to D^\mu \to D^\mu_A$, with $D^\mu_A = g^{\mu\nu} D^{-A}_\nu$. The reason for this definition is that it extends the property of the null-gauged Laplace-Beltrami operator as the functional derivative

$$\frac{1}{2}\frac{\delta}{\delta u}\int g^{\mu\nu}(\xi) D_\mu u(\xi) D_\nu u(\xi) D\xi = -\Delta_g u$$

to the case involving a non-trivial gauge field A_μ:

$$\frac{1}{2}\frac{\delta}{\delta u}\int g^{\mu\nu}(\xi) D^A_\mu u(\xi) D^A_\nu u(\xi) D\xi = -\Delta^A_g u.$$

Combined with reparametrization of the codomain we find the generalization of Eq. (9.16):

$$\partial_t S_t - \Delta^A_g S_t - \mu \|\nabla^A S_t\|^2_g = 0, \tag{9.19}$$

with $\|\nabla^A u\|^2_g = g^{\mu\nu} D^A_\mu u D^A_\nu u = D^\mu_{-A} u D^A_\mu u$. This is our main result. The conjecture is that it underlies many instances of multi-resolution systems in physics, biology, and engineering sciences.

9.2.5 Examples

9.2.5.1 Human Vision

In the following example we take $m = n = 2$. (There will consequently be no need to distinguish between Greek and Latin spatial indices, since embedded surface and embedding space coincide.)

Elsewhere it has been argued that the retino-cortical pathway of the human visual system can be understood from a differential geometric point of view in terms of a certain conformal metric transform that reflects the foveal properties of the human eye [152, 153]. The proposed metric takes the following form in polar coordinates,[5] $(\xi^1, \xi^2) = (r, \theta)$:

$$g_{\mu\nu} = \left(\frac{r_0}{r}\right)^2 \eta_{\mu\nu}, \tag{9.20}$$

in which r_0 is a biological size parameter characteristic of the central fovea, and $\eta_{\mu\nu}$ the Euclidean metric,

$$\eta_{\mu\nu} = \begin{pmatrix} 1 & 0 \\ 0 & r^2 \end{pmatrix}. \tag{9.21}$$

[5] At this level of rigor we ignore the singularity at the origin, but cf. [153].

This conformal metric reflects the fact that, to reasonable approximation, the human visual system has a foveal bias (objects near the foveal point are magnified inversely proportional to eccentricity), but is otherwise isotropic (angular relations are not affected). The corresponding Laplace-Beltrami operator Δ_g turns out to be proportional to the Euclidean Laplacian Δ_η, which should be regarded as a peculiarity of a conformal metric in $n = 2$ dimensions (it no longer holds when $n > 2$):

$$\Delta_g = \left(\frac{r}{r_0}\right)^2 \Delta_\eta. \tag{9.22}$$

Consequently the generating equation, Eq. (9.19) based on metric Eq. (9.20) and with $\gamma = \mathrm{id}$ and $A_\mu = 0$, is given by

$$\partial_t S_t - \left(\frac{r}{r_0}\right)^2 \Delta_\eta S_t = 0. \tag{9.23}$$

A natural scale reparametrization presents itself, viz.

$$\left(\frac{r}{r_0}\right)^2 t = s, \tag{9.24}$$

after which the reparametrized system appears homogeneous.[6] Recall that t and s are quadratic scales, thus one expects biological receptive field sizes to scale linearly with eccentricity r/r_0. (Resolution limitations are not taken into account here.) There exists ample evidence in support of this. The structure of the human retina and its retinotopic mapping onto the striate cortex (also known as cortical area V1) reflects the above-sketched geometric structure remarkably well, and so does ample psychophysical evidence [38, 359].

Let us now briefly return to the parametrization issue, recall Sect. 9.2.3. The Weber-Fechner law is a well-known psychophysical law that applies to various perceptual modalities. It states that the intensity of the percept is a logarithmic function of the intensity of the physical stimulus. Here we consider the relation between apparent brightness U_t and retinal irradiation S_t. The Weber-Fechner law is a direct consequence of Eq. (9.16) if we take μ to be a positive constant. Indeed, in this case one readily finds $\gamma(U_t)$ to be an exponential function (defined up to a pair of integration constants). If we impose the boundary conditions $\gamma(-\infty) = 0$ and $\gamma(\infty) = \infty$ (so that S_t is positive definite), and set $\gamma(0) = \gamma_0$ for some constant $\gamma_0 > 0$, we obtain

$$U_t = \frac{1}{\mu} \ln \frac{S_t}{\gamma_0}.$$

[6]Homegeneity in fact holds almost everywhere in the sense that space remains flat *except* at the foveal centre, at which the Ricci curvature tensor degenerates. This singularity can be removed at the expense of introducing global curvature with appreciable magnitude in the fovea centralis [153].

A least noticeable perceptual difference dU_t then corresponds to a logarithmic increment $\mu^{-1} dS_t/S_t$ of photon flux. One could interpret μ^{-1} as a psychophysical unit of dimension for the quantity U_t. The constant γ_0 may be adaptive to ambient light conditions.

The Weber-Fechner law holds only within an interval of physical photon fluxes of a few orders of magnitude. In general one will need to account for threshold and saturation phenomena. A way to achieve this is to replace the unbounded mapping γ above by $\gamma_\chi = \gamma \circ \chi$ for some suitably chosen psychophysical function χ, such that χ^{-1} is bounded and monotonic, say $\chi^{-1} : \mathbb{R} \to (0, 1)$. This amounts to a replacement of the nonlinearity coefficient $\mu \to \mu_\chi = \mu\chi' + \chi''/\chi'$.

Finally, it may be worth exploring the feasibility of the general equation, Eq. (9.19), as a generic model for "early vision", accounting for retino-cortical mapping (via the stipulated conformal metric transform), nonlinear signal transduction (via suitable codomain reparametrization), and illusionary "lightness", or perceived brightness, in relation to the actual physical stimulus (via adaptation mediated by some gauge field). As far as the latter aspect is concerned we find ourselves still largely in the dark. Most likely the gauge is fixed by some dynamic, *global* adaptation mechanism (point entities such as they occur in Eq. (9.17) are physically void). This supports the intriguing conjecture, raised by Koenderink [262], that the brain can be understood as a "geometry engine".

9.2.5.2 High Angular Resolution Diffusion Imaging

In the following example we take $m = 2$, $n = 3$, and identify Ω^2 with the unit sphere embedded in Euclidean 3-space, $\mathbb{R}^3$. This case is of considerable interest in high angular resolution diffusion imaging (HARDI), a non-invasive magnetic resonance imaging technique for mapping local water diffusivity profiles in vivo [26, 27, 277]. It is conjectured that such profiles convey important information on the architecture of fibrous tissues, such as brain white matter (axons) and muscles, since water diffusion is facilitated in the direction of the underlying fibres.

A typical HARDI acquisition entails a large number of signal attenuation measurements in different directions. Various representations have been proposed in the literature to handle such multi-directional data. Typically these take the form of (square-integrable) scalar functions on the unit sphere.

Let therefore $S : \Omega^2 \to \mathbb{R}$ denote a given raw HARDI signal confined to the unit sphere $\Omega^2 : \|x\| = 1$, $x \in \mathbb{R}^3$. Ω may be parametrized using two coordinates, ξ^μ, $\mu = 1, 2$, say. An obvious choice would be the polar angles $\xi = (\xi^1, \xi^2) = (\theta, \phi) \in [0, \pi] \times [0, 2\pi)$, given in terms of 3D Cartesian coordinates x^i, $i = 1, 2, 3$, by

$$\begin{cases} x^1 = \sin\theta\cos\phi, \\ x^2 = \sin\theta\sin\phi, \\ x^3 = \cos\theta. \end{cases} \tag{9.25}$$

The components of the Riemannian metric for the unit sphere embedded in Euclidean 3-space $\mathbb{R}^3$ are given by Eq. (9.1). Using polar and Cartesian coordinates for Ω^2 and $\mathbb{R}^3$, respectively, we obtain the familiar matrix representations

$$g_{\mu\nu} = \begin{pmatrix} 1 & 0 \\ 0 & \sin^2\theta \end{pmatrix} \quad \text{resp. } \eta_{ij} = \begin{pmatrix} 1 & 0 & 0 \\ 0 & 1 & 0 \\ 0 & 0 & 1 \end{pmatrix}. \tag{9.26}$$

In particular, $D\xi = \sqrt{g}\, d\xi^1 d\xi^2 = \sin\theta\, d\theta d\phi$ is the invariant volume element on Ω^2.

We may think of Ω^2 as a "fibre" over the Euclidean space $\mathbb{R}^3$ at some implicit base point $x \in \mathbb{R}^3$. More specifically, in the context of HARDI, Ω is the domain of definition of the diffusive attenuation signal S or diffusion coefficient D related via the Stejskal-Tanner formula [324, 392], recall Chap. 4, notably Sect. 4.5:

$$S(\theta, \phi) = S_0 \exp(-bD(\theta, \phi)). \tag{9.27}$$

In what follows we shall consider the scale space extension of S, but this may be equally well replaced by any other $L^2(\Omega^2)$-function, notably D, or by derived probabilistic quantities such Tuch's orientation distribution function (ODF) [415], the higher order diffusion tensor model and the diffusion orientation transform (DOT) by Özarslan et al. [324, 325], and the diffusion tensor distribution model by Jian et al. [238]. See also Descoteaux et al. [103, 104] and Hess et al. [226].

Consider the scale space representation S_t of S derived in Sect. 9.2.2. The (dimensionless) parameter $t \in \mathbb{R}^+$ now controls (inverse) *angular resolution*. Expressed in terms of spherical coordinates it is natural to decompose the functions S_t and S relative to the orthonormal basis of spherical harmonics, since this diagonalizes the Laplace-Beltrami operator:

$$\Delta_g Y_\ell^m = -\ell(\ell+1) Y_\ell^m, \tag{9.28}$$

for any $\ell \in \mathbb{Z}_0^+$ and $m \in \{-\ell, -\ell+1, \ldots, \ell-1, \ell\}$. We employ the following convention for spherical harmonics, in which P_ℓ^m are the associated Legendre polynomials:[7]

$$Y_\ell^m(\theta, \phi) = \sqrt{\frac{(2\ell+1)(\ell-m)!}{4\pi(\ell+m)!}}\, e^{im\phi} P_\ell^m(\cos\theta), \tag{9.29}$$

$$P_\ell^m(z) = \frac{(-1)^m}{2^\ell \ell!}(1-z^2)^{\frac{m}{2}} \frac{d^{\ell+m}}{dz^{\ell+m}}(z^2-1)^\ell, \tag{9.30}$$

with $-1 \le z \le 1$. Setting

$$S(\theta, \phi) = \sum_{\ell,m}^{*} c_{\ell m}(0) Y_\ell^m(\theta, \phi), \tag{9.31}$$

[7] Cf. functions.wolfram.com for further properties of Y_ℓ^m and P_ℓ^m.

in which the asterisk indicates summation over the effective indices $\ell \in \mathbb{Z}_0^+$ and $m \in \{-\ell, -\ell+1, \ldots, \ell-1, \ell\}$, yields

$$S_t(\theta, \phi) = \sum_{\ell,m}^{*} c_{\ell m}(t) Y_\ell^m(\theta, \phi), \tag{9.32}$$

with

$$c_{\ell m}(t) = e^{-t\ell(\ell+1)} c_{\ell m}(0). \tag{9.33}$$

Notice the analogy with the $e^{-t\|\omega\|^2}$-attenuation of high frequency components of scalar images in the Euclidean plane under Gaussian blurring.

The coefficients in Eq. (9.31) are determined as follows:

$$c_{\ell m}(0) = \int_0^{2\pi} \int_0^{\pi} S(\theta, \phi) Y_\ell^{-m}(\theta, \phi) \sin\theta \, d\theta \, d\phi, \tag{9.34}$$

by virtue of $Y_\ell^{-m} = (Y_\ell^m)^*$ and orthonormality:

$$\int_0^{2\pi} \int_0^{\pi} Y_\ell^m(\theta, \phi)^* Y_{\ell'}^{m'}(\theta, \phi) \sin\theta \, d\theta \, d\phi = \delta_{mm'} \delta_{\ell\ell'}. \tag{9.35}$$

Equations (9.32–9.34) summarize the operational scheme for extending a raw HARDI signal into a continuous scale space family defined implicitly via minimization of Eq. (9.8). The required integrals over the sphere can be numerically approximated in the standard way, e.g. by employing a regular sampling of directions on vertices of a tessellated icosahedron with a corresponding numerical integration measure induced by the areas of the polygons defined by the dual tessellation.

The asymptotic cases are as expected:

$$\lim_{t \to 0^+} S_t(\theta, \phi) = S(\theta, \phi), \tag{9.36}$$

$$\lim_{t \to \infty} S_t(\theta, \phi) = \frac{1}{4\pi} \int_0^{2\pi} \int_0^{\pi} S(\theta, \phi) \sin\theta \, d\theta \, d\phi, \tag{9.37}$$

i.e., in the hypothetical limit of infinite angular resolution Eqs. (9.32–9.34) reproduce the raw signal, whereas the limit of vanishing angular resolution corresponds to a complete averaging of this raw signal over the sphere. Similar Tikhonov regularization procedures of unit sphere HARDI functions have been proposed in the literature by Descoteaux et al. [104] and by Hess et al. [226]. Figure 9.3 illustrates Eq. (9.32) on a test image.

9.3 Conclusion

We have adapted the linear scale space paradigm to the local geometry of the image domain (base manifold) as well as to the potentially local nature of codomain gauge

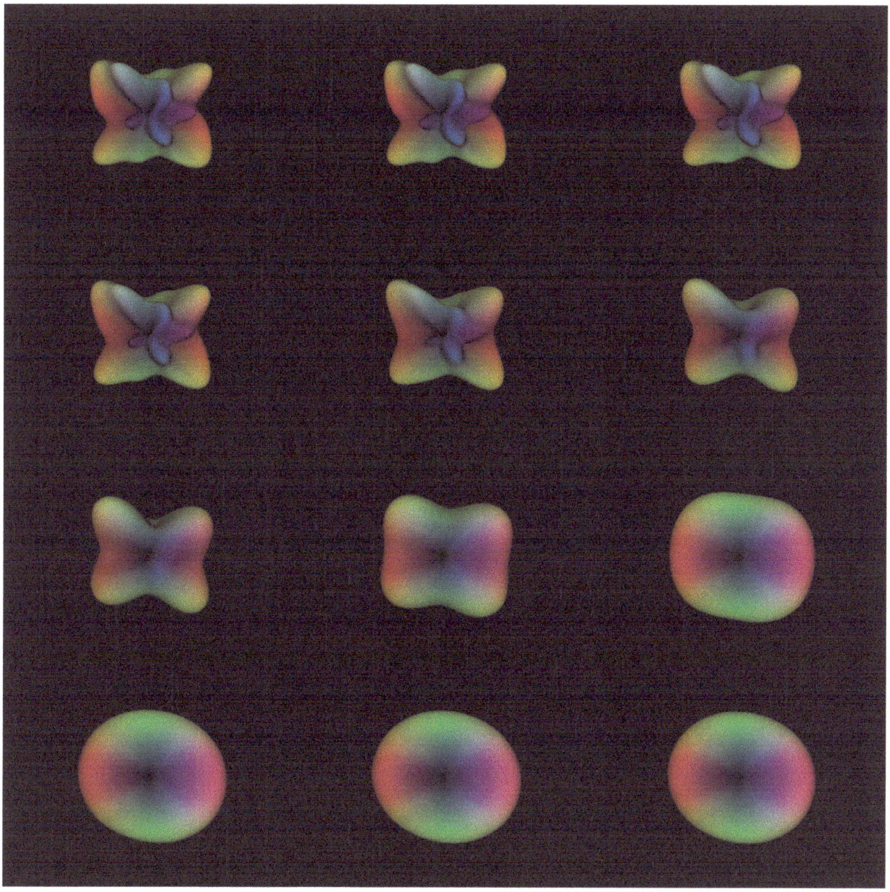

Fig. 9.3 A scalar function on the unit sphere, with intensity displayed as radial distince from the origin, viewed at (exponentially) increasing values of the inverse angular resolution parameter t

(fibre bundle over the base manifold). The theory has been illustrated with two examples, viz. regularization of high angular resolution diffusion imaging (spherical base manifold geometry), and modeling of the retino-cortical pathway in human vision (conformal metric transform). In the latter case we have pointed at the potential relevance of a fibre bundle construct, but detailed mechanisms are hitherto unknown. Much research will be needed to understand the role of local adaptation mechanisms in the psychophysics of human perception.

Acknowledgements The Netherlands Organisation for Scientific Research (NWO) is gratefully acknowledged for financial support.

Chapter 10
An A Priori Model of Line Propagation

Markus van Almsick

Abstract For the robust detection of lines in 2-dimensional images, it is feasible to have a general, a priori model reflecting the properties of lines. Based on three simple and generally applicable assumptions we introduce a stochastic line propagation model with its resulting Fokker-Planck equation and Green's function. The line model implies a line diffusion scheme that is not simply another anisotropic diffusion of scalar-valued luminosity functions, but a mechanism for the anisotropic diffusion of oriented line segments in a 3-dimensional space that encodes position and orientation.

10.1 Introduction

Diffusion schemes are well established in image processing. They constitute a mechanism to exchange image information between pixels. A short introduction to diffusion and diffusion-like methods is given in Chap. 1.

The main unwanted diffusion effect is the blurring of interesting features, like contours and lines. Anisotropic and non-linear, edge-preserving diffusion schemes help to preserve contours. Coherence enhancing diffusion conserves line structures to a certain degree. In most cases one applies diffusion directly to the luminosity function of a gray-scaled image or to the luminosity functions of several (color) channels. Only a few diffusion schemes, so far, deal with the diffusion of more complex image features, such as matrices and tensors in the contributions in Chaps. 3 and 5.

We take a quite general approach by considering the diffusion, or rather the stochastic propagation of anisotropic filter responses. These filter responses are usually obtained by translating and rotating a filter across an image. Each filter response is, thus, parameterized by a translation vector and a rotation angle. Consequently, these filter responses constitute a function on the Euclidean group manifold and we therefore turn to stochastic processes and diffusion on the Euclidean group mani-

M. van Almsick (✉)
Department of Biomedical Engineering, Eindhoven University of Technology, Den Dolech 2, 5600 MB Eindhoven, The Netherlands
e-mail: M.v.Almsick@tue.nl

L. Florack et al. (eds.), *Mathematical Methods for Signal and Image Analysis and Representation*, Computational Imaging and Vision 41,
DOI 10.1007/978-1-4471-2353-8_10,

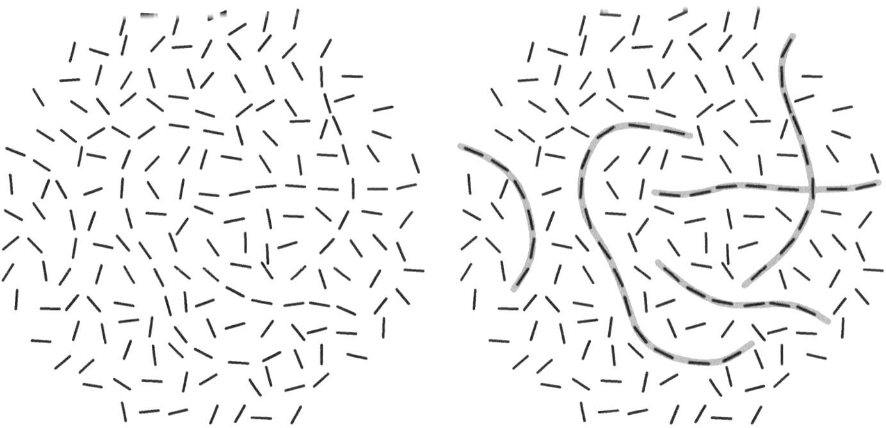

Fig. 10.1 *The line segments in the left image* are grouped in the visual system of the observer according to the Gestalt law of good continuations as indicated in *the right image*

fold. Other group manifolds (e.g. translation and scaling group as in wavelet theory) are of course possible as well, but not considered here.

In this chapter we introduce a diffusion scheme of line and contour segments by deriving a stochastic propagation for lines and contours. Our model does not rely on a specific line or contour generation process that may be found in nature (e.g. branches of trees, coast lines, outlines of clouds, etc.). Instead, we rely on generally applicable and basic first principles that apply to almost any line independent of its source. In this sense, we introduce an *a priori model* for lines and contours. A more general, encompassing diffusion scheme without an axiomatic line modeling approach is presented by Duits et al. [112–115]. A different form of line and contour diffusion is given by Franken et al. [169].

The motivation and original application of our line propagation scheme has been the regularization of noisy image data with faint lines as encountered in fluoroscopy images during an electrophysiology intervention on the heart. Another field of application is the regularization of diffusion tensor images (DTI) and high-angular resolution diffusion images (HARDI) prior to fiber tracking.

Our approach is inspired by the human visual system in two ways. First, the Gestalt laws [265] that are based on psychophysical experiments imply several heuristics to group image features. These heuristics are: proximity, similarity, continuity, closure, and common fate in time. To group the detection of line segments into lines and contours, we utilize the laws of proximity and continuity (see Fig. 10.1).

Second, the physiology of the visual system implies an orientation space. The visual cortex V1 exhibits a cortical map of the visual field that not only encodes the position of every visual input, but that also encodes the orientation of the receptive fields in the retina from where anisotropic visual stimuli are conveyed to the brain. Furthermore, lateral neural connections in V1 are orientation specific [47]. They prefer to connect neurons that encode the same orientation and that are spatially grouped along their orientation axes.

We assume in our line and contour model that these lateral neural connections mutually enforce oriented visual stimuli, which in most cases are line or contour segments. Hence, we try to derive from first principles the probability that two line or contour segments mutually join to a common line or contour and we attempt to provide the most probable connection between the two segments. To achieve this, we proceed as follows.

We begin in Sect. 10.2 with the space, in which we construct our line and contour model. It is the Euclidean group manifold and the function space upon it. Then, in Sect. 10.3, we utilize three axioms and determine with these a stochastic differential equation (SDE) that governs the probabilistic propagation of lines and contours on the Euclidean group manifold. We solve the SDE in two different ways. First, in Sect. 10.4, we convert the SDE into a Lagrangian density and determine the extremal path via the corresponding Euler-Lagrange equation. This way, we obtain the most probable line or contour between two segments. Second, in Sect. 10.5, we convert the SDE into the corresponding Fokker-Planck (or Chapman-Kolmogorov) equation and obtain a diffusion equation for line and contour segments. The Green's function of this partial differential equation is the transition probability for a line that starts at a given point and orientation and that extends to a second point and orientation. In analogy to the Green's function of a diffusing scalar point, which in some cases is referred to as the *point-spread-function*, we refer to our Green's function as the *line-spread-function* (LSF). With regard to high energy physics, one may also call the Green's function the *line propagator*. The LSF provides us with the probability that two line or contour segments do connect. In order to apply the LSF to all line segment responses in an image, one has to perform a Euclidean group convolution of the LSF with the orientation space of an image. In the last Sect. 10.6 we bring all the results together and show how to apply these to our initial example in Fig. 10.1.

10.2 Orientation Encoding

In image analysis one usually probes an image via a filter kernel ψ. The filter kernel acts like a receptive field of neurons on the retina. It collects the incoming light stimuli in an excitatory or inhibitory manner over a small region of the visual field. These receptive fields are located everywhere in the retina, in all orientations (if anisotropic), and also in all sizes. Hence, it is necessary to move a filter kernel ψ to all locations in an image, to rotate it into all directions, and to resize it. If we denote the transformations by g and the representations of g acting on the image domain Ω by U_g, we obtain the linear filter responses W_ψ by an inner product of the filter kernel ψ with the luminosity function L over the image domain Ω.

$$W_\psi(g) := \int_\Omega \psi^*(U_{g^{-1}}\boldsymbol{x})L(\boldsymbol{x})\,d\boldsymbol{x}.$$

Obviously, the filter response W_ψ is parameterized by transformation g. This is often overlooked, if one only considers translations of the filter kernel. The parameter

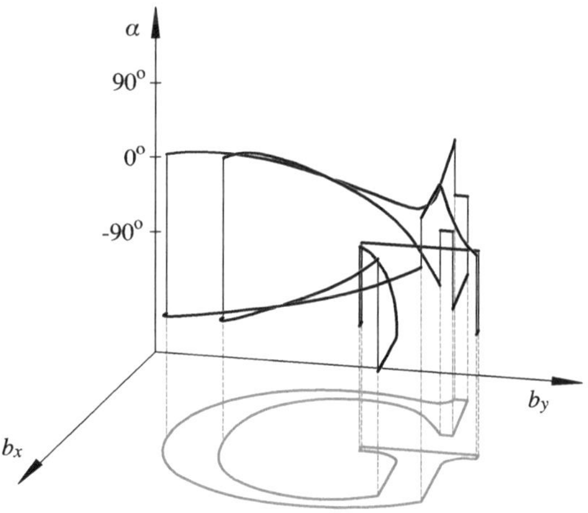

Fig. 10.2 The outline of the letter G is lifted into the third, angular dimension of the Euclidean group manifold that encodes the receptive field parameters location and orientation

manifold of translations is isomorphic to the image domain Ω. As a consequence, $W_\psi(g)$ appears to be defined on the image domain and is often viewed as an image $W_\psi(\mathbf{x})$ itself, which is not quite correct.

The detection of line or contour segments is achieved by anisotropic filters ψ. Hence, we need to consider at least translations and rotations to move W_ψ to all locations and into all directions. In this article we omit rescaling, since we neglect the thickness and, thus, the size/scale of lines or contours. The resulting group of transformations for 2-dimensional images is the Euclidean group of two dimensions. The above equation in this specific case is

$$W_\psi(\mathbf{b}, \alpha) := \int_{\mathbb{R}^2} \psi^*(\mathscr{R}_\alpha^{-1}(\mathbf{x} - \mathbf{b})) L(\mathbf{x})\, d^2\mathbf{x}$$

with translations specified by the translation vectors $\mathbf{b}$ and rotations given by the 2×2-dimensional rotation matrices $\mathscr{R}_\alpha$. The image domain is assumed to be a subset of $\mathbb{R}^2$ and we include the possibility of complex-valued filters by considering the hermitian inner product with the complex-conjugate filter kernel ψ^*.

The linear response W_ψ is a complex- or real-valued function on the 3-dimensional Euclidean group manifold consisting of the continuous parameters $\mathbf{b}$ that denote the 2-dimensional translation vector and the rotation angle α ranging from 0 to 2π. As an example, Fig. 10.2 depicts the idealized response W_ψ to the outline of the letter G. Of course, the linear response of a filter kernel does not render such well-defined lines in the Euclidean manifold. Figure 10.2 is only a schematic drawing.

To detect a line or contour segment, the filter ψ should resemble locally the luminosity function L of a line or a contour. Hence, a good choice is a second-order Gaussian derivative or gradient filter of appropriate scale, but other filters are just as suitable. To obtain a measure-preserving, invertible mapping from the luminosity function L to the linear response W_ψ on the Euclidean manifold, one can also consider, in analogy to wavelet theory, so-called admissible filters ψ that fulfill the admissibility constraint $\int_0^{2\pi} \hat{\psi}^*(\rho,\varphi)\hat{\psi}(\rho,\varphi)\,d\phi = \frac{1}{4\pi^2}$ where $\hat{\psi}(\rho,\varphi)$ denotes the Fourier-transformed filter of ψ in polar coordinates of the frequency space. These admissible filters lead to the inverse mapping

$$L(\mathbf{x}) = \int_0^{2\pi} \int_{\mathbb{R}^2} \psi(\mathscr{R}_\alpha^{-1}(\mathbf{x}-\mathbf{b}))W_\psi(\mathbf{b},\alpha)\,d^2\mathbf{b}\,d\alpha. \tag{10.1}$$

The linear response W_ψ is only the first step in a line or contour detection scheme. In the subsequent step, one needs to convert the linear responses W_ψ into a line/contour-probability or indicator field $p(g)$ that reflects the probability to find a line/contour segment at the given location (translation) and orientation (rotation) g. The main difficulty in this processing step is the fact that a linear filter ψ renders a response not only exactly at the location and orientation of a line/contour segment, but also, to a lesser degree, at deviating positions and orientations. Furthermore, other image features besides lines and contours may trigger a linear response W_ψ. Hence, it is practical to only consider the local maxima of W_ψ perpendicular to the line/contour orientation as a good line or contour indication $p(g)$. This approach resembles the well-known Canny edge detector [66] without the final hysteresis step. Another approach via so-called logical/linear operators has been introduced by Iverson and Zucker [234]. Here, one combines several differential properties of lines or contours by Boolean-like logical/linear operators. The result is a non-negative, quasi linear response to lines or contours that one can interpret as a line or contour probability function $p(g)$.

The line and contour detection up to this point only takes local line and contour characteristics into account. A decisive advantage of the orientation space is the capacity to measure and encode more than one direction in a single position. Thus, it is possible to deal with line crossings and bifurcations. Consequently, the orientation space, that is $W_\psi(g)$ or $p(g)$ on the Euclidean motion group manifold, constitutes an ideal state space for a line or contour model that incorporates the non-local, spatial properties. We address such a line/contour model in the next section.

10.3 Stochastic Line and Contour Propagation

To cope with the spatial properties of lines and contours, we model their propagation across an image. Initially this is a process in time t, but we are going to eliminate time t in due course. We commence with a drawing process as depicted in Fig. 10.3 that places the ink of a line or contour onto an image. This appears to be a rather artificial ansatz, since most of the lines and contours in images have a completely different cause. However, our ansatz can be based on three general and basic axioms that qualify our line and contour model as an a priori model for any line or contour.

The three axioms of our line and contour model are:

- *Invariance* under symmetry operations, such as spatial and temporal translation and spatial rotation.

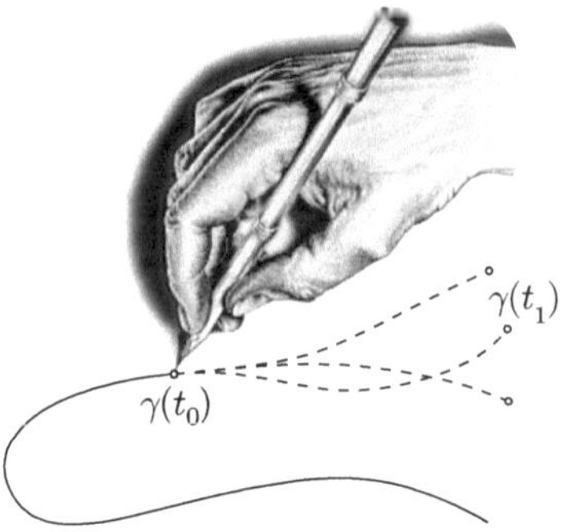

Fig 10.3 The stochastic process emulates the drawing of *a line* or *contour* $\boldsymbol{\gamma}$ starting at time t_0 at location $\boldsymbol{\gamma}(t_0)$ and extrapolating the *line* or *contour* until t_1

- *Smoothness* for the sake of good continuation. We assume that the lines and contours are continuous and differentiable at least up to first order, thereby ensuring the notions of line direction.
- *Extensibility*, which implies that if a line or contour satises the model so do all the parts of any line or contour partition.

The axiom of extensibility allows us to formulate the line propagation in an infinitesimal form. We formulate the line drawing process $\boldsymbol{\gamma}(t) = (\mathbf{x}(t), \alpha(t))$ as a general stochastic differentiable equation.

$$\partial_t \begin{pmatrix} \mathbf{x}(t) \\ \alpha(t) \end{pmatrix} = \boldsymbol{a}(\mathbf{x}(t), \alpha(t), t) + \mathbf{B}(\mathbf{x}(t), \alpha(t), t) \cdot \boldsymbol{\eta}(t). \tag{10.2}$$

The term $\boldsymbol{a}(\mathbf{x}(t), \alpha(t), t)$ models the deterministic portion of the process and $\mathbf{B}(\mathbf{x}(t), \alpha(t), t) \cdot \boldsymbol{\eta}(t)$ stands for the probabilistic portion, where the random variable $\boldsymbol{\eta}(t)$ is coupled to the process via a coupling matrix $\mathbf{B}(\mathbf{x}(t), \alpha(t), t)$. Note, that the functions $\boldsymbol{a}$ and $\mathbf{B}$ in (10.2) are so far completely undetermined. The random variable $\boldsymbol{\eta}(t)$ renders white noise of unit-variance as specified in the cumulant equations (10.3) and (10.4). Hence, all probabilistic properties of the Markov process (10.2) depend on the coupling matrix $\mathbf{B}$.

$$\langle\!\langle \boldsymbol{\eta}(t) \rangle\!\rangle = 0, \tag{10.3}$$

$$\langle\!\langle \boldsymbol{\eta}(t_0)\, \boldsymbol{\eta}(t_1) \rangle\!\rangle = \delta(t_1 - t_0). \tag{10.4}$$

In order to obtain more specific functions for $\boldsymbol{a}$ and $\mathbf{B}$, we transcribe the stochastic differential (10.2) into the corresponding Fokker-Planck respectively Chapman-Kolmogorov equation. These partial differential equations determine the conditional probabilistic distribution $p(\mathbf{x}, \alpha, t \,|\, \mathbf{x}_0, \alpha_0, t_0)$, that denotes the transition probability of a line or contour starting at time t_0 at location $\mathbf{x}_0$ and orientation α_0 to end up at time t in location $\mathbf{x}$ with orientation α:

$$\begin{aligned} &\partial_t\, p(\mathbf{x}, \alpha, t \,|\, \mathbf{x}_0, \alpha_0, t_0) \\ &= - \sum_{j=\{x,y,\alpha\}} \partial_j (a_j(\mathbf{x}(t), \alpha(t), t) p(\mathbf{x}, \alpha, t \,|\, \mathbf{x}_0, \alpha_0, t_0)) \end{aligned}$$

$$+\frac{1}{2}\sum_{j,k=\{x,y,\alpha\}}\partial_j\,\partial_k(D_{jk}(\mathbf{x}(t),\alpha(t),t)p(\mathbf{x},\alpha,t\,|\,\mathbf{x}_0,\alpha_0,t_0)),$$

$$\text{with } \mathbf{D}(\mathbf{x}(t),\alpha(t),t) := \mathbf{B}(\mathbf{x}(t),\alpha(t),t)\cdot\mathbf{B}^t(\mathbf{x}(t),\alpha(t),t). \tag{10.5}$$

For convenience we rewrite the above equation by introducing the differential operator $\mathscr{L}$ that represents all the differential terms in (10.5).

$$\partial_t\,p(\mathbf{x},\alpha,t\,|\,\mathbf{x}_0,\alpha_0,t_0) = \mathscr{L}(\mathbf{x}(t),\alpha(t),t)p(\mathbf{x},\alpha,t\,|\,\mathbf{x}_0,\alpha_0,t_0). \tag{10.6}$$

It is now time to discard time t. A line or contour in an image has been drawn. It consists of all the ink that has been deposited in the drawing process. Hence, we need to integrate the probability $p(\mathbf{x},\alpha,t\,|\,\mathbf{x}_0,\alpha_0,t_0)$ and Eq. (10.5) respectively Eq. (10.6) over time t from t_0 to infinity.

Before we do so, a few decisive considerations regarding the time-dependence of a drawing process are due. First, the drawing process should be time-invariant, since the shape of the resulting line or contour should not depend on the date and time it was drawn. Hence, we can omit the explicit t-dependence of differential operator $\mathscr{L}$. Second, the Fokker-Planck equation (10.5) conserves the conditional probability $p(\mathbf{x},\alpha,t\,|\,\mathbf{x}_0,\alpha_0,t_0)$ over time t as if the lines and contours go on for ever. This scenario, however, is counter productive. If one keeps drawing, one eventually obtains a 'black' image densely covered with one infinite line all over. We consequently must introduce a term that models the termination of lines or the closure of contours. Obviously, the only way to do so in a temporally and spatially invariant manner is the addition of a constant term $-\lambda\,I$ to operator $\mathscr{L}$ in Eq. (10.6), which will lead to an exponential decay of the conditional probability $p(\mathbf{x},\alpha,t\,|\,\mathbf{x}_0,\alpha_0,t_0)$ in time t.

We formally solve (10.6) as follows: Any function $f(t)$ in time t can locally be expanded in a first order Taylor expansion (10.7).

$$f(t+\tau)\approx\; f(t)+\tau\,\partial_t f(t) = (I+\tau\,\partial_t)f(t). \tag{10.7}$$

With expansion (10.7) we can extrapolate function $f(t)$ by a time step τ into the future. However, the larger τ gets the less accurate the linear Taylor expansion becomes. It is therefore better to divide the extrapolation into arbitrary small time-steps τ/N. The concatenation of N of these arbitrary small time-steps via N linear extrapolations renders

$$f(t+\tau) = \lim_{N\to\infty}\left(I+\frac{\tau}{N}\partial_t\right)^N f(t) = e^{\tau\,\partial_t}f(t). \tag{10.8}$$

Hence, replacing ∂_t by $\mathscr{L}$ in (10.6) and the constant decay term $-\lambda\,I$ mentioned above, we can formally write

$$p(\mathbf{x},\alpha,t\,|\,\mathbf{x}_0,\alpha_0,t_0) = e^{(t-t_0)\,(\mathscr{L}(\mathbf{x}(t),\alpha(t))-\lambda I)}\,p(\mathbf{x},\alpha,t_0\,|\,\mathbf{x}_0,\alpha_0,t_0). \tag{10.9}$$

With Eq. (10.9) we have a formal solution for the line/contour drawing process in time. As mentioned above, the drawing process is just a mathematical construct to derive our model. In the next step we integrate over time t,

$$p(\mathbf{x}, \alpha \,|\, \mathbf{x}_0, \alpha_0) := \int_{t_0}^{\infty} p(\mathbf{x}, \alpha, t \,|\, \mathbf{x}_0, \alpha_0, t_0)\, dt, \tag{10.10}$$

to obtain the conditional probabilities $p(\mathbf{x}, \alpha \,|\, \mathbf{x}_0, \alpha_0)$ of the lines and contours themselves, which are independent of the drawing process. Placing equation (10.9) into (10.10) we obtain

$$\begin{aligned} p(\mathbf{x}, \alpha \,|\, \mathbf{x}_0, \alpha_0) &= \int_{t_0}^{\infty} e^{(t-t_0)(\mathscr{L}(\mathbf{x}(t),\alpha(t))-\lambda I)}\, dt\; p(\mathbf{x}, \alpha, t_0 \,|\, \mathbf{x}_0, \alpha_0, t_0) \\ &= -\big(\mathscr{L}(\mathbf{x}(t), \alpha(t)) - \lambda I\big)^{-1} p(\mathbf{x}, \alpha, t_0 \,|\, \mathbf{x}_0, \alpha_0, t_0). \end{aligned} \tag{10.11}$$

A fixed and well localized starting position and direction of a line or contour propagation demands a Dirac δ-distribution as the initial condition.

$$p(\mathbf{x}, \alpha, t_0 \,|\, \mathbf{x}_0, \alpha_0, t_0) = \delta(\mathbf{x} - \mathbf{x}_0, \alpha - \alpha_0), \tag{10.12}$$

Thus, we finally obtain for the line and contour propagation the following partial differential equation

$$-\big(\mathscr{L}(\mathbf{x}(t), \alpha(t)) - \lambda I\big) p(\mathbf{x}, \alpha \,|\, \mathbf{x}_0, \alpha_0) = \delta(\mathbf{x} - \mathbf{x}_0, \alpha - \alpha_0). \tag{10.13}$$

To completely determine the conditional probability $p(\mathbf{x}, \alpha \,|\, \mathbf{x}_0, \alpha_0)$, we also need to specify the boundary conditions in $\mathbf{x}$ and α. The conditional line/contour probability decays exponentially. Consequently, lines/contours are not arbitrarily long and will vanish at infinity. And, of course, the direction angle α is periodic in 2π. Hence, natural boundary conditions are

$$\lim_{|\mathbf{x}-\mathbf{x}_0|\to\infty} p(\mathbf{x}, \alpha \,|\, \mathbf{x}_0, \alpha_0) = 0, \tag{10.14}$$

$$p(\mathbf{x}, 0 \,|\, \mathbf{x}_0, \alpha_0) = p(\mathbf{x}, 2\pi \,|\, \mathbf{x}_0, \alpha_0). \tag{10.15}$$

Equation (10.13) is still too general to render explicit and useful results. At this point we recall the set of axioms that form the basis of our line/contour model. We utilize the first axiom, the invariance under spatial translation and rotation, to further constrain the differential operator $\mathscr{L}(\mathbf{x}(t), \alpha(t)) - \lambda I$ in (10.13). To do so, we express the translations and rotations, or rather the infinitesimal generators of these symmetry transformations as differential operators:

$$\mathscr{T}_x = -\partial_x, \qquad \mathscr{T}_y = -\partial_y, \qquad \mathscr{S} = -x\partial_y + y\partial_x - \partial_\alpha. \tag{10.16}$$

The differential generators $\mathscr{T}_x$ and $\mathscr{T}_y$ for translation in x and y-direction are well known. The infinitesimal generator for rotation in the orientation space needs some explanation. The differential generator for rotation in a 2-dimensional Euclidean space is $-x\partial_y + y\partial_x$. However, during a rotation one changes direction. It is therefore necessary to also translate an orientation space in α, which is done by adding the translation generator $-\partial_\alpha$. We refer to this combined generator as the *shift-twist-generator* $\mathscr{S}$.

To achieve invariance under translation and rotation, one simply has to ensure that the differential operator $\mathscr{L}(\mathbf{x}(t), \alpha(t)) - \lambda I$, the generator of line/contour propagation in (10.6), commutes with the differential generators $\mathscr{T}_x$, $\mathscr{T}_y$ and $\mathscr{S}$ of the symmetry transformations. If they commute, also the corresponding rotation and translation transformations commute with the line and contour model and, as a consequence, leave the model invariant. Note, that the decay term $-\lambda I$ is constant and, thus, it commutes with any other spatial or angular differential operator. So, we are left with the commutation relations

$$[\mathscr{T}_x, \mathscr{L}] = 0, \qquad [\mathscr{T}_y, \mathscr{L}] = 0, \qquad [\mathscr{S}, \mathscr{L}] = 0. \tag{10.17}$$

The first two commutation relations (10.17) with the translation generators $\mathscr{T}_x$ and $\mathscr{T}_y$ are 0, if and only if the gradient of the drift term $\nabla \boldsymbol{a}(\mathbf{x}(t), \alpha(t))$ and the gradient $\nabla \mathbf{B}(\mathbf{x}(t), \alpha(t))$ vanish. Obviously, this is the case when $\boldsymbol{a}$ and $\mathbf{B}$ are constant and, thus, independent of $\mathbf{x}$. Consequently, we continue to work with $\boldsymbol{a}(\alpha(t))$ and $\mathbf{B}(\alpha(t))$. The t- and $\mathbf{x}$-dependence is obsolete.

The last commutation relation (10.17) with shift-twist-generator $\mathscr{S}$ amounts to nine ordinary differential equations for $\boldsymbol{a}$ and $\mathbf{B}$, one for each coefficient in front of the resulting partial differentials ∂_x, ∂_y, ∂_α, ∂_x^2, ∂_y^2, ∂_α^2, ∂_{xy}, $\partial_{x\alpha}$, and $\partial_{y\alpha}$.

One can explicitly solve these ODEs, but a more intuitive and more elegant approach is the use of a left-invariant tangent vector basis [113, 115]

$$\begin{aligned} \boldsymbol{e}_\parallel &:= \cos\alpha\, \boldsymbol{e}_x - \sin\alpha \boldsymbol{e}_y, \\ \boldsymbol{e}_\perp &:= \sin\alpha \boldsymbol{e}_x + \cos\alpha \boldsymbol{e}_y, \\ \boldsymbol{e}_\alpha &= \boldsymbol{e}_\alpha. \end{aligned} \tag{10.18}$$

With respect to this basis, the ODEs require the components of $\boldsymbol{a}$ and $\mathbf{D}$ to be constant. As a result, we are left with nine constant parameters $a_\parallel$, $a_\perp$, a_α, $D_{\parallel\parallel}, \ldots$. Note that matrix $\mathbf{D}$ is symmetric. In addition, we have the constant λ determining the decay in line probability.

A model with ten parameters is not desirable, but fortunately most of these parameters turn out to be fixed. We can resort to three conditions that will reduce the number of parameters to three.

First, note that the origin of the α-axis can be chosen freely. We determine $\alpha = 0$ such, that the line propagation runs parallel to direction α. This is intuitive for line modeling and would imply $a_\perp = 0$. One could also chose α to point into the direction of a contour gradient. This would imply $a_\parallel = 0$, but we will follow the *line convention*. Furthermore, the parameterization of lines and contours by time variable t represents and internal degree of freedom that the line/contour model does not need. The speed of drawing is irrelevant. Hence, we set the drawing speed to unit speed, which renders $a_{\|} = 1$ (or $a_{\perp} = 1$ for the *contour convention*). Furthermore, we can now replace the time-parametrization t by the line/contour path-length-parametrization s. The last component of the drift term $\boldsymbol{a}(s)$, constant a_α, remains a true parameter that cannot be fixed. It describes a drift in the line/contour orientation, which we denote by $\bar{\kappa}$. $\bar{\kappa}$ is not the actual curvature of a line or contour, but the overall mean curvature. At unit speed, $\bar{\kappa}$ stands for the average change of orientation angle per unit length.

The axiom regarding the smoothness and, thus, differentiability of a line or contour determines almost all constant components of $\mathbf{D}$. Note that a line or contour would become in-differentiable, if the random variable $\boldsymbol{\eta}(s)$ in SDE (10.2) would couple directly to a spatial component $x(s)$ or $y(s)$. Hence, all the components of the first two rows of matrix $\mathbf{B}$ have to be 0 and, consequently, only $D_{\alpha\alpha}$ is unequal 0. $D_{\alpha\alpha}$ denotes the diffusion constant square σ^2 of the angular α-coordinate. Finally, we are left with three constants: λ, σ, and $\bar{\kappa}$. It is now possible to give an explicit form of Eq. (10.13).

$$\left(\cos\alpha\,\partial_x + \sin\alpha\,\partial_y - \frac{\sigma^2}{2}\partial_\alpha^2 + \bar{\kappa}\,\partial_\alpha + \lambda\right) p(x, y, \alpha | x_0, y_0, \alpha_0)$$
$$= \delta(x - x_0, y - y_0, \alpha - \alpha_0). \tag{10.19}$$

The corresponding, explicit SDE for line and contour propagation is

$$\partial_s \begin{pmatrix} x(s) \\ y(s) \\ \alpha(s) \end{pmatrix} = \begin{pmatrix} \cos\alpha(s) \\ \sin\alpha(s) \\ \bar{\kappa} \end{pmatrix} + \begin{pmatrix} 0 & 0 & 0 \\ 0 & 0 & 0 \\ 0 & 0 & \sigma \end{pmatrix} \cdot \boldsymbol{\eta}(s). \tag{10.20}$$

The above derivation of the line model given by Eqs. (10.19) and (10.20) is not quite stringent. Besides the three axioms *smoothness*, *invariance*, and *extensibility*, it contains the assumption that the line model can adequately be described in an orientation space. This assumption entered our derivation by the ansatz that a Markov process in $x(t)$, $y(t)$, and $\alpha(t)$ is sufficient. Higher derivatives, such as curvature $\kappa(t)$ were not considered in the Markov process, but these more complex line models are, in principle, possible. Nevertheless, our model is the simplest line and contour model that satisfies the three axioms and, furthermore, it is possible to empirically verify our ansatz as follows.

If we extract the α-component from the SDE (10.20), we obtain a Wiener process for the direction variable $\alpha(s)$:

$$\partial_s \alpha(s) = \bar{\kappa} + \sigma\boldsymbol{\eta}(s). \tag{10.21}$$

Given the cumulants of the random variable $\boldsymbol{\eta}(s)$ in (10.3) and (10.4), we can readily deduce the stochastic properties of the curvature $\kappa(s) = \partial_s\,\alpha(s)$ and direction $\alpha(s)$. We should observe for the line or contour curvature $\kappa(s)$:

$$\begin{aligned}\langle\!\langle \kappa(s) \rangle\!\rangle &= \bar{\kappa}, \\ \langle\!\langle \kappa^2(s) \rangle\!\rangle &= \sigma^2, \\ \langle\!\langle \kappa(s_0)\,\kappa(s_1) \rangle\!\rangle &= \sigma^2 \delta(s_1 - s_0),\end{aligned} \tag{10.22}$$

and for the line and contour direction $\alpha(s)$:

$$\begin{aligned}\langle\!\langle \alpha(s) \rangle\!\rangle &= \bar{\kappa} s, \\ \langle\!\langle \alpha^2(s) \rangle\!\rangle &= \sigma^2 s, \\ \langle\!\langle \alpha(s_0)\,\alpha(s_1) \rangle\!\rangle &= \sigma^2 \min(s_0, s_1).\end{aligned} \tag{10.23}$$

One can empirically verify the variances and correlations in Eqs. (10.22) and (10.23). Most line and contours in natural images closely adhere to these equations, but it may be prudent to verify these cumulant equations before applying the line and contour model to a specific task.

We have derived the equations that define our line and contour model. In the next step, we have to solve these equations. We do so in two ways. The SDE (10.20) can be transcribed into a Lagrangian density. The extremal paths of this density represent the optimal connecting lines or contours between two given line/contour segments. We derive an explicit solution of the extremal path in the next section. Furthermore, it is possible to solve (10.19). The result is the transition probability $p(x, y, \alpha | x_0, y_0, \alpha_0)$ for a line or contour to connect two segments at (x, y, α) and (x_0, y_0, α_0). We sketch the derivation of this transition probability in the subsequent section.

10.4 Extremal Path

The Fokker-Planck equation of the Wiener process (10.21) is well-known. Here we include the decay term $-\lambda I$.

$$\partial_s p(\alpha, s | \alpha_0, s_0) = \left(\frac{\sigma^2}{2} \partial_\alpha^2 - \bar{\kappa}\,\partial_\alpha - \lambda \right) p(\alpha, s | \alpha_0, s_0). \tag{10.24}$$

The corresponding solution of the transition probability is

$$\begin{aligned}&p(\alpha_{j+1}, s_j + \Delta s \,|\, \alpha_j, s_j) \\ &\quad = \frac{1}{\sqrt{2\pi\,\Delta s}\,\sigma} \exp\left(-\frac{(\alpha_{j+1} - \alpha_j - \bar{\kappa}\,\Delta s)^2}{2\sigma^2 \Delta s} - \lambda \Delta s \right).\end{aligned} \tag{10.25}$$

We can approximate the directional trajectory $\alpha(s)$ of path $\boldsymbol{\gamma}(s)$ by arbitrarily small, linear steps Δs. The overall probability $p(\alpha_0, \alpha_1, \alpha_2, \ldots)$ of such an approximate path is given by the probability product of all its segments.

$$p(\alpha_0, \alpha_1\alpha_2, \ldots) \propto \prod_{j=0,1,\ldots} p(\alpha_{j+1}, s_j + \Delta s \,|\, \alpha_j, s_j) \tag{10.26}$$

$$= \prod_{j=0,1,\ldots} \exp\left(-\frac{(\alpha_{j+1} - \alpha_j - \bar{\kappa}\,\Delta s)^2}{2\sigma^2 \Delta s} - \lambda \Delta s \right) \tag{10.27}$$

$$= \exp\left(\sum_{j=0,1,\ldots} -\frac{(\alpha_{j+1} - \alpha_j - \bar{\kappa}\,\Delta s)^2}{2\sigma^2 \Delta s} - \lambda \Delta s \right). \tag{10.28}$$

Taking the limit $\Delta s \to 0$, we obtain the actual probability $p(\alpha(s))$ of the path. Here we neglect the normalization factor to avoid mathematical complications:

$$p(\alpha(s)) \propto \lim_{\Delta s \to 0} \exp\left(-\sum_{j=0,1,\ldots} \left(\frac{(\frac{\alpha_{j+1}-\alpha_j}{\Delta s} - \bar{\kappa})^2}{2\sigma^2} + \lambda \right) \Delta s \right)$$

$$= \exp\Bigg(\underbrace{-\int \left(\frac{(\partial_s \alpha(s) - \bar{\kappa})^2}{2\sigma^2} + \lambda \right) ds}_{\mathscr{L}[\kappa(s)]} \Bigg). \tag{10.29}$$

Our aim is to find the extremal path $\boldsymbol{\gamma}(s)$ of a line or contour. Note that the path $\boldsymbol{\gamma}(s)$ is uniquely determined by the directional trajectory $\alpha(s)$ via integration in s. As a consequence, the probability of $p(\alpha(s))$ is the same as $p(\boldsymbol{\gamma}(s))$. Hence, to maximize $p(\boldsymbol{\gamma})$ we have to minimize the Lagrangian $\mathscr{L}[\kappa(s)]$ given in Eq. (10.29). The Lagrangian is defined for the line or contour curvature $\kappa(s)$:

$$\mathscr{L}[\kappa(s)] = \int \left(\frac{(\kappa(s) - \bar{\kappa})^2}{2\sigma^2} + \lambda \right) ds, \tag{10.30}$$

$$\text{with } \kappa(s) = \partial_s\, \alpha(s). \tag{10.31}$$

A look at the Lagrangian $\mathscr{L}[\kappa(s)]$ reveals the effects that the parameters λ, σ, and $\bar{\kappa}$ have. Parameter λ determines the degree, by which the path length s is taken into account. The larger λ, the higher the *cost* for a long path $\boldsymbol{\gamma}$ and the shorter paths tend to be. Parameter $\bar{\kappa}$ determines the average path curvature that is favored. Hence, for $\bar{\kappa} = 0$, the line and contour does not have a bias to turn left or right. Parameter σ determines the cost, which a curvature deviation from $\bar{\kappa}$ will inflict. The smaller σ, the less $\kappa(s)$ tends to deviate from $\bar{\kappa}$. Larger σ values make directional changes more likely. Thus, a line or contour behaves like an elastic rod whose stiffness is controlled by σ.

The corresponding Euler-Lagrange equation for $\mathscr{L}$ is

$$\partial_s^2\, \kappa(s) + \frac{\kappa^3(s)}{2} - \left(\lambda\, \sigma^2 + \frac{\bar{\kappa}^2}{2} \right) \kappa(s) = 0. \tag{10.32}$$

For $\bar{\kappa} = 0$ the above equation is in fact Euler's elastica equation. The solution $\kappa(s)$ of Euler's elastica equation is the Jacobi elliptic sin-function with arbitrary real integration constants c and s_Δ:

$$\kappa(s) = 2c\sqrt{-m}\,\mathrm{sn}(c(s - s_\Delta)\,|\,m), \tag{10.33}$$

with $m := -1 - \frac{1}{c^2}(\lambda\sigma^2 + \frac{\bar{\kappa}^2}{2}) \le -1$. Twice the integration with respect to s yields $\alpha(s)$ and $\boldsymbol{\gamma}(s)$ and we obtain

$$\begin{aligned}\boldsymbol{\gamma}(s) = {} & \begin{pmatrix} x_0 \\ y_0 \end{pmatrix} + \frac{4}{c(m-1)^2}\begin{pmatrix} \cos\alpha_0 & -\sin\alpha_0 \\ \sin\alpha_0 & \cos\alpha_0 \end{pmatrix} \\ & \times \begin{pmatrix} \mathrm{dn}^2(cs_\Delta\,|\,m) + \frac{m-1}{2} & -\sqrt{-m}\,\mathrm{cn}(cs_\Delta\,|\,m)\,\mathrm{dn}(cs_\Delta\,|\,m) \\ \sqrt{-m}\,\mathrm{cn}(cs_\Delta\,|\,m)\,\mathrm{dn}(cs_\Delta\,|\,m) & \mathrm{dn}^2(cs_\Delta\,|\,m) + \frac{m-1}{2} \end{pmatrix} \\ & \times \begin{pmatrix} \frac{m-1}{2}cs + E(\mathrm{am}(c(s - s_\Delta)\,|\,m)|\,m) + E(\mathrm{am}(cs_\Delta\,|\,m)\,|\,m) \\ -\sqrt{-m}(\mathrm{sn}(c(s - s_\Delta)\,|\,m) + \mathrm{sn}(cs_\Delta\,|\,m)) \end{pmatrix}. \end{aligned} \tag{10.34}$$

In the above explicit formula for the extremal path the following elliptic integrals and functions are used. The incomplete elliptic integral of the first kind with modulus m is

$$F(\phi\,|\,m) := \int_0^\phi \frac{1}{\sqrt{1 - m\sin^2(\varphi)}}\,d\varphi. \tag{10.35}$$

The incomplete elliptic integral of the second kind with modulus m is

$$E(\phi\,|\,m) := \int_0^\phi \sqrt{1 - m\sin^2(\varphi)}\,d\varphi. \tag{10.36}$$

The Jacobi amplitude $\mathrm{am}(z\,|\,m)$ is the inverse function of the incomplete elliptic integral $F(\phi\,|\,m)$ of the first kind. All the remaining Jacobi elliptic functions are defined via the Jacobi amplitude $\mathrm{am}(z\,|\,m)$:

$$\mathrm{dn}(z\,|\,m) := \sqrt{1 - m\sin^2(\mathrm{am}(z\,|\,m))} = \partial_z\,\mathrm{am}(z\,|\,m) \tag{10.37}$$

with

$$\mathrm{sn}(z\,|\,m) := \sin(\mathrm{am}(z\,|\,m)) \quad \text{and} \quad \mathrm{cn}(z\,|\,m) := \cos(\mathrm{am}(z\,|\,m)). \tag{10.38}$$

Integration constants c, s_Δ, α_0, x_0 and y_0 need to be adjusted according to the initial and final side constraints of the extremal curve $\boldsymbol{\gamma}(s)$. Constant c affects the scale and shape, the path length offset s_Δ changes the initial curvature, α_0 is the initial direction, and (x_0, y_0) the initial path position. Examples with $\tilde{\lambda} = 0$ and $\tilde{\lambda} = 1/\sigma^2$ are displayed in Fig. 10.4.

Given two line or contour segments, we are now able to determine the optimal connecting line/contour. The optimal connecting line, however, does not provide the probability, with which such a connection may occur. It is of course possible

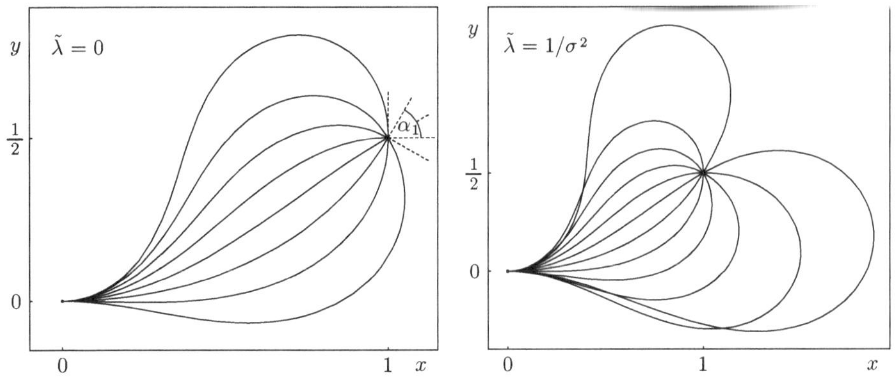

Fig. 10.4 *The figure to the left* depicts the extremal curves with $\tilde{\lambda} := \lambda\sigma^2 + \bar{\kappa}^2/2 = 0$ running from the origin (0, 0) with direction $\alpha_0 = 0°$ to the endpoint (1, 1/2) with angles of incidence α_1 at the endpoint ranging from $-90°$ to $120°$ in steps of 30°. *The figure on the right* depicts the extremal curves with $\tilde{\lambda} = 1/\sigma^2$ with angles of incidence α_1 ranging from $-150°$ to $210°$ in steps of 30°

to take (10.29) and calculate the probability of the optimal line. But this is just the probability of the optimal line and not the overall probability of all possible connections between two segment. For that purpose, we need to solve Eq. (10.19).

10.5 Line Spread Function

Solving (10.19) to obtain the transition probability $p(x, y, \alpha | x_0, y_0, \alpha_0)$ is a non-trivial task. A numerical method for cyclic spatial boundary condition has been proposed by August Jonas [16]. Here, we provide an explicit solution in Fourier space, that we have derived in collaboration with Remco Duits. It is beyond the scope of this article to provide a detailed derivation, which you can find in [110, 419]. Here, we only sketch the necessary steps.

The Fourier transformation in spatial coordinates of (10.19) and a subsequent transformation into polar coordinates renders an ordinary differential equation in α

$$\left(-i\omega_r \cos(\alpha - \omega_\phi) - \frac{\sigma^2}{2}\partial_\alpha^2 + \bar{\kappa}\,\partial_\alpha + \lambda\right)\hat{p}(\omega_r \cos\omega_\phi, \omega_r \sin\omega_\phi, \alpha)$$
$$= \frac{\delta(\alpha - \alpha_0)}{2\pi} \tag{10.39}$$

with

$$\hat{p}(\omega_r \cos\omega_\phi, \omega_r \sin\omega_\phi, \alpha)$$
$$:= \frac{1}{2\pi}\iint p(x, y, \alpha) \,|\, 0, 0, \alpha_0) e^{i(\omega_x x + \omega_y y)}\, dx\, dy \tag{10.40}$$

and polar coordinates

$$\omega_r := \sqrt{\omega_x^2 + \omega_y^2} \quad \text{and} \quad \omega_\phi := \arctan \frac{\omega_y}{\omega_x}. \tag{10.41}$$

The homogeneous part of Eq. (10.39) is, with the adequate substitutions (10.43), the Mathieu equation, Eq. (10.42):

$$(\partial_z^2 + a - 2q\cos(2z))G(a,q,z) = -\frac{2}{\pi\sigma^2}\delta(z - z_0), \tag{10.42}$$

$$\begin{aligned} &\text{with } z := \frac{1}{2}(\alpha - \omega_\phi), \; z_0 := \frac{1}{2}(\alpha_0 - \omega_\phi), \\ &a := -\frac{8}{\sigma^2}\lambda - \frac{4}{\sigma^4}\bar{\kappa}^2, \; q := -i\frac{4}{\sigma^2}\omega_r, \\ &G(a,q,z) := e^{-\frac{\bar{\kappa}}{\sigma^2}(\alpha-\alpha_0)}\hat{p}(\omega_r\cos\omega_\phi, \omega_r\sin\omega_\phi, \alpha). \end{aligned} \tag{10.43}$$

The Mathieu ODE, the homogeneous part of (10.42), defines the following Mathieu functions as solutions. We denote the even elliptic cosine by $\mathrm{Ce}(a,q,z)$ and the odd elliptic sine by $\mathrm{Se}(a,q,z)$. $\mathrm{Ce}(a,q,z)$ and $\mathrm{Se}(a,q,z)$ solve the Mathieu equation (10.42) for $z \in \mathbb{R}$ and are determined up to a complex-valued normalization constant. We can, thus, assemble the Fourier-transformed line spread function $\hat{p}(\omega_x, \omega_y, \alpha)$ piecewise for $\alpha < \alpha_0$ and $\alpha > \alpha_0$ by a linear combination of $\mathrm{Ce}(a,q,z)$ and $\mathrm{Se}(a,q,z)$. This, however, only renders a solution with boundary conditions at $z \to \pm\infty$. To obtain a solution with a 2π-periodic boundary condition in α, we wrap the previous solution of infinite extent around a 2π circular domain, which results in an infinite sum that can be solved like a geometric series due to the Floquet [150] theorem. The result is

$$\hat{p}(\omega_x, \omega_y, \alpha) = \frac{e^{2\bar{\kappa}(z-z_0)/\sigma^2}}{\pi\sigma^2\chi\,\mathrm{Se}_z(a,q,0)\,\mathrm{Ce}(a,q,0)} \times \begin{cases} \frac{(\mathrm{Ce}(a,q,z_0)-\chi\,\mathrm{Se}(a,q,z_0))(\mathrm{Ce}(a,q,z)+\chi\,\mathrm{Se}(a,q,z))}{1-e^{-\pi(2\bar{\kappa}/\sigma^2+i\nu)}} \\ \quad -\frac{(\mathrm{Ce}(a,q,z_0)+\chi\,\mathrm{Se}(a,q,z_0))(\mathrm{Ce}(a,q,z)-\chi\,\mathrm{Se}(a,q,z))}{1-e^{-\pi(2\bar{\kappa}/\sigma^2-i\nu)}} & \text{if } z < z_0, \\ \frac{\sinh(i\nu\pi)(\mathrm{Ce}^2(a,q,z_0)-\chi^2\,\mathrm{Se}^2(a,q,z_0))}{\cosh(i\nu\pi)-\cosh(2\pi\bar{\kappa}/\sigma^2)} & \text{if } z = z_0, \\ \frac{(\mathrm{Ce}(a,q,z_0)+\chi\,\mathrm{Se}(a,q,z_0))(\mathrm{Ce}(a,q,z)-\chi\,\mathrm{Se}(a,q,z))}{1-e^{\pi(2\bar{\kappa}/\sigma^2-i\nu)}} \\ \quad -\frac{(\mathrm{Ce}(a,q,z_0)-\chi\,\mathrm{Se}(a,q,z_0))(\mathrm{Ce}(a,q,z)+\chi\,\mathrm{Se}(a,q,z))}{1-e^{\pi(2\bar{\kappa}/\sigma^2+i\nu)}} & \text{if } z > z_0. \end{cases} \tag{10.44}$$

A more detailed derivation can be found in [419] and [110].

To obtain the transition probability $p(x, y, \alpha \,|\, 0, 0, \alpha_0)$ in position space an inverse Fourier transformation is needed. A typical result of the transition probability $p(x, y, \alpha \,|\, 0, 0, \alpha_0)$ is depicted in Fig. 10.5.

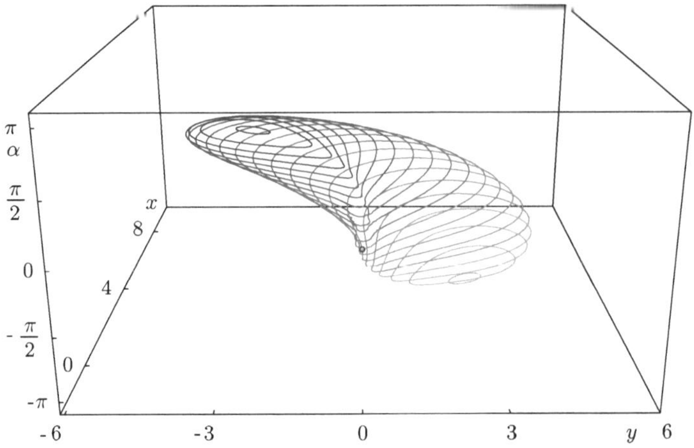

Fig. 10.5 The equi-probability contours of the transition probability $p(x, y, \alpha \mid 0, 0, \alpha_0) = 0.02$ in an orientation space at different levels of α, each $\pi/16$ apart. The parameters are $\sigma = 1/2$, $\lambda = 1/10$, $\alpha_0 = 0$, and $\bar{\kappa} = 0$

The transition probability $p(x, y, \alpha \mid 0, 0, \alpha_0)$ provides a measure for the affinity of two line or contour segments to connect. It is the Green's function or *propagator* kernel of line/contour propagation. At first sight it may appear to be a disadvantage that we only provide a closed solution of the line/contour transition probability in Fourier space and not in position space. However, in practice one needs to apply the line/contour propagation kernel to a dense distribution of line/contour segment measurements. This is achieved via a convolution, a group convolution on the Euclidean motion group (10.45).

$$(\Phi *_{\mathscr{E}_2} L)(\mathbf{b}, \alpha) := \int_0^{2\pi} \int_{\mathbb{R}^2} \Phi(\mathscr{R}_{\alpha'}^{-1}(\mathbf{b} - \mathbf{b}'), \alpha - \alpha') L(\mathbf{b}', \alpha')\, d^2\mathbf{b}'\, d\alpha',$$
$$\text{with } \Phi(x, y, \alpha) := p(x, y, \alpha \mid 0, 0, \alpha_0). \tag{10.45}$$

We can apply the Fourier theorem to the spatial part of the above convolution and obtain

$$\widehat{(\Phi *_{\mathscr{E}_2} L)}(\omega, \alpha) := \int_0^{2\pi} \hat{\Phi}(\mathscr{R}_{\alpha'}^{-1}(\omega - \omega'), \alpha - \alpha') \hat{L}(\omega', \alpha')\, d\alpha'. \tag{10.46}$$

This equation can be implemented quite efficiently, utilizing the Fourier transformed transition probability $\hat{p}(\omega_x, \omega_y, \alpha \mid 0, 0, \alpha_0)$.

10.6 Application and Conclusion

Based on just three general axioms we have obtained for the orientation space an elegant model for lines and contours. The model exhibits only three parameters

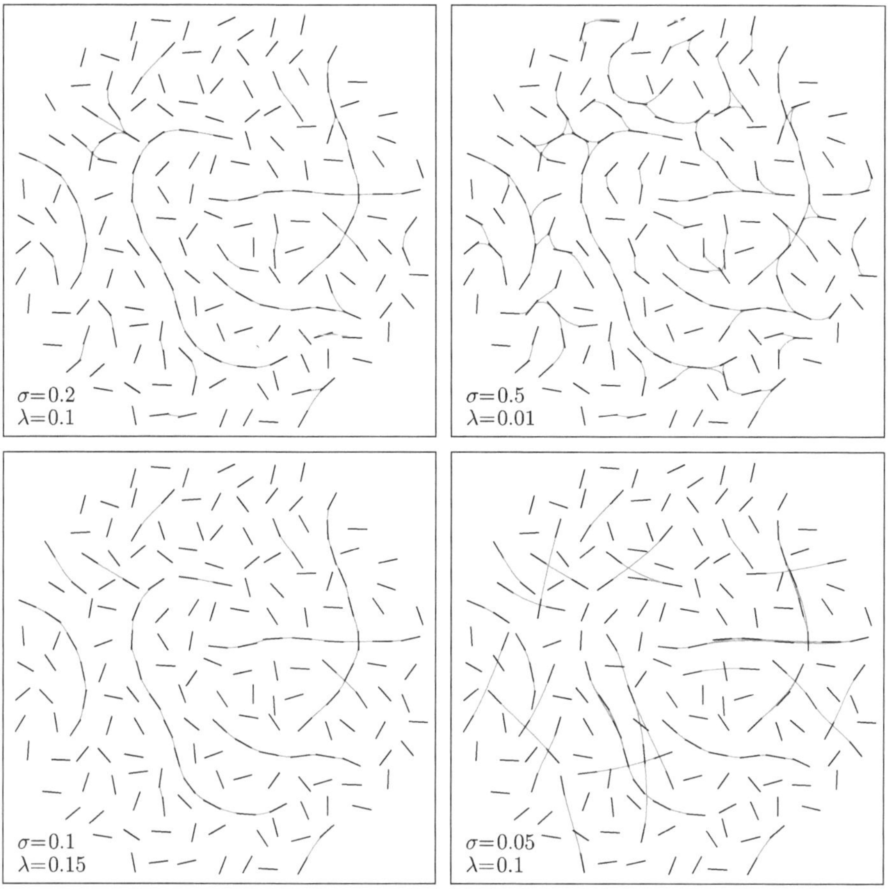

Fig. 10.6 Applying the theory to Fig. 10.1 with different σ and λ values. Small values for σ favor *straight lines*, large values allow curves with large curvatures. Large λ values lead to short connections, small λ values permit long range connections

specifying the line/contour length via $\lambda > 0$, the stiffness with respect to turning or bending via $\sigma > 0$, and the bias for curving left or right via $\bar{\kappa}$. Given two line/contour segments, we can determine the probability, with which these two segments may connect, and we can provide the most probable connection. How can we apply this knowledge to image analysis?

There are two types of application scenarios: sparse or dense line/contour data. Figure 10.1 in the introduction is an example of sparse line or contour data. At discrete points we are given the measurements of line or contour segments. Most images, however, lead to a continuous and, thus, dense line/contour probability distribution in the orientation space. The application of our line model in both scenarios is basically the same, but technically slightly different.

In the case of sparse line/contour data, one can apply to each data pair Eq. (10.34) to render the probability for a line/contour-connection and Eq. (10.44) to draw the

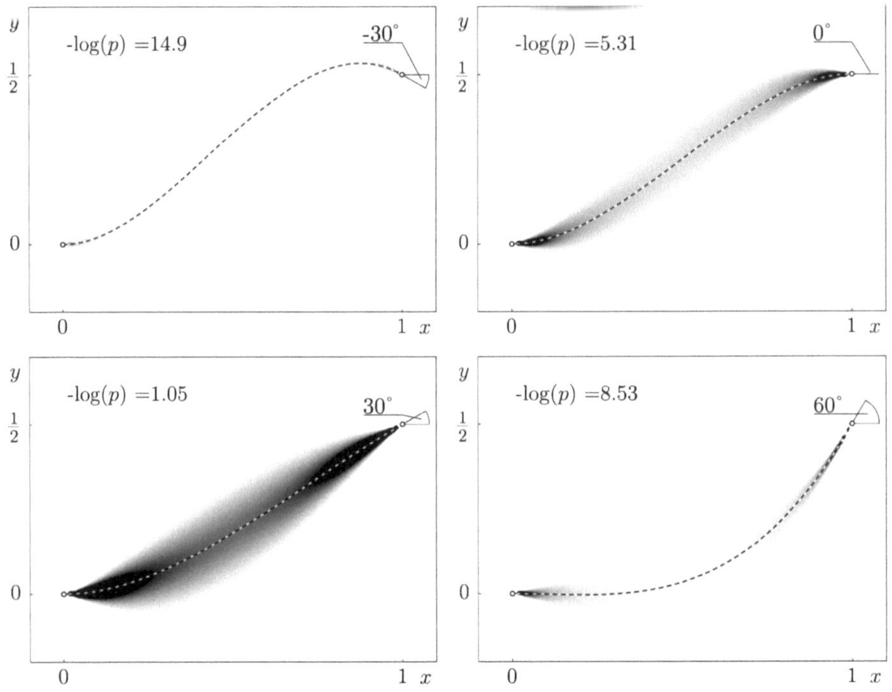

Fig. 10.7 All four completion fields have the source at the origin with an initial direction of 0° and the sink at location $(1, 1/2)$. The parameters are set to $\sigma = \frac{1}{2}$, $\lambda = \frac{1}{10}$, and $\bar{\kappa} = 0$. The final direction at the sink varies from $-30°$ to $60°$. The completion field is stronger when the source and sink are roughly aligned resulting in a stronger affinity between the source and sink line segment. The logarithmic probability of $p(\mathbf{x}_1, \alpha_1 | \mathbf{x}_0, \alpha_0)$ for a connecting line is given in *the top left corner*. *The dashed line* is the extremal path

ideal connecting line/contour. As an example, we provide the line grouping results of the introductory Gestalt law sample for several model parameter λ and σ in Fig. 10.6.

Furthermore, one can generate a so-called *completion field* [384, 446, 447, 468], that represents the probability distribution for any possible connecting line/contour. If we refer to the starting segment of a connecting line/contour as the *source* at (x_0, y_0, α_0) and if we refer to the ending segment as the *sink* at (x_1, y_1, α_1), then any intermediate segment of the connecting line at an arbitrary location and direction (x, y, α) has to connect to both segments, to the *source* and the *sink*. The probability of these intermediate line/contour segments is therefore proportional to the product of $p(x, y, \alpha \,|\, x_0, y_0, \alpha_0)$ and $p(x_1, y_1, \alpha_1 \,|\, x, y, \alpha)$, the probability to connect to the source as well as to the sink. Hence, the distribution given by

$$C(\mathbf{x}, \alpha) := p(\mathbf{x}_1, \alpha_1 | \mathbf{x}, \alpha) \; p(\mathbf{x}, \alpha | \mathbf{x}_0, \alpha_0) \tag{10.47}$$

is the completion field. Four examples of completion fields are shown in Fig. 10.7.

Now we turn to the dense data scenario. In this case, it is futile to consider the connections of all possible data point pairs. Instead we return to our initial consideration regarding the diffusion of lines. The Green's function $p(x, y, \alpha \mid x_0, y_0, \alpha_0)$ is essentially the diffusion kernel of lines or contours. Recall that the convolution of a Gaussian kernel and an image yields ordinary diffusion. Accordingly, the Euclidean group convolution (10.45) of the Green's function $p(x, y, \alpha \mid x_0, y_0, \alpha_0)$ with an orientation space of an image yields the directed diffusion of oriented line segments in the orientation space. Hence, almost all the image analysis techniques that rely on diffusion can now be extended to the diffusion of lines in an orientation space. This amounts to a lot of research and engineering opportunities that await us.

Chapter 11
Local Statistics on Shape Diffeomorphisms Using a Depth Potential Function

Maxime Boucher and Alan Evans

Abstract Shape diffeomorphisms are used along with statistical models to localize shape differences within a set of similar surfaces. Diffeomorphisms in images are found by looking at similarities within the intensity function of the image. Surfaces do not enjoy the availability of such an intensity function and one therefore needs to choose which function will give the most accurate match between similar features. Thus, it is important to choose an intensity function that faithfully represents features of the surface as accurately as possible. In this paper, we present a fast method which uses a curvature scale space to capture surface features over a wide range of scales. We recall the relationship between Poisson equation and scale space representations, and we use this relationship to formulate a potential function which integrates curvature information over a wide range of scales. The resulting map is a global function of bending (as measured by curvature) that is used to find diffeomorphic maps between surfaces. The potential function showed to be more accurate than a depth map computed on the surface.

As an application of the depth potential function, we use it to analyze diffeomorphic maps of the human brain through a statistical analysis of surface deformation. We used surfaces extracted from 92 subjects affected with very mild to mild dementia and 97 healthy subjects. Using T^2-statistics on the diffeomorphic map between the groups of demented and non-demented subjects, we were able to detect the atrophy of the interior and exterior temporal lobe due to the presence of mild dementia.

M. Boucher (✉)
School of Computer Science, McGill University, Montreal, Canada
e-mail: boucher@bic.mni.mcgill.ca

A. Evans
McConnell Brain Imaging Center, Montreal Neurological Institute, McGill University, Montreal, Canada
e-mail: alan@bic.mni.mcgill.ca

L. Florack et al. (eds.), *Mathematical Methods for Signal and Image Analysis and Representation*, Computational Imaging and Vision 41,
DOI 10.1007/978-1-4471-2353-8_11,

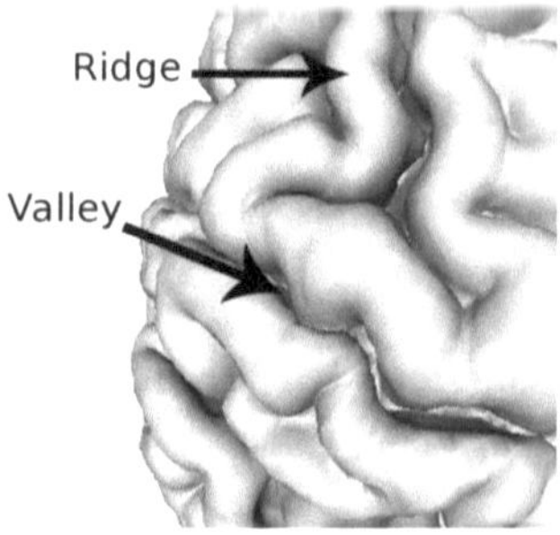

Fig. 11.1 Ridges and valleys on a human brain cortical surface

11.1 Introduction

In Chaps. 1 to 9 the authors explain how diffusion processes are used to find salient features in images and tensor fields. Using these methods, a single image can be enhanced using a diffusion model. In this chapter, we look at the situation where multiple images of similar objects are available to build a statistical description of that object. For example, let us suppose that we have several images of livers. A statistical description of the liver would represent the different livers observed within the set. Moreover, let us suppose that some images come from patients with a liver-related disease. The purpose of statistical shape analysis is then to determine which regions of the liver are most affected by the disease.

In this chapter, the problem of building a statistical description of the cortical surface is studied. More specifically we estimate the impact of a disease on the local shape of the cortical surface by an appropriate diffeomorphic shape model. In a diffeomorphic shape model, shape differences are represented as a deformation of a template surface. In general, the parameters of the model are unknown and can only be determined by finding accurate mappings between the shape of several observations of diseased and healthy organs. This is the so-called registration problem [50, 147, 296, 407, 411, 422]. In the case of images and volumes, a gray-level intensity function is a convenient choice to find diffeomorphic maps between different shapes. However, for some shapes like the human cerebral cortex (e.g. Fig. 11.1), folds make volume registration prone to getting trapped in local mis-registration minima. A fold can be described in loose terms as the juxtaposition of a ridge and a valley, as shown on Fig. 11.1. Thus in the case of cortical surfaces, a surface representation better describes the geometry of the features to be matched. Surfaces do not enjoy the availability of gray-level images for registration purposes, one therefore has to choose which "intensity" function will give an accurate diffeomorphic map. This chapter proposes a scalar field that can be used to register cortical surfaces. Mean curvature is used as a local measure of bending of a surface. A diffusion process is then applied onto the mean curvature scalar field of a surface to capture the shape of a surface at different scales. As was explained in [156], it is possible to integrate the output of an isotropic diffusion process very efficiently using a screened Poisson equation. Using this method, Boucher et al. [50] proposes a bending measure of the surface (as measured by mean-curvature) to register surfaces. The bending measure obtained through an integration of a diffusion process allows it to capture the local bending of the surface over a large range of scales. The resulting scalar field was named a depth potential function. For surface registration, the concern is more efficient computations and accurate registration. On this side, the screened Poisson equation is computationally fast to solve, which makes the depth potential function a very efficient choice to find local similarities between surfaces. Accuracy for surface registration was evaluated using surface labels, and we found that the depth potential function outperformed the geodesic depth map of [358].

As an application to surface registration with the depth potential function, we studied shape similarities and differences among a group of 97 subjects with mild cognitive impairment against a group of healthy subjects. We use the Log-Euclidean tensor probabilistic model proposed in [14, 284] and then used in [48] to analyze surface deformations between the two groups of surfaces (healthy control subjects and patients with mild cognitive impairment). The statistical shape model uses T2-statistics on the matrix logarithm of the deformation tensor. Using this framework, we found the presence of a significant deformation in the interior temporal lobe due to the presence of mild dementia. These deformations are mostly due to atrophy in the temporal lobe, as is highlighted in [49].

11.2 Local Bending of the Surface

The choice of surface image is of prime importance for registration purposes, because it has an impact on the accuracy and the quality of the alignment of similar geometric features across surfaces. Let us put the problem formally: let $I_i : \mathcal{M}_i \to \mathbb{R}, i = 1, \ldots, N$, be a real-valued function over surface $\mathcal{M}_i$; this is the so-called surface image. We suppose each surface has a Riemannian metric g induced by the imbedding in $\mathbb{R}^3$. Surface registration aims at finding a correspondence map ϕ_i between a surface $\mathcal{M}_i$ and a template surface $\bar{\mathcal{M}}$ such that ϕ_i is diffeomorphic and is a global minimum of a registration energy functional:

$$\phi_i = \text{argmin}_{\phi_i^*} E_{distance}(I_i \circ \phi_i^*, \bar{I}) + E_{smoothness}(\phi_i), \tag{11.1}$$

where $\bar{I}$ is the surface image of the template. We observe that Eq. (11.1) leaves out almost entirely any shape information from either $\mathcal{M}_i$ and $\bar{\mathcal{M}}$ to find an optimal map. Thus it is very important to choose I_i and $\bar{I}$ such that features on $\mathcal{M}_i$ and $\bar{\mathcal{M}}$ are well represented.

Section 11.2.1 presents the proposed surface image and Sect. 11.2.2 relates the surface image to an approximation of a locally adapted depth map for surfaces. Finally, we provide some validation experiments in Sect. 11.2.3.

11.2.1 Curvature Scale Space

The size of folds on different subjects varies spatially and across subject and a good surface image I_i should identify and match folds of all scales. We define a scale space representation of a surface shape through a curvature scale space, which we describe shortly. Let H be the mean curvature of the surface defined as

$$\Delta_{\mathcal{M}_i}\mathcal{M}_i = 2H\mathbf{n}, \tag{11.2}$$

where $\Delta_{\mathcal{M}_i}$ is the Laplace-Beltrami operator for surface $\mathcal{M}_i$ and $\mathbf{n}$ is the normal vector to the surface. A positive mean curvature indicates that the sum of the principal curvatures of the surface is positive, and thus it is locally bent outward, while a negative H indicates that the surface is locally bent inward. However, mean curvature is a very local shape descriptor that often fails to capture more global shape characteristic. For example, a region of positive mean curvature may either be a small bump at the bottom of a fold or at the top of a gyrus. Thus, scale space analysis is used to decompose mean curvature over different scales. A scale space representation of a function $f : \mathcal{M} \to \mathbb{R}$ is obtained by solving a heat-diffusion Equation of the form

$$\frac{\partial \hat{f}}{\partial t} = -\Delta_{\mathcal{M}}\hat{f} \tag{11.3}$$

with initial value condition $\hat{f}(\mathcal{M}, 0) = f(\mathcal{M})$. For simpler notation, let $G_{t,i}$ the diffusion operator which, to each function f, associates the parametrized function $\hat{f}(\mathcal{M}_i, t)$ by solving Eq. (11.3). In other word, $G_{t,i}(f)$ is the diffusion operator which diffuses a function f on surface $\mathcal{M}_i$ for an interval of time t.

A mean curvature scale space is given as $G_{t,i}(H_i)$. As said previously, a good surface image I_i allows optimally to identify and match folds of all scales. In [50], it was proposed to obtain such a surface image by summing the curvature scale space over a broad range of scales

$$I_{\alpha,i}(\mathcal{M}_i) = \int_{\mathbb{R}^+} e^{-\frac{\alpha}{2}t} G_{t,i}(H_i)\,dt \tag{11.4}$$

with α a parameter that controls how much emphasis to put on finer scales, cf. also [156]. As such, this approach is not conceptually different from the approach of [147] where they integrate a signed traveled distance throughout an Euler characteristic-preserving flow. However, what makes Eq. (11.4) particularly compelling is that it does not need to be solved by explicitly integrating curvature over a wide range of scale. It was shown in [156] that solving Eq. (11.4) is equivalent to solving

$$\frac{1}{2}(\alpha\mathbb{I} + \Delta_{\mathcal{M}_i})I_{\alpha,i} = H_i \tag{11.5}$$

where $\mathbb{I}$ is the identity operator. Equation (11.5) is a screened Poisson equation, which does not need to solve any costly Euler-characteristic preserving flows and can be solved very efficiently using numerical methods (see suggested numerical method in [50]). We can finally observe that Eq. (11.4) does not converge for $\alpha = 0$ for closed surfaces. We correct this situation by adding a corrector term as

$$H_i' = H_i - \frac{\int_{\mathcal{M}_i} H_i\,ds}{\int_{\mathcal{M}_i} ds}, \tag{11.6}$$

where ds is the volume form. We see that H_i' integrates to 0 and that $\lim_{t\to\infty} G_{t,i}(H_i') \to 0$ everywhere. The resulting equation is

$$\frac{1}{2}(\alpha\mathbb{I} + \Delta_{\mathcal{M}_i})I_{\alpha,i} = H_i'. \tag{11.7}$$

11.2.2 Local Bending of the Surface and Depth Approximation

One of the advantages of Euler-characteristic preserving flows is that the signed traveled distance corresponds to a geometric measure on the surface: how far inward or outward has the surface to travel until it is considered flattened. This interpretation, along with other natural Euclidean interpretation of depth maps, is lost while using Eq. (11.7). In this section, we want to present a variational approximation of depth on surfaces, which will later allow us to show that Eq. (11.5) is an approximation of depth on surfaces. To explain this claim, we first give the variational formulation for depth on surfaces.

Let $\mathbf{e}_i, i = 1, 2, 3$, be an orthonormal basis of $\mathbb{R}^3$. Let **up** denote a unit-length vector which is used to measure depth on a surface. The **up** vector is unknown a priori. Let $f : \mathcal{M} \to \mathbb{R}$ be a function that describes depth on a surface. Let (u_1, u_2) be a coordinate system on $\mathcal{M}$ for an open subset $\Omega \subset \mathbb{R}^2$. The first intuition about what should be a depth f of a surface $\mathcal{M}$ is that it should allow us to write $\mathcal{M}$ parametrically over a large Ω as a graph:

$$\mathcal{M}(u_1, u_2) = u_1\mathbf{e}_1 + u_2\mathbf{e}_2 + f(u_1, u_2)\mathbf{e}_3. \tag{11.8}$$

Suppose in Eq. (11.8) that $\mathbf{up} = \mathbf{e}_3$. Let D be the differential operator $D = (\partial/\partial u_1, \partial/\partial u_2)$ and let $\nabla_{\mathcal{M}}$ be the gradient operator on $\mathcal{M}$. For a function f, which describes depth as in Eq. (11.8), we have

$$D\mathcal{M}(\nabla_{\mathcal{M}} f) = \mathbf{up} - (\mathbf{up}^t.\mathbf{n})\mathbf{n}. \tag{11.9}$$

where $\mathbf{n}$ is the normal vector of the surface. Inspired by Eq. (11.9), we pose the following variational problem for a function f representing depth on a surface:

$$f = \operatorname{argmin}_{f'} \int_{\mathcal{M}(\Omega)} \|Df' - (\mathbf{up}^t - (\mathbf{up}^t.\mathbf{n})\mathbf{n}^t).D\mathcal{M}\|_g^2 \, ds, \tag{11.10}$$

with given boundary conditions on $f(\mathcal{M}(\partial\Omega))$ and where ds is the volume form on $\mathcal{M}$. Unless $\mathcal{M}$ is a particular case, it is not possible to find a function f such that $\mathcal{M}$ can be parametrically written as the graph of f for all of $\mathcal{M}$. Since the depth function represents a distance to a projection plane with respect to the **up** direction, an easy method to adapt the depth function is to penalize large deviations from zero:

$$E_{regularized}(f_\alpha) = \operatorname{argmin}_{f_\alpha'} \int_{\mathcal{M}} \|Df_\alpha' - \mathbf{up}D\mathcal{M}\|^2 + \alpha\|f_\alpha'\|^2 \, ds, \tag{11.11}$$

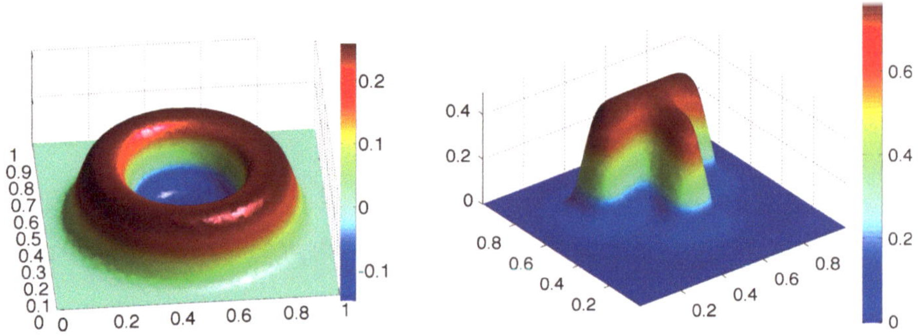

Fig. 11.2 Depth potential function without any damping ($\alpha = 0$) on example surfaces. The value of the depth potential function is shown as a color map and the scale of the color map is with respect to the units shown on the axis. *Left*: A circle relief. *Right*: A "T" relief

where we used the orthonormality between $\mathbf{n}$ and $D\mathcal{M}$ to simplify the equation. Equations (11.11) allows many solutions which mostly depend on our choice of **up**. The solution to obtain a good depth map of the relief on a surface is to pose that **up** is constant over $\mathcal{M}(\Omega)$, in which case the solution is given by

$$(\alpha\mathbb{I} + \Delta_{\mathcal{M}}) f_\alpha = \Delta_{\mathcal{M}} \mathcal{M}^t.\mathbf{up} \tag{11.12}$$

with an extra term to describe the relationship to the boundary condition on $\mathcal{M}(\partial\Omega)$. Looking at Eqs. (11.7) and (11.9), we see that if we pose

$$\mathbf{up}^t\mathbf{n} = 1 - \mathcal{O}(\|D\mathcal{M}.\nabla_{\mathcal{M}} f\|^2) \tag{11.13}$$

then Eq. (11.12) becomes equivalent to Eq. (11.5). This is a rough approximation to a true depth map representation of the surface. For surface registration, we can expect that an approximation to the true geometry will yield good result. Also, the approximation given in Eq. (11.13) is mostly accurate for points of extremal depth (minimum, maximum and saddle points: where $\mathbf{n} = \mathbf{up}$).

11.2.3 Illustration and Examples

We provide in this section some illustrations and examples of depth potential functions computed using Eq. (11.7). The first two examples are relief shown in Fig. 11.2. We solved Eq. (11.7) on both these surfaces while imposing as boundary conditions $f(\partial\mathcal{M}) = 0$ without any damping ($\alpha = 0$). The results on Fig. 11.2 shows that the depth potential function correctly identifies the presence of relief as well as their elevation. It should be noted that the maximal height of the "T" relief is 0.5 while the depth potential function gives a maximum difference in elevation of 0.78. Thus the depth potential function overestimates differences in height, which is expected from Sect. 11.2.2. However, for surface registration, this is not a concern because we are mostly interested at how the depth potential function correlates with the shape of the relief.

11.3 Validation of the Depth Potential Function Using Cortical Surfaces

In this section, we present validation experiments and results that were performed using surfaces that were extracted from in vivo magnetic resonance images (MRIs). The CIVET pipeline [249] was used to obtain surfaces from T1-weighted MRI from the ICBM database [304] and the OASIS brain database [307]. The MRIs from the ICBM database were taken from healthy subjects between 20 to 28 years of age. The OASIS database consists of MRIs of subjects of 60 years old and above. Within the OASIS database, 87 subjects are healthy and 82 are diagnosed with very mild or mild dementia (Clinical Dementia Ratio ≤ 1). The CIVET pipeline extracts two pairs of surfaces from any given MRI image, one pair for each hemisphere. One pair being the interface between the gray matter and white matter (the White surface, one surface per hemisphere), the other pair of surfaces being the interface between the gray matter and the cerebro-spinal fluid (the Gray surface). The surface extraction pipeline functions such that the Gray surface is constrained to be diffeomorphic to the White surface through a constrained Laplacian technique [297]. In the end, each vertex on the Gray surface corresponds to a vertex on the White surface. A mid surface representation of the cortex is then obtained by averaging the coordinate of corresponding vertices on the Gray and the White surface. The resulting surface is called the mid surface.

Figure 11.3 shows an example of a mid surface along with the depth potential function with $\alpha = 0.0015$ and $\alpha = 0$. Damping α removes coarse features from the I_α as is expected from Eq. (11.4).

11.3.1 Correlation Between Fold Depth and the Depth Potential Function

For surface registration, it is important to use a surface image that correlates well with the surface pattern to be studied. Thus, we evaluated the correlation between the depth potential function and fold depth on a set of surfaces where the depth of the left central sulcus was artificially reduced (shown on Fig. 11.5). We first asked a neuroanatomist to label the region corresponding to the left central sulcus on 20 surfaces from the ICBM database. Let L_i be the set of points labeled as being part of the left central sulcus and χ_{L_i} the indicator function on $\mathscr{M}_i$. We created a smooth transition between the labeled and non-labeled regions by smoothing χ_{L_i} using $G_{12.5}(\chi_{L_i})$. Each surfaces was also smoothed using an area-minimizing flow given as

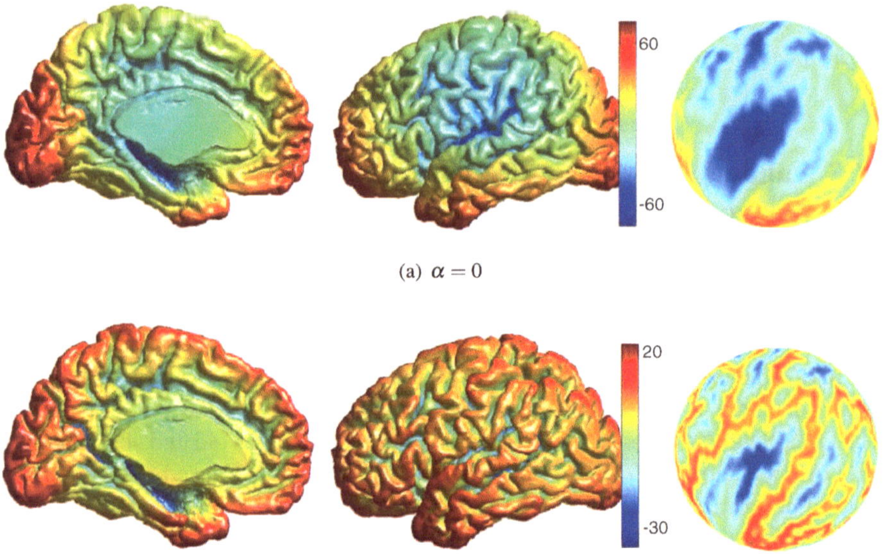

(a) $\alpha = 0$

(b) $\alpha = 0.0015$

Fig. 11.3 Examples of a depth potential function for a cortical surface. The spherical map shown *in the third column* is obtained during the surface extraction process [249]. Coarse features such as the global, ellipsoidal shape of the brain are mostly visible with $\alpha = 0$ (**a**). However, if some screening is added ($\alpha = 0.0015$), the folding pattern is better represented (**b**)

$$\frac{\partial \hat{\mathscr{M}}_i}{\partial t} = -\Delta_{\hat{\mathscr{M}}_i} \hat{\mathscr{M}}_i \qquad (11.14)$$

with initial value condition $\hat{\mathscr{M}}_i(\mathscr{M}_i, 0) = \mathscr{M}_i$. With a slight abuse of notation, let us note $\hat{\mathscr{M}}_i(\mathscr{M}_i, t) = G_{t,i}(\mathscr{M}_i)$. The set of surfaces with a left central sulcus of reduced depth is given as:

$$\mathscr{M}_{i,smoothed}(\mathscr{M}_i) = (1 - G_{12.5}(\chi_{L_i}))\mathscr{M}_i + G_{12.5}(\chi_{L_i})G_{12.5}(\mathscr{M}_i). \qquad (11.15)$$

The second group of surfaces consists of 20 surfaces extracted directly from the ICBM data set, without any smoothing. All surfaces were registered onto the same template surface.

We evaluated the impact of fold depth on the depth potential function by measuring the local standard deviation of the depth potential function on each surface. Local standard deviation σ_t is expressed as

$$\sigma_t^2 = G_t((f_\alpha - G_t(f_\alpha))^2). \qquad (11.16)$$

For comparison purposes, we used σ_{70} and $\alpha = 0.0015$. To illustrate the ability of the method to localize a correlation between fold depth and the depth potential function, we first compared two identical surfaces before and after smoothing the left central sulcus. The results are shown in the left plot of Fig. 11.4. In this figure,

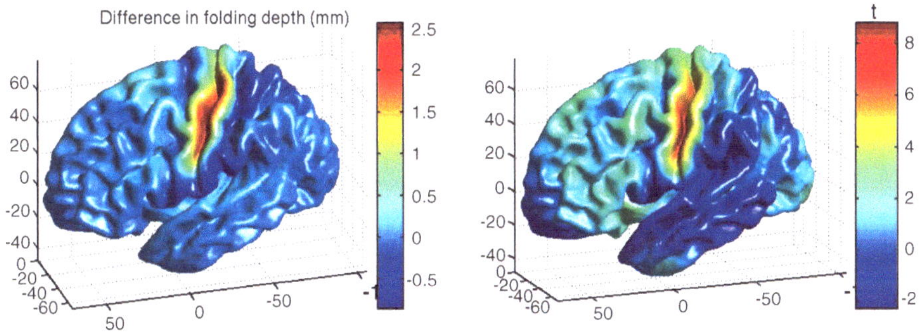

Fig. 11.4 Correlation between fold depth and the depth potential function. *Left*: Difference in local standard deviation before and after smoothing. *Right*: Statistical difference between the two groups of surfaces. The left central sulcus is correctly identified as being shallower. If no smoothing is applied to the left central sulcus, all T-values are $|T| < 3$

we can see that a correlation is detected around the left central sulcus, which corresponds to the labeled region. We then used the following generalized linear model [305] to detect differences in local standard deviation

$$\sigma_t \sim \mu + cS_i + \varepsilon_i \tag{11.17}$$

where ε_i is a normally distributed random variable and S_i is equal to 1 if surface $\mathscr{M}_i$ is member of the smoothed set and -1 if it isn't. The regression coefficient c was determined through linear regression and the probability that $c = 0$ was evaluated through a standard T-test. The result of the analysis is shown in the right plot of Fig. 11.4. This figure shows that for this example and in spite of statistical variability, the depth potential function is well correlated with fold depth.

11.4 Surface Registration Results

If the depth potential function, given in Eq. (11.7) is a good surface image, then the mapping that minimizes Eq. (11.1) should provide quantifiably better result than another surface image. The surface registration algorithm of [358] was used to quantify surface registration accuracy and validate if a depth potential function offers any improvement over a geodesic depth map based on the painting algorithm [357, 358].

First, we determined if surface registration is consistent. Let $\mathscr{M}_i$ and $\mathscr{M}_j$ be two surfaces and $\phi_{i,j}$ the mapping that minimizes Eq. (11.1). We define a lack of consistency as a difference between $\phi_{i,j}$ and $\phi_{j,i}^{-1}$. To measure this lack of consistency, we approximate the geodesic distance on surfaces through a spherical mapping $S_i : \mathscr{M}_i \rightarrow \mathbb{S}^2$. The distance between $\phi_{i,j}$ and $\phi_{j,i}^{-1}$ is then measured as

$$d_{sphere}(S_i \circ \mathscr{M}_i, S_i \circ \mathscr{M}_i \circ \phi_{i,j} \circ \phi_{j,i}), \tag{11.18}$$

where $d_{sphere}(\cdot, \cdot)$ is the geodesic distance on $\mathbb{S}^2$.

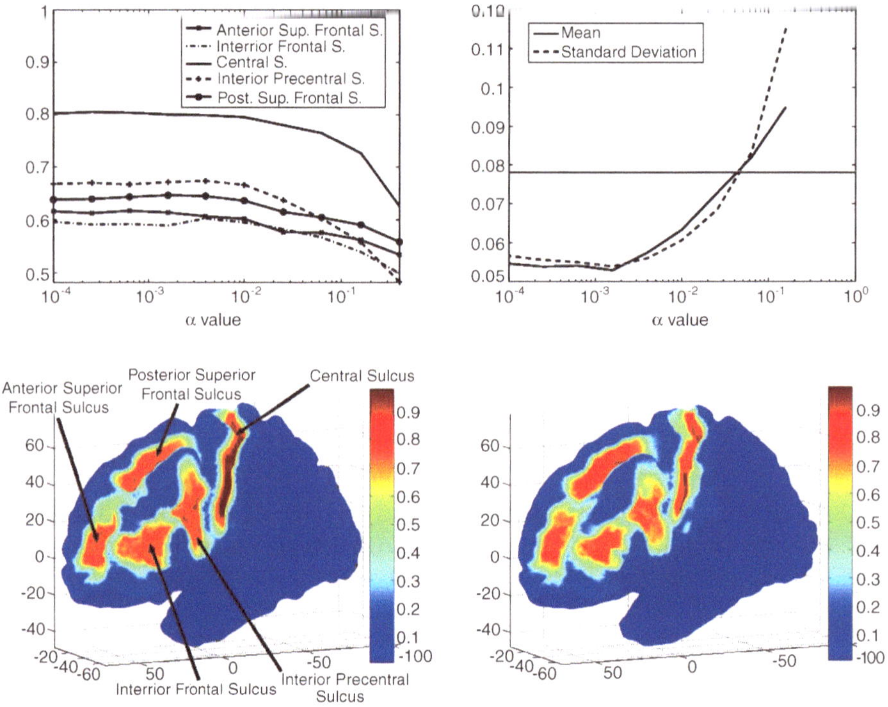

Fig. 11.5 *Top*: Accuracy (*left*) and consistency (*right*) of registration using a depth potential function with different α values. The most consistent result is obtained for $\alpha = 0.0015$ and registration of each individual fold is accurate for a wide range of α values. *Bottom*: Accuracy of registration when all subjects are registered on the same target using a depth potential function ($\alpha = 0.0015$) (*left*) and a geodesic depth map (*right*). Boundary of each sulcus is better defined when using the depth potential function

We used 40 surfaces extracted from the ICBM data set to evaluate consistency of surface registration. Twenty pairs of surfaces were formed and registration was performed between each pairs using a depth potential function and a depth map based on a painting algorithm [357, 358]. Then, d_{sphere} was measured for every pair of surfaces. For the geodesic depth map of [358], we found that lack of consistency d_{sphere} was on average 0.078 over the entire surface with a standard deviation of 0.076 (this is shown as thin horizontal lines in the top right of Fig. 11.5). We tested consistency of the depth potential function for several values of α and the results are plotted in Fig. 11.5. This figure shows that for a wide range of α, consistency of the depth potential function is better than the geodesic depth map of [358]. Across different pairs of surfaces, a minimum is reached around $\alpha = 0.0015$. Hence, in this case, the depth potential function improves registration consistency over a map produced with a painting algorithm.

Accuracy of surface registration is measured with labels of major sulci. A neuroanatomist labeled five major sulci of the frontal lobes, which are shown in Fig. 11.5 (Anterior Superior Frontal Sulcus, Posterior Superior Sulcus, Inferior

Table 11.1 Registration results for different regions of the cortical surface

Sulcus	Sulcal depth map	Depth potential function
Anterior sup. frontal	0.56	0.62
Inferior frontal	0.5	0.59
Central	0.73	0.8
Inferior precentral	0.59	0.67
Posterior sup. frontal	0.59	0.62

Frontal Sulcus, Inferior Precentral Sulcus, Central Sulcus). The sulci were manually labeled on 40 native MRIs and the labels were mapped onto the mid-surface by intersecting the surface with the labeled regions. We measured accuracy as the overlap between labels on surface $\mathscr{M}_i$ and $\mathscr{M}_j$. Formally, let L_k be the set of points which has label k and let $V(L)$ be the volume of the set L. Then

$$accuracy(\mathscr{M}_i, \mathscr{M}_j, k) = \frac{V(\mathscr{M}_j(L_{k,i}) \cap \mathscr{M}_i(L_{k,i}) \circ \phi_{i,j})}{V(\mathscr{M}_j \circ L_{k,j})}. \tag{11.19}$$

Twenty pairs of surfaces were formed and accuracy was measured for $\phi_{i,j}$ and $\phi_{j,i}$. The average accuracy of registration is shown in the top left plot in Fig. 11.5. Table 11.1 compares accuracy of surface registration using a depth potential function ($\alpha = 0.0015$) and a geodesic depth map. As we can see in this table, the DPF constantly outperforms the sulcal depth map of [358]. On average, the improvement in overlap is 11%.

The improvement in accuracy is illustrated in the bottom of Fig. 11.5 where all surfaces were registered onto the same target using a depth potential function with $\alpha = 0.015$ and a geodesic depth map. On this figure, we can see that the boundary of each sulcus (most notably of the central sulcus) is better defined for the depth potential function.

11.5 Application: Detecting Shape Differences Through Local Statistics on Shape Diffeomorphisms

As an application, we propose to use shape diffeomorphisms to find statistical differences between two groups of surfaces. The framework we describe in this section was first proposed in [14, 284] for volume diffeomorphism. It was implemented for surfaces in [48] and we propose to use the expected Euler characteristic of random fields [451, 454] to determine if locally a deformation is unlikely to have been generated from the shape model.

We used the unbiased surface template $\bar{\mathscr{M}}$ of [296] to find surface diffeomorphisms $\phi_i : \bar{\mathscr{M}} \to \mathscr{M}_i$ between each surface. The deformation tensor is given by

$$A_i = \nabla\phi_i . D\mathscr{M}_i . (\nabla\phi_i . D\mathscr{M}_i)^t, \tag{11.20}$$

where A_i is 2×2 symmetric positive definite matrix field. Here, we suppose that A_i is a smooth field. Let $A = \sum_j \lambda_j r_j r_j^t$ be the diagonalization of A and let Log be the matrix log operator defined as

$$\mathrm{Log}(A) = \sum \log(\lambda_j) r_j r_j^t. \tag{11.21}$$

The matrix logarithm is a diffeomorphism between the space of symmetric positive matrix field and the space of symmetric matrices. Let $vec(A) = (A_{11}, A_{22}, \sqrt{2}A_{12})$ be the operator which maps a symmetric matrix onto a column vector. The Log-Euclidean framework of [14] uses the matrix logarithm Log to model tensor variability as

$$vec(\mathrm{Log}(A_i)) \sim \mathcal{N}(\mu, \Sigma), \tag{11.22}$$

where μ is the expected value of $vec(\mathrm{Log}(A_i))$ and Σ describes the variability of the model.

In practice, the model parameters μ and Σ are unknown and need to be estimated. We use a least square estimator to estimate μ as

$$\hat{\mu} = \frac{1}{N} \sum_i vec(\mathrm{Log}(A_i)). \tag{11.23}$$

The unbiased covariance estimator $\hat{\Sigma}$ is given as

$$\hat{\Sigma} = \frac{1}{N-1} \sum_i (vec(\mathrm{Log}(A_i)) - \hat{\mu}) \otimes (vec(\mathrm{Log}(A_i)) - \hat{\mu}). \tag{11.24}$$

Let the Mahalanobis distance be given as

$$d_{Mahalanobis}(\mathrm{Log}(A))^2 = (vec(\mathrm{Log}(A)) - \hat{\mu})\hat{\Sigma}^{-1}(vec(\mathrm{Log}(A)) - \hat{\mu}). \tag{11.25}$$

There is a relationship between the Mahalanobis distance as expressed in Eq. (11.25) and T^2 distributed random variables:

$$N d_{Mahalanobis}(\mathrm{Log}(A))^2 \sim T^2_{3,N-1}. \tag{11.26}$$

Equation (11.26) is a T^2 random field. The probability that Eq. (11.26) goes above a preset bound c, $d_{Mahalanobis}(\mathrm{Log}(A)) > c$ is evaluated through the expected Euler characteristic of the random field [451, 454] at that value c.

Chapter 1, Sect. 1.5.2, and Chap. 5 discussed the problem of tensor field regularization. In statistical shape analysis, regularization is used to increase the detection power of linear models. In our case, we used a simple diffusion kernel as explained in [49] and we diffused for up to $t = 50$.

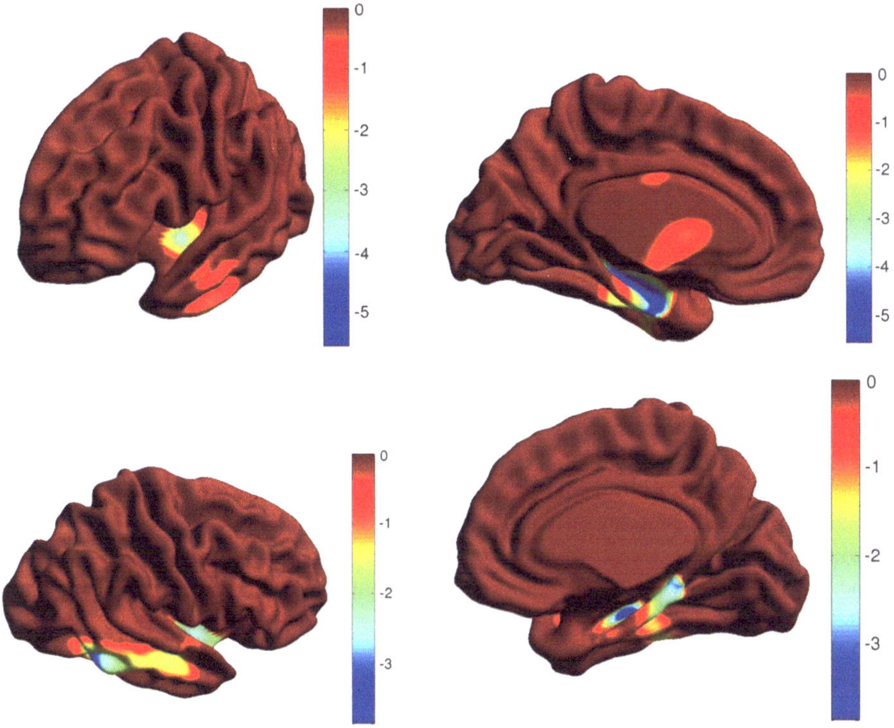

Fig. 11.6 Difference between patients with mild dementia and elderly healthy subject. *Color bars* show logarithmic probabilities: anything below $\log(0.05) = -1.3$ is significant. The Sylvian fissure and the temporal saggital lobe shows significant difference in the group with mild dementia

11.5.1 Results on a Population of Cortical Surfaces

The method described in Sect. 11.5 was applied on the OASIS database. We used the following linear model

$$Y_i \sim c_0 + c_1 Gender_i + c_2 Age_i + c_3 CDR_i + \varepsilon_i, \tag{11.27}$$

where Y_i is the measured deformation tensor on the surface for subject i, $Gender_i$ is the gender of the subject, Age_i the age, CDR_i the clinical dementia ratio and ε_i is a normally distributed random variable. We tested the null hypothesis that $c_3 \approx 0$ using the expected Euler characteristic of the T^2 random field [451, 454]. The results are shown in Fig. 11.6. The color bar on these figures shows logarithmic probabilities to emphasize regions of low probability. On this scale, $\log(0.01) = -2$ and anything smaller than $\log(0.05) = -1.3$ is significant. In Fig. 11.6, we can see the Sylvian fissure and the temporal saggital lobe show significant difference in the group with mild dementia. Other differences between the healthy subjects and patients with mild dementia are observed on the right hemisphere. It is known in the literature that there is an atrophy in the temporal lobe of patients with Alzheimer's disease [408]. However, in our results, we found that the atrophy is spread over wider regions in the right hemisphere. This is in opposition to other results, which have shown a larger atrophy of the cortical thickness in the left hemisphere. It should be noted that our study does not look at cortical thickness but rather cortical shape differences.

11.6 Conclusion

We presented in this chapter a surface image to register surfaces. The surface image integrates mean curvature information over a wide range of scale, thus creating a faithful representation of the surface. The use of our surface image showed to improve surface registration (Sect. 11.4). We then presented a statistical shape model for surface morphometry in Sect. 11.5. We used the shape model on the OASIS database, which contains patients with Alzheimer's Disease and elderly subjects. We used this shape model on the OASIS database and found that the presence of Alzheimer's disease was correlated with several shape differences in the temporal lobe of both hemispheres. Results look promising, and we are investigating their exact biological meaning.

Chapter 12
Preserving Time Structures While Denoising a Dynamical Image

Yves Rozenholc and Markus Reiß

Abstract In restoration or denoising of a movie, the classical procedures often do not take into account the full information provided by the movie. These procedures are either applied spatially "image per image" or locally on some neighborhood combining both closeness in one image and closeness in time. The local information is then combined homogeneously in order to realize the treatment. There is one type of movie where both approaches fail to provide a relevant treatment. Such a movie, called *dynamical image*, represents the same scene along the time with only variations, not in the positions of the objects in the scene but, in their gray levels, colors or contrasts. Hence, at each point of the image, one observes some regular temporal dynamics. This is the typical output using Dynamic Contrast Enhanced Computed Tomography (DCE-CT) or Dynamic Contrast Enhanced Magnetic Resonance Imaging (DCE-MRI) where at each spatial location (called voxel) of the image a time series is recorded.

In such a case, in order to preserve the full temporal information in the image, a proper denoising procedure should be based on averaging over spatial neighborhoods, but using the full dynamics of all pixels (or voxels) within the neighborhood. It is thus necessary to search homogeneous spatial neighborhoods with respect to the full dynamical information.

We introduce a new algorithm which meets these requirements. It is based on two main tools: a multiple hypothesis testing for the zero mean of a vector and an adaptive construction of homogeneous neighborhoods. For both tools, results of mathematical statistics ensure the quality of the global procedure. Illustrations from medical image sequences show a very good practical performance.

Y. Rozenholc (✉)
MAP5, UMR CNRS 8145, University Paris Descartes, Paris, France
e-mail: yves.rozenholc@parisdescartes.fr

M. Reiß
Institute of Mathematics, Humboldt University Berlin, Berlin, Germany
e-mail: mreiss@math.hu-berlin.de

L. Florack et al. (eds.), *Mathematical Methods for Signal and Image Analysis and Representation*, Computational Imaging and Vision 41,
DOI 10.1007/978-1-4471-2353-8_12,

12.1 Introduction

Classically, the restoration or the denoising of a movie involve 2D techniques for smoothing [56, 287], enhancing (see contributions on structure enhancement and use of line diffusion), filtering (see contribution of Michael Felsberg and references within), partitioning (see [339] or contributions of J. Polzehl and K. Tabelow), etc. applied image per image. If some methods use instead of one image a small number of images to take into account the time domain, all methods ignore the full time structure. A natural mathematical reason, outside the computational cost, is that the information contained in one image is considered to be less and less connected to the next images as the time increases. In some sense, the time regular structures existing in a movie do not have a long range effect and disappear quickly as the time distance between two images increases. One exception to this last rule is given by movies made from one *dynamical image* where at each 2D location (pixel or voxel) one observes not only a gray level or a color, but a full dynamical process represented by a time series or a vector. This is typically the case when the objects in the picture do not have changes in their position, but have changes only in their brightness, contrast or color. Medical examples of interest are sequences like DCE-CT or DCE-MRI which register the variations of a contrast medium in the internal tissues of the body in response to a dynamical injection.

Our aim is to develop a new technique, adapted to these *dynamical images*, where each pixel presents a time series. In order to denoise by local averaging, we combine two approaches: statistical multiple hypothesis testing in the spirit of the works [22, 120] to compare the time series between two pixels, together with a spatial growth of locally homogeneous neighborhoods closely related to [285, 287]. The originality of our method is not only to use the full dynamics to compare pixels, but also the way of sequentially growing the neighborhood. While usual paradigms to grow such neighborhoods compare the estimates built on the nested neighborhoods as in [287], our method compares, at each step, independent estimates to stop or not the recursive growth. In the context of noise with heavy tails like Laplace noise for example, this comparison using independent estimates and robust statistics like the median offers a clear benefit in reducing the error.

Because the test procedure is based on the comparison to zero of the difference between two noisy vectors, i.e. two noisy time series, no modeling assumption is used. As we use an adaptive multiple hypothesis testing procedure, neither regularization in time is needed nor made. From a theoretical point of view, we only need the dynamics to be not too wild in the sense that the enhancement differences—viewed as a function of time—should remain in some Hölder regularity ball.

The article is organized as follows: In Sect. 12.2, we introduce the statistical framework of *dynamical images* and summarize our method. The two main statistical tools used to construct this method, multiple testing and neighborhood/ring growth, are respectively studied in Sects. 12.3 and in 12.4. In these sections, results from mathematical statistics are provided to guarantee the behavior of our method. Section 12.5 is devoted to the comparison of our estimation technique with estimates introduced in [287] and in [339]. Finally in Sect. 12.6, we briefly describe an application to DCE-CT of renal metastasis in the liver, an extract from the specific work [363].

12.2 Denoising a Dynamical Image

We consider the following model called *dynamical image* indexed by the two quantities t and x representing time and spatial location, respectively:

$$I = \{I_x(t), x \in \mathscr{X}, t \in \{t_1, t_2, \dots, t_K\}\}.$$

Here $I_x(\cdot)$ denotes a noisy time series at location x. The notation I_x will denote the vector

$$(I_x(t_1), \dots, I_x(t_K)),$$

of the discrete observation at times $t_1, \dots, t_K$ of this time series. In our setting, $\mathscr{X}$ denotes the finite grid of pixels.

We assume that the noisy observations $I_x(t_k)$ may be written as

$$I_x(t_k) = i_x(t_k) + \sigma \varepsilon_k^x, \tag{12.1}$$

where $i_x(t)$ denotes an unobservable true gray level at time t, where ε_k^x denotes a standardized noise with fixed distribution and where σ denotes the level of noise. We assume that the noise variables ε_k^x are independent and identically distributed with respect to both spatial location x and (time index) k. It is not necessary to assume that σ and the distribution of ε^x are known, but we will illustrate our construction using the simple setting σ known and ε^x having a Gaussian distribution.

In order to compare the full time series, we use multiple hypothesis tests to decide whenever the vectors I_x and I_y at spatial locations x and y are statistically similar or not. For this purpose, we ask if the expectation of difference vector $I_x - I_y$ is the zero vector or not at a given level α which could be considered as a natural tuning parameter. Such a test is presented in Sect. 12.3.

Given a set V of pixels, let us define the estimated dynamics using V as the coordinate-wise empirical mean

$$\hat{I}_V = \frac{1}{|V|} \sum_{y \in V} I_y, \tag{12.2}$$

where $|A|$ denotes the cardinality of a set A. Our aim is to construct at each spatial location x, a spatial neighborhood $\mathscr{V}_x$ of x made from voxels y such that the difference $I_x - I_y$ does not deviate significantly from the zero vector and such that the bias of the estimated dynamics obtained by replacing V in (12.2) by $\mathscr{V}_x$ remains under control. Controlling the statistical bias means that we aim at using a procedure which ensures that the principal goal of nonparametric statistics, which is the trade-off between bias and variance, is achieved. This method is introduced in Sect. 12.4 and can be summarized as follows:

Let a geometrically increasing sequence (n_i) of positive integers be given with $n_1 = 1$. For a fixed location $x \in \mathscr{X}$, we consider the set $\mathscr{W}_x$ of all pixels y such that, at a level $\alpha \in (0, 1)$, $I_x - I_y$ is accepted to have a zero mean by the previous test. We set $i = 0$ as well as $V_0 = \emptyset$ and $W_0 = \{x\}$. Our method repeats the following steps until one test, at least, rejects the null hypothesis:

- Increment i.
- Define $V_i = V_{i-1} \cup W_{i-1}$ and build the set W_i of the n_i closest y in $\mathscr{W}_x \setminus V_i$.
- For $j = 1, \dots, i$, test the $i - 1$ null hypotheses $\mathbb{E}(\hat{I}_{W_i} - \hat{I}_{V_j}) = 0$ for $j = 1, \dots, i - 1$ against a non-zero expectation.

The final estimate is $\hat{I}_{V_{i-1}}$. The algorithm realizes a pointwise adaptive selection of homogeneous dynamics. Because the last step involves a multiple hypothesis test, in order to ensure that its level is α, each individual test has to be calibrated with respect to α and i by a Bonferroni correction.

12.3 Multiple Testing

Given two spatial locations x and y, we present the statistical test used to compare to the zero vector the expectation of the difference vector Z with components

$$Z_k := I_y(t_k) - I_x(t_k), \quad k = 1 \dots K.$$

Such a test of comparison to the zero vector derives from the theoretical works [22, 120] which consider a more general framework where σ is unknown and/or where ε is not necessarily assumed to be Gaussian. The theoretical study relies on large K asymptotics, which in our case means that we observe the continuous time signal more and more frequently. For the sake of simplicity, we write $Z = f + \sqrt{2}\sigma\varepsilon$ and we introduce these tests in their simplified version which is the Gaussian case with σ known. We aim to test that the mean vector f is zero or not and hence consider the hypotheses

$$\mathscr{H}_0 : \text{“}f = 0\text{”} \quad \text{versus } \mathscr{H}_1 : \text{“}f \neq 0\text{”}.$$

To enlighten the notations, we suppose that K, the number of time indices, is of the form $K = 2^d$ and we consider the regular dyadic decomposition of the observation times $t_1 \dots t_{2^d}$. For $j = 0, \dots, d - 1$, we denote by $T_1^j \dots T_{2^j}^j$ the 2^j intervals of time indexes

$$T_l^j = \{t_k : k = 2^{d-j}(l - 1) + (1 \dots 2^{d-j})\}.$$

The set T_l^j contains 2^{d-j} time indices. Given j in $0, \dots, d - 1$, let us introduce the projection of the observation Z onto the space spanned by the vectors with same coordinates on each time index T_l^j:

$$\Pi_j Z = (\underbrace{m_1^j, \dots, m_1^j}_{2^{d-j} \text{ times}}, \dots, \underbrace{m_{2^j}^j, \dots, m_{2^j}^j}_{2^{d-j} \text{ times}})$$

with the local mean over T_l^j

$$m_l^j = \frac{1}{2^{d-j}} \sum_{t \in T_l^j} Z_t.$$

The test is based on the size of the squared Euclidean norm $\|\Pi_j Z\|_K^2$ which is:

$$2^{d-j} \sum_{l=1}^{2^j} (m_l^j)^2 = \frac{1}{2^{d-j}} \sum_{l=1}^{2^j} \Big(\sum_{t \subset T_l^j} Z_t \Big)^2.$$

Under $\mathscr{H}_0$, the difference vector $Z = I_y - I_x$ is a centered Gaussian vector with covariance matrix $2\sigma^2 \mathrm{Id}_K$, where Id_K denotes the identity matrix in $\mathbb{R}^K$. Hence, under $\mathscr{H}_0$,

$$\|\Pi_j Z\|_n^2 / 2\sigma^2 \sim \chi^2(2^j)$$

which defines a distribution-free test statistic. Let us denote by Ψ_D^{-1} the inverse cumulative distribution function of a χ^2 with D degrees of freedom, using a Bonferroni correction, our test procedure works as follows:

$$\text{Reject } \mathscr{H}_0 \text{ at level } \alpha \text{ if for any } j = 0, \dots, d-1, \ \|\Pi_j Z\|_K^2 / 2\sigma^2 > \Psi_{2^j}^{-1}(\alpha/d).$$

If σ is unknown, instead of a χ^2-test based procedure, [22] proposes a Fisher like construction and considers the following procedure:

$$\text{Reject } \mathscr{H}_0 \text{ at level } \alpha \text{ if for any } j = 0, \dots, d-1,$$
$$\frac{\|\Pi_j Z\|_K^2 / 2^j}{\|Z - \Pi_j Z\|_K^2 / 2^{K-j}} > \mathscr{F}_{2^j, 2^{K-j}}^{-1}(\alpha/d),$$

where $\mathscr{F}_{2^j, 2^{K-j}}^{-1}$ denotes the inverse cumulative distribution function of a Fisher distribution with 2^j and 2^{K-j} degrees of freedom. Moreover, if the distribution of the noise variables ε_k^x is unknown, as $\varepsilon := (\varepsilon^x - \varepsilon^y)/\sqrt{2}$ is by construction standardized and symmetrical, under the weak condition that the common distribution of the ε does not put mass in 0, we can follow the construction proposed in [120]:

Consider p Rademacher vectors $R_1, \dots, R_p$ of length K. Rademacher vectors have independent coordinates R_i with $P(R_i = +1) = P(R_i = -1) = 1/2$. Introduce the vectors $N_j = Z \otimes R_j$ where $\otimes$ denotes the coordinate-wise product. The previous constructions can be generalized by considering the following procedure:

$$\text{Reject } \mathscr{H}_0 \text{ at level } \alpha \text{ if for any } j = 0, \dots, d-1,$$
$$\|\Pi_j Z\|_K^2 > Q^{-1}[\|\Pi_j Z \otimes R_1\|_K^2, \dots, \|\Pi_j Z \otimes R_p\|_K^2](\alpha/d),$$

where $Q^{-1}[X_1, \dots, X_K]$ denotes the inverse cumulative distribution of the empirical distribution of the X_i, $i = 1, \dots, K$.

Introducing the continuous-time signal F defined by $F(t) := \iota_x(t) - \iota_y(t)$, the difference vector $Z = I_x - I_y$ follows the fixed-design regression model

$$Z_k = f_k + \sqrt{2}\varepsilon_k = F(t_k) + \sqrt{2}\sigma\,\varepsilon_k.$$

In this framework, the introduced multiple hypothesis tests are known to have a power which may be controlled on balls of the form $\|f\|_K \geq \rho(K)$ where $\rho(K)$ is a radius decreasing with K, the number of components of f (= number of observation times in the *dynamical image*). Moreover, these multiple hypothesis tests are adaptive with respect to the unknown Hölder regularity s of the unknown function F: at a fixed level α and for a given fixed power $1 - \beta$, this test automatically achieves (up to logarithmic factors) the minimax rate of testing $\rho_s(K)$ obtained in [180] for all regularities $s > 1/4$. More precisely, we can derive following [22, Theorem 1] and [120, Theorem 4]) the following control of the power:

Theorem 12.1 *If the distribution of* ε_1 *does not put mass on* 0, *assuming that*: (*Bernstein's conditions*) $\max_{1\leq i\leq n} E(\varepsilon_i^{2p}) \leq \gamma p! \mu^{p-2}$ *holds for all integers* $p \geq 1$ *and some positive* γ *and* μ, *then there are* $A_1, \ldots, A_4 > 0$ *such that the multi-test procedure is of power* $1 - \beta$ *outside the ball defined by*

$$\|f\|_K^2 \leq \inf_j \Delta(J_j, f, A_1, \ldots, A_5)$$

where

$$\begin{aligned}
&\Delta(J_j, f, A_1, \ldots, A_5)\\
&\quad = A_1\|f - \Pi_{J_j} f\|_K^2 + A_2\sqrt{\frac{\gamma}{\beta}}2^{(j-1)/2}\\
&\qquad + A_3\Big(\sqrt{\gamma} + \mu + \max_{k=1..K}(f_k^2)\Big)\left[1 + \frac{j\log 2 + \log(1/\beta)}{2}\right]\log\left(\frac{2d}{\beta\alpha}\right)\\
&\qquad + A_4\Big(\sqrt{\gamma} + \mu + \max_{k=1..K}(f_k^2)\Big)2^{(j-1)/2}\sqrt{\left[1 + \frac{j\log 2 + \log(1/\beta)}{2}\right]\log\left(\frac{2d}{\beta\alpha}\right)}.
\end{aligned}$$

Corollary 12.1 *If* $f_k = F(t_k)$ *then the test is adaptative w.r.t. the Hölder regularity* s *of* F *and achieves the optimal minimax rate for* $s > 1/4$.

This theorem is the exact translation of [120, Theorem 4], using our simple setting with $K = 2^d$ and dyadic partitions. It ensures a good power of the test (up to logarithmic terms) if the true unknown function to be tested is, at least for one dyadic partition, at a squared distance from 0 large enough to ensure that a noisy version has also its squared distance from 0 large enough. Let us recall that the squared distance of a noisy version is of the order of the sum of the variance term given by the number of pieces of the partition and the bias term $\|f - \Pi_{J_j} f\|_K^2$ which measures how well f can be approximated by a piecewise constant vector on the index partition.

We have now at hand the tool to compare dynamics at two locations x and y with respect to the known noise level σ. We denote "$I_y \equiv^{\alpha}_{\sigma^2} I_x$" when the enhancement difference average vector $i_x - i_y$ is accepted to be the zero vector at level α with respect to a noise variance $2\sigma^2$ following the above construction.

12.4 Spatial Neighborhood Growth

We now present the construction of the spatial neighborhood $\mathscr{V}_x$ for a fixed location x. In all that follows, the distance between spatial locations x in $\mathscr{X}$ can be measured by any metric; usually the Euclidean metric will be taken.

The main idea of the sequential neighborhood growth can be explained in terms of testing. Suppose that neighborhoods $V_1 \subset V_2 \subset \cdots \subset V_M$ of x are given. Then the stochastic fluctuations in the mean $\hat{I}_{V_i}$ of the observed signal over V_i, as measured by the variances, decrease with i. On the other hand, the bias terms usually increase with i (we smooth with a larger bandwidth). We can assume that the bias in $\hat{I}_{V_1}$ is negligible (or we just set $V_1 = \{x\}$). Considering next the empirical mean $\hat{I}_{V_2 \setminus V_1}$ over the observations in the set difference $V_2 \setminus V_1$, a test is constructed to decide on the hypothesis that $\mathbb{E}(\hat{I}_{V_2 \setminus V_1} - \hat{I}_{V_1}) = 0$ (recall that $\mathbb{E}Z$ denotes the coordinate-wise expectation of a vector Z). If the test rejects, we conclude that the bias of the new observations is already too large and we take V_1 as neighborhood, thus keeping $\hat{I}_{V_1}$ as estimator. Otherwise, we accept V_2 as new neighborhood and consider $V_3 \setminus V_2$. If $\hat{I}_{V_3 \setminus V_2}$ does not deviate significantly from $\hat{I}_{V_1}$ and from $\hat{I}_{V_2}$, we continue. Otherwise, we stop and select V_2 as neighborhood, $\hat{I}_{V_2}$ as estimator. This way we continue growing the neighborhoods until the first time the new observations differ statistically significantly from any of the estimators on the previously built neighborhoods or until we reach the last index M.

This simple idea is reminiscent of Lepski's method [286, 287] in mathematical statistics with the important difference that at the ith step we do not test whether the new estimator $\hat{I}_{V_{i+1}}$ is statistically homogeneous with all previous $\hat{I}_{V_j}$, $j \leq i$, but we compare only the average $\hat{I}_{V_{i+1} \setminus V_i}$ over the new observations with the previous estimates. This has the clear advantage that the underlying test statistics are based on independent quantities and that in practice the number of too large grown neighborhoods is reduced because changes in the signal are detected earlier.

In order to correctly define our method, it is, of course, of great importance how to define precisely the tests and how to preselect the potential neighborhoods (V_i). In the static case (i.e. we only observe a scalar, not a time series at each point) this is described in full detail in [356], here we describe the main points. Let us first indicate how two averages $\hat{I}_V$ and $\hat{I}_{V'}$ over disjoint sets V and V' are compared. The test statistic is always based on the size of the difference $Z_{V,V'} = \hat{I}_V - \hat{I}_{V'}$. In the static case and for Gaussian noise, we can standardize $Z^2_{V,V'}$ by its expectation under the null, which is $\rho(V, V') = |V|^{-1} + |V'|^{-1}$, and compare it with some critical value $z_{V,V'}$. This critical value can be specified by a Bonferroni correction of the corresponding χ^2-quantile, taking into account the multiple testing. A more refined and powerful, but also computationally more demanding calibration of the quantiles in this case is proposed in [356]. Together with the prescription of geometrically growing neighborhoods (V_i), it is proved there that this results in estimators that optimally adapt to the local regularity of the signal (in the asymptotic minimax sense over Hölder balls). We thus achieve an unsupervised estimation procedure which is pointwise adaptive.

Here, the focus is on dynamic images and we need to compare the mean of time series over different spatial locations V and V'. For this we use exactly the adaptive multiple testing procedure presented in the preceding section. A second feature is that an a priori prescription of potentially good neighborhoods (V_i) around each pixel or voxel x is not feasible. The potential neighborhoods should rather be preselected based on the data observed in order to adapt to possibly heterogeneous geometric structures like borders or bands where a priori defined balls or squares are obviously not adequate. In a first preprocessing step we therefore exclude pixels that carry with high probability a significantly different time series. On the pixels kept after preprocessing we apply our neighborhood growth, which eventually results in the following algorithm.

We consider the set

$$\mathscr{W}_x = \{y \in \mathscr{X} \text{ s.t. } y \neq x \text{ and } y \equiv^{\alpha}_{\sigma^2} x\}$$

of the spatial locations for which the time series are statistically similar to those of x with respect to the multi-test introduced in Sect. 12.3 above. Let us emphasize that $\mathscr{W}_x$ is neither necessarily contiguous with $\{x\}$ nor connected.

Our algorithm works as follows:

1. Start with $i = 0$ and set $V_0 = \{x\}$ and $\hat{I}_0 = I_x$ as the first neighborhood and its associated denoised time series.
2. Find W_i the subset of the $C_0 2^i$ closest points to V_i in $\mathscr{W}_x \setminus V_i$. We call this set the "ring" around V_i, alluding to the case $\mathscr{W}_x = \mathscr{X}$ with the Euclidean distance.
3. Compute $\hat{J}_i$ the estimated time series using locations in the ring W_i.
4. Test if $\hat{J}_i$ is statistically not to be distinguished from all previously constructed estimates $\hat{I}_j$ for $j = 0, \dots, i$. See hereafter for precision on this test.
5. If equality is accepted, consider

$$V_{i+1} = V_i \cup W_i$$

   and define the estimate $\hat{I}_{i+1}$ using locations in V_{i+1}. Return to (2) with $i = i + 1$.
6. Else stop, define $\mathscr{V}_x = V_i$ and use $\hat{I}_x = \hat{I}_i$ as denoised time series in x.

To compare $\hat{J}_i$ with the previously constructed estimates $\hat{I}_0, \dots, \hat{I}_i$, we test the hypothesis

$$\mathscr{H}_0 : \text{“}\mathbb{E}(\hat{J}_i - \hat{I}_j) = 0, \quad \text{for all } j = 0, \dots, i\text{”},$$

against

$$\mathscr{H}_1 : \text{“}\mathbb{E}(\hat{J}_i - \hat{I}_j) \neq 0, \quad \text{for at least one } j\text{”},$$

using a generalization of our test introduced in Sect. 12.3. We accept $\mathscr{H}_0$ if

$$\hat{J}_i \equiv^{\alpha/(i+1)}_{\sigma^2 \rho(|V_j|,|W_i|)} \hat{I}_j.$$

The use of $\alpha/(i+1)$ is a Bonferroni correction due to multiple hypothesis tests. It would be possible to use more clever corrections as proposed in [30] for example. The correction factor ρ in the noise level takes into account that the estimates come from independent samples with respective size $|V_j|$ and $|W_i|$, which in the Gaussian case is just the variance $\rho(|V_j|, |W_i|) = |V_j|^{-1} + |W_i|^{-1}$.

12.5 Static Toy Examples

Not yet exploring the full scope of the method for treating dynamical images, we present first how the denoising is accomplished for one- and two-dimensional signals (images) without an attached time series structure.

12.5.1 1D Example

We propose a simple 1D example to compare the use of rings combined with balls with the algorithm introduced in [287] based only on balls. Given an increasing sequence of neighborhoods V_j, indexed by $j = 1, \dots, j_{\max}$, the selected index in [287] is defined by

$$\hat{j} = \inf\left\{ j \geq 0 \text{ s.t. } \exists \ell \leq j,\ \frac{\|\hat{I}_{V_j} - \hat{I}_{V_\ell}\|^2}{\mathbb{E}[\|\hat{I}_{V_j} - \hat{I}_{V_\ell}\|^2]} > z_\ell^2 \right\} \wedge j_{\max}$$

while for our method the index is selected as follows:

$$\hat{j} = \inf\left\{ j \geq 0 \text{ s.t. } \exists \ell \leq j,\ \frac{\|\hat{I}_{V_j \setminus V_\ell} - \hat{I}_{V_\ell}\|^2}{\mathbb{E}[\|\hat{I}_{V_j \setminus V_\ell} - \hat{I}_{V_\ell}\|^2]} > z_\ell^2 \right\} \wedge j_{\max}.$$

Both methods achieve the same rate of convergence for α-Hölder classes of signals in $\mathbb{R}^d$:

$$\mathbb{E}[\|\hat{I}_{V_{\hat{j}}} - i\|^2] \leq c\left(\frac{\log n}{n}\right)^{2\alpha/(2\alpha+d)},$$

where n is the size of the neighborhood. This rate is known to be optimal in a minimax sense over classes of Hölder functions, already in the classical mean regression model with known errors (see [286]). The logarithmic term has to be paid in order to be adaptive. Note that the dimension d, driven by the dimension of the neighborhood, is two for classical images, the length K of the time series itself does not change the rate, but will certainly influence the finite sample performance: more data in time increases the signal to noise ratio.

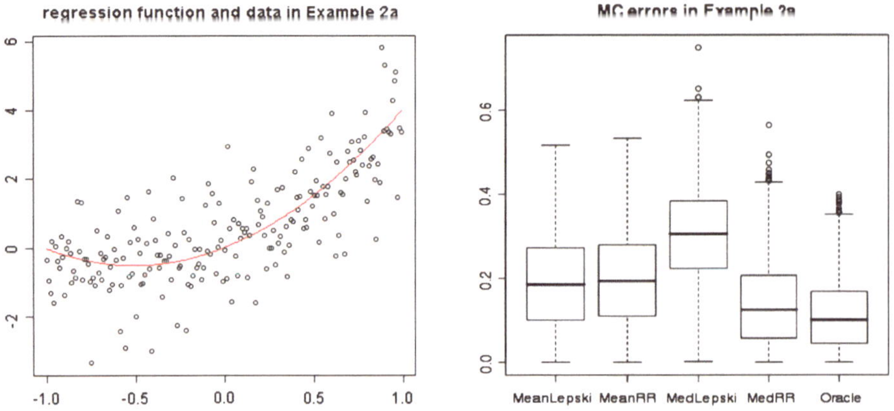

Fig. 12.1 A simple example with a parabolic signal. Estimation point is $x = 0$ and noise is Laplace distributed. Errors from 1000 Monte Carlo replications

In the left part of Fig. 12.1 the true signal is presented together with a typical sample of 200 data points obtained by adding iid Laplace distributed noise. The estimation point is at $x = 0$ in the center. As possible balls, i.e. intervals, around x we take $V_k := [-a_j, +a_j]$ with $a_j = 4 \cdot (5/4)^j/200$ with $j = 1, 2, \ldots$ We compare the following methods: (1) balls as in Lepski [287] using the local mean; (2) balls/ring using the local mean; (3) balls as in Lepski [287] using the local median; (4) balls/ring using the local median; all these methods are compared to the (5) oracle where the median is taken over the (a posteriori optimal) interval $[-0.39, +0.39]$, which is slightly larger than V_{10}. We have used the empirical median in addition to the empirical mean (12.2) since for Laplace distributed noise the median is more efficient and robust. The calibration of the methods follows always the same algorithm as recommended in [356]. The boxplots present the errors computed for the same 1000 samples with each method. The boxplots in Fig. 12.1 show a comparable performance for our method in the mean case with the benchmark Lepski approach while in the median case a clear advantage for our ball/ring method can be seen. It is remarkable that the oracle error is almost achieved.

12.5.2 2D Example

Figure 12.2 compares our method to AWS introduced in [339]. The left subfigure shows a noisy image obtained using a pattern with two values and a Laplace noise with noise level equal to the difference of these two values. The middle subfigure shows the denoised version using AWS, a typical benchmark procedure for off-the-shelf spatially adaptive denoising. We have used the R package AWS version 1.3–3.1 developed by the authors of [339] available on the web site of the CRAN using the default 2D settings to produce this subfigure. Finally, the right subfigure shows the denoised image obtained by our method. While, at the used default setting, AWS

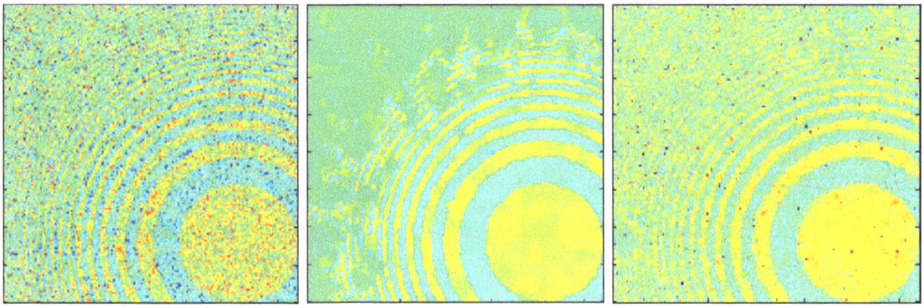

Fig. 12.2 *Left*: Noisy data. *Middle*: AWS. *Right*: Balls/Rings

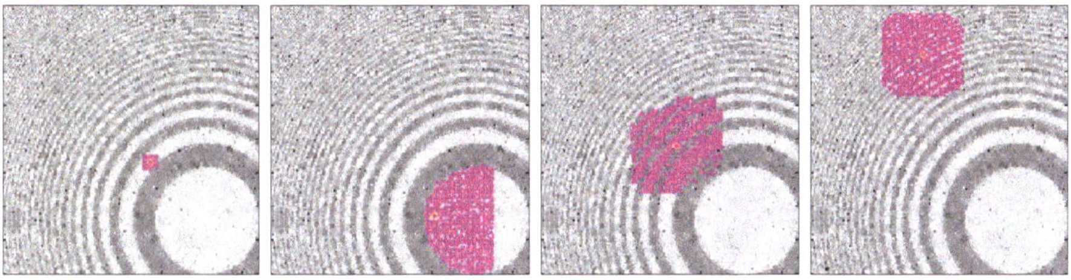

Fig. 12.3 Examples of neighborhood (*pink crosses*) found for 4 selected points of interest (*red circle*)

offers a very clean picture in the right lower corner, all details are lost in the upper left corner. Using our method, the signal is less smoothed out, but allows to recover details even in the last part of the upper corner. Clearly more details can be found in this last image even if one could complain about the remaining noise due the noise level used.

In Fig. 12.3, four examples of selected neighborhoods are presented. One can remark the special effect of the first selection of the set $\mathscr{W}_x$ in the presented examples: the neighborhood follows the borders and can even be not connected, allowing for rich geometrical structures in the signal.

12.6 Practical Results

A complete study, using our technique in the framework of DCE-CT, has been conducted with radiologists at the European George Pompidou Hospital (HEGP), Paris, and the complete results are reported in [363]. In order to provide a flavor of how our method works on real data, we have extracted from this study a few illustrations. The precise setting of the medical experiments as well as the way the tuning parameters like σ and α are fixed is left to that work.

Due to the limited irradiation dose used during the sequential acquisition, the *dynamical images* in this study suffer from a poor signal-to-noise ratio. Figure 12.4

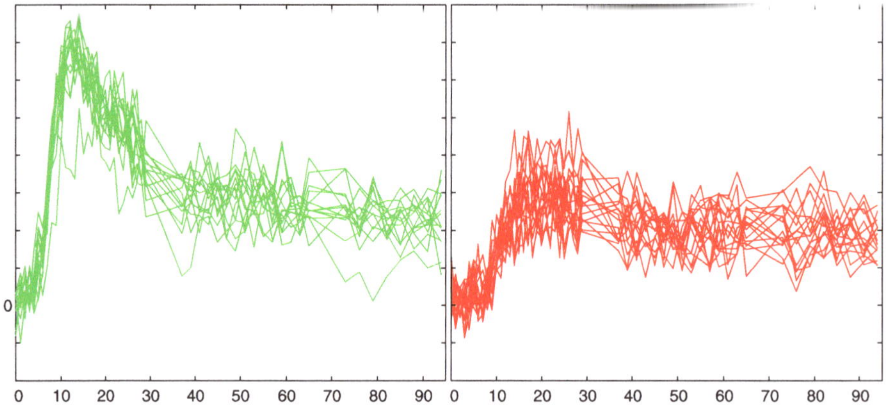

Fig. 12.4 Typical curves (time, enhancement) of individual voxels in ROI manually drawn within the Aorta (*left*) and the tumor (*right*)

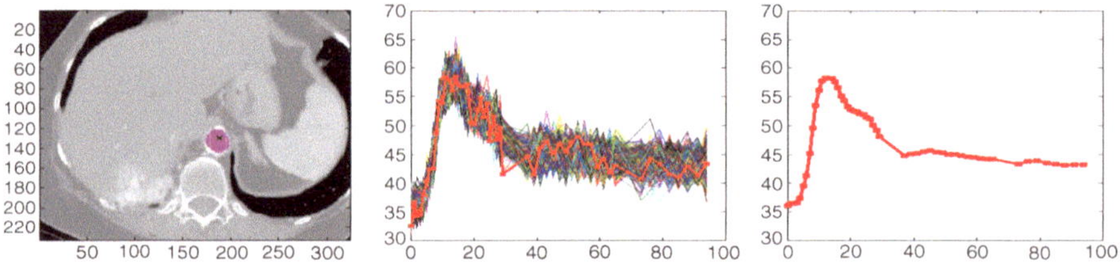

Fig. 12.5 Voxel within the aorta

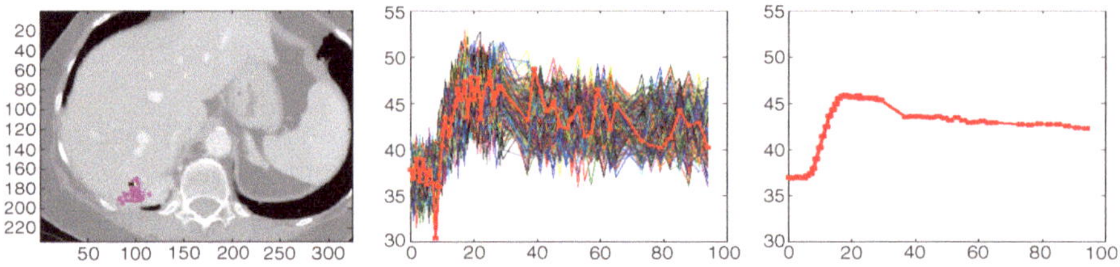

Fig. 12.6 Voxel within the tumor

shows typical time series obtained from voxels in manually selected Regions of Interest (of size $\simeq$15 voxels) within the Aorta (left curves) and a tumor (right curves).

The Figs. 12.5 and 12.6 present the result of our method applied at two different voxels, one inside the aorta and the other inside the tumor.

Each figure is divided into 3 sub-figures. Left: the denoised slice at a specific time, with the selected voxel x (black cross) and the voxels from its neighborhood $\mathscr{V}_x$ (pink dots). Center: the original enhancement vector I_x (red thick curve) and the associated enhancement vectors I_y for $y \in \mathscr{V}_x$ (background curves). Right: the estimated enhancement using a generalized median as a robust center of the selected neighborhood (see Sect. 12.4).

12.7 Conclusion

Using a two-step procedure, each step based on multiple hypothesis testing, we have introduced a novel algorithm to denoise *dynamical images* where each point of the picture exhibits a complete time series. Based on a comparison of the time series dynamics, this algorithm preserves the full structure of the time series by averaging spatially over adaptively grown neighborhoods. The efficiency of our algorithm is shown on artificial examples and on real medical images. The quality of the denoised dynamical image is shown by the remarkable reconstruction of the details in these spatially heterogeneous images.

Acknowledgements This work was supported in 2008 by a grant "Bonus Qualité Recherche" from the University Paris Descartes for the project "Cancer Angiogénèse et Outils Mathématiques".

Chapter 13
Interacting Adaptive Filters for Multiple Objects Detection

Xavier Descombes

Abstract In this chapter, we consider a marked point process framework for analyzing high resolution images, which can be interpreted as an extension of the Markov random field modelling (see Chaps. 14 and 15). The targeted applications concern object detection. Similarly to Chap. 10, we assume that the information embedded in the image consists of a configuration of objects rather than a set of pixels. We focus on a collection of objects having similar shapes in the image. We define a model applied in a configuration space consisting of an unknown number of parametric objects. A density, composed of a prior and a data term, is described. The prior contains information on the object shape and relative position in the image. The data term is constructed from local filters matching the object shape. Two algorithms for optimizing such a model are described. Finally, two applications, concerning counting of a given population, are detailed. The first application concerns small lesions in the brain whereas the second aims at counting individuals in a flamingo colony.

13.1 Introduction

Statistical physics and probabilistic approaches have been brought in image analysis starting with the seminal paper by Besag in 1974 [33]. Ten years later, the papers, based on Gibbs modeling, either in texture analysis [86] or in image restoration [183], mark the beginning of a new field in image processing: the Markov Random Field (MRF) modeling. Since then, Gibbs fields methods have been intensively developed in the Bayesian framework. These early works generated an explosion of applications of Gibbs fields for modeling high-dimensional inverse problems of image processing such as restoration, denoising, deblurring, classification, segmentation, feature extraction, surface reconstruction, stereo matching, etc. Some examples of advanced Markovian modeling are given in Chaps. 14 and 15.

X. Descombes (✉)
Laboratoire d'Informatique, Signaux et Systèmes de Sophia-Antipolis I3S, UMR6070, UNS CNRS 2000, route des Lucioles, Les Algorithmes, bât. Euclide B, BP 121, 06903 Sophia Antipolis Cedex, France
e-mail: xavier.descombes@inria.fr

L. Florack et al. (eds.), *Mathematical Methods for Signal and Image Analysis and Representation*, Computational Imaging and Vision 41,
DOI 10.1007/978-1-4471-2353-8_13,

The basic characteristic of the Gibss distributions is their decomposition as a product of factors depending only on few variables. Moreover, the distributions involve usually only few types of factors. One of them arises from the observable image (the data term) and has the form of an external field term. Other factors are due to generic or prior knowledge on the structure of images. Prior terms in the distribution function are specified by potentials associated with local interactions defined on finite sets of neighboring variables. Thus, each variable directly depends only on its neighborhood, although from a global point of view, all variables are mutually dependent through the combination of successive local interactions. Therefore, the defined prior permits to include constraints on the regularity or the smoothness of the solution. However, the MRF machinery is not well suited for introducing geometric constraints. The actual shape of objects or segments in the image are hardly modeled by local interactions between neighboring pixels. For example, as shown in Chap. 10, a Gestalt law promotes good continuation in linear structures based on object interactions rather than on pixel values. This limit becomes a major drawback when dealing with high resolution images. In such images, the geometry of objects is well described and actually represents the most relevant information. To analyze these images, it is therefore necessary to generalize MRFs. We present in this chapter such a generalization by considering marked point processes [89, 90, 423]. The main idea is to define a probabilistic model on a configuration space defined by objects instead of pixels. We do not consider general shapes, such as described in Chap. 11, but parametric objects, such as discs, ellipses, rectangles, or segments, allowing optimization of the model within reasonable computation time. The point process theory offers the possibility to work with an unknown number of variables, which is particulary adapted to our context as we do not know the number of objects in the image.

In this chapter, we focus on the different terms involved in the definition of a marked point process. We propose a general model based on local filters to extract the information from the data at an object level. We then describe two optimization algorithms and present two applications.

13.2 Point Processes of Local Filters

In this section, we define a general model for detecting a configuration of objects, geometrically specified, in an image. After having very briefly reviewed the marked point process approach in image processing, we present a general model based on the definition of spatially adaptative local filters.

13.2.1 Marked Point Processes in Image Processing

Marked point processes have been introduced in image processing in the early nineties [18, 365]. These models can be interprated as a generalization of Markov

Random Fields (MRFs), widely spread in the image processing community, cf. Chaps. 14 and 15. The main idea is to mix information estimated on the data and prior knowledge on the solution. MRFs hardly embed geometric information, such as features on object shapes. One can address this problem by considering a MRF on a graph, where the nodes represent an underlying object [418]. However, the approach requires the specification of the graph, which means the knowledge of the objects and their spatial relations. During last years, the increasing number of high resolution data, especially in satellite imagery, has motivated the development of new algorithms for analyzing images by taking into account the geometrical information. In this context, marked point processes have proven their efficiency to solve real applications such as road detection [269, 393] or building footprint extraction [270, 323]. We now give the main definitions and ingredients to define a marked point process. For a detailed construction of point processes, the reader can refer to [423].

Let $K \subset \mathbb{R}^p$ be a subset of Euclidean space and $L \subset \mathbb{Z}^p \cap K$ the corresponding discrete lattice. Usually $p = 2$ or 3 for image processing applications. The image is denoted by $I = (i_s)_{s \in L}$, where $i_s \in \Lambda$ is the grey level (or color) of pixel s. Consider $\Omega_n = \{\mathbf{x} = \{x_1, \dots, x_n\}, x_1, \dots, x_n \in K\}$ the set of unordered configurations of n points in K. We define the configuration space as $\Omega = \bigcup_{n \in \mathbb{N}} \Omega_n$. We consider a reference measure π on Ω defined as follows:

$$d\pi(\mathbf{x}) = f(\mathbf{x}) d\nu(\mathbf{x}), \tag{13.1}$$

where $\nu(\mathbf{x})$ is the measure of the Poisson process of intensity $\lambda(u), u \in K$ and $f(\mathbf{x})$ is a density (Radon-Nykodim derivative of π with respect to ν).

Let us consider now a space of marks $M = C \times N$, where $C \in \mathbb{R}^c$ and $N \in \mathbb{N}^m$. A mark $m \in M$ is associated to each point in the configuration $\mathbf{x}$. These marks can be continous, for example the radius of a disk, or discrete, for example a label selecting a shape (disk, rectangle, ...). An object is then associated to the couple $o_i = (x_i, m_i) \in K \times M$. We consider the Lebesgue measure μ_c on C and the counting measure ρ on N.

A marked point process is then defined by a density $h(\cdot)$ with respect to the product measure $\pi(\cdot)\mu_c(\cdot)\rho(\cdot)$:

$$\begin{aligned} &\forall \mathbf{o} = \{(x_1, m_1), \dots, (x_n, m_n)\} \in (K \times M)^n, \\ &dp(\mathbf{o}) = h(o) d\pi(\mathbf{x}) d\mu_c(\mathbf{m}^1) d\rho(\mathbf{m}^2), \end{aligned} \tag{13.2}$$

where $\mathbf{x} = \{x_1, \dots, x_n\}$, $\mathbf{m}^1 = \{m_1^1, \dots, m_n^1\}$ corresponds to the continuous marks, defined in C and $\mathbf{m}^2 = \{m_1^2, \dots, m_n^2\}$ corresponds to the discrete marks, defined in N.

The density $h(.)$ is usually written as a product of a prior term, reflecting prior knowledge on the solution, and a data term, reflecting the radiometric properties of the image. By writing h as a Gibbs density, we obtain:

$$h(\mathbf{o}) = \frac{1}{Z} \exp - [U_p(\mathbf{o}) + U_d(\mathbf{o})], \tag{13.3}$$

where $U_d(\mathbf{o})$ depends on the data I and $U_p(\mathbf{o})$ is a prior energy, Z is the partition function which is usually unknown.

Once the model is defined, the solution is obtained by maximizing the density:

$$\hat{\mathbf{o}} = \operatorname{argmax} h(\mathbf{o}). \tag{13.4}$$

To estimate this solution, we consider a simulated annealing scheme, which consists in iteratively simulating:

$$(h(\mathbf{o}))^{\frac{1}{T}} d\pi(\mathbf{o}), \tag{13.5}$$

where T is a temperature parameter slowly decreasing to zero during iterations.

13.2.2 Prior Energy

The goal of the prior energy is to model some constraints on the object configuration. These constraints aim at favoring or penalizing some local configurations depending on the knowledge we have on the solution apart from the data themselves. We only consider, in this subsection, pairwise interactions. Let us consider a set of symmetric relations $\sim_i$ in $K \times M$. We consider that two objects o and o' interact if and only if there exists a relation such that $o \sim_i o'$. Two classical examples of such relations are given by:

Proximity:

$$\forall \{o = (x, m), o' = (x', m')\} \in (K \times M)^2, \quad o \sim_p o' \quad \Leftrightarrow \quad d(x, x') \leq R. \tag{13.6}$$

Overlapping:

$$\forall \{o = (x, m), o' = (x', m')\} \in (K \times M)^2, \quad o \sim_o o' \quad \Leftrightarrow \quad o \cap o' \neq \emptyset. \tag{13.7}$$

The prior energy is then decomposed on the different relations as follows:

$$U_p(\mathbf{o}) = \sum_i \sum_{\{o,o'\} \subset \mathbf{o} : o \sim_i o'} V_i(o, o'). \tag{13.8}$$

For example, penalizing close or overlapping objects can be obtained by:

Proximity:

$$\forall \{o, o'\} \in (K \times M)^2 : o \sim_p o', \qquad V_p(o, o') = C > 0. \tag{13.9}$$

Overlapping:

$$\begin{aligned} &\forall \{o, o'\} \in (K \times M)^2 : o \sim_o o', \\ &V_o(o, o') = C \frac{\mu(o \cap o')}{\min(\mu(o), \mu(o'))}, \quad C > 0. \end{aligned} \tag{13.10}$$

where $\mu(x)$ is the surface of x.

More sophisticated interactions will be described in Sect. 13.4.

13.2.3 Local Filters Versus Likelihood

The data term $U_d(\mathbf{o})$ is classically obtained by modeling the likelihood of the data:

$$p(I|\mathbf{o}) = \frac{1}{Z_l}\exp(-[U_d(\mathbf{o})]). \tag{13.11}$$

This leads to the Bayesian approach which consists in modeling the posterior:

$$p(\mathbf{o}|I) \propto p(\mathbf{o})p(I|\mathbf{o}), \tag{13.12}$$

where $p(\mathbf{o}) = \frac{1}{Z_p}\exp(-[U_p(\mathbf{o})])$ is the prior.

Let consider that the image is composed of a background, where the pixel grey levels follow a Gaussian distribution $\mathcal{N}(\mu_b, \sigma_b^2)$ and some objects, for which the pixel grey levels follow a Gaussian distribution $\mathcal{N}(\mu_o, \sigma_o^2)$. Consider a configuration of objects $\mathbf{o}$. The silhouette $S(\mathbf{o})$ of $\mathbf{o}$ is defined as follows:

$$S(\mathbf{o}) = \{s \in L : \exists u \in \mathbf{o}, s \in u\}. \tag{13.13}$$

In other words, the silhouette of $\mathbf{o}$ is the projection of the union of the objects composing $\mathbf{o}$ on the lattice L. The likelihood of the image I given the object configuration is then given by:

$$\begin{aligned} p(I|\mathbf{o}) = & \prod_{s\in S(\mathbf{o})} \frac{1}{\sqrt{2\pi\sigma_o^2}} \exp\left(-\left[\frac{(i_s-\mu_o)^2}{2\sigma_o^2}\right]\right) \\ & \times \prod_{s\notin S(\mathbf{o})} \frac{1}{\sqrt{2\pi\sigma_b^2}} \exp\left(-\left[\frac{(i_s-\mu_b)^2}{2\sigma_b^2}\right]\right). \end{aligned} \tag{13.14}$$

Defining such a likelihood requires a model of the pixel radiometry for both objects and background. In this chapter, we consider an object approach, specially adapted for high resolution images. In this case, a radiometric model is usually not available, at least at the pixel level. One can imagine to specify the radiometry of the targeted objects, but the background is usually composed of different structures having highly variable radiometry. For example, let us consider the problem of tree detection. The background may be composed of roads, buildings, lakes,... Therefore, a background radiometric model does not make sense. The object class is also not always completely specified by the radiometry. Some vegetation index can be computed allowing the definition of the likelihood term for trees. However, vegetation is not restricted to trees. Therefore, using a likelihood to define the data term in the detection model will lead to false alarms in meadows. To overcome this drawback, we address an object based approach to define the data term. We consider an energy, written as a sum on the objects in the configuration:

$$p_I(\mathbf{o}) = \frac{1}{Z}\exp(-[U_p(\mathbf{o}) + U_d(\mathbf{o})]), \tag{13.15}$$

where

$$U_d(\mathbf{o}) = \sum_{o \in \mathbf{o}} u_d(o). \tag{13.16}$$

The term $u_d(o)$ is the output of a local filter, assessing, from the data point of view, the relevance of object o. Note that the object contains information both on its location and on its shape. The data term can thus be interpreted as an adaptative local filter by selecting, or more precisely favoring, a specific shape of the object depending locally on the data.

Consider an example, based on a distance between pixels inside and outside the object. Let o be an object and $S(o) \subset L$ its silhouette. We define the discrete ρ-neighborhood of o as follows:

$$\rho(o) = \{l \in L/\{S(o)\}, \exists t \in S(o) : d(l,t) \leq \rho\}, \tag{13.17}$$

where $d(.,.)$ is the Euclidean distance. The local filter is based on a distance between the radiometries inside and outside the object as follows:

$$f(o) = \frac{(\mu_{in} - \mu_{out})^2}{4\sqrt{\sigma_{in}^2 + \sigma_{out}^2}} - \frac{1}{2}\log\frac{2\sigma_{in}^2\sigma_{out}^2}{\sigma_{in}^2 + \sigma_{out}^2}, \tag{13.18}$$

where μ_{in} and σ_{in} (resp. μ_{out} and σ_{out}) are the mean and the standard deviation of pixels inside $S(o)$ (resp. in the ρ-neighborhood of o). This local filter is then mapped to $[-1, 1]$, to get the data term, as follows:

$$u_d(o) = \begin{cases} 1 - \frac{f(o)}{d_0} & \text{if } f(o) < d_0, \\ \exp(-[\frac{f(o)-d_0}{c}]) - 1 & \text{if } f(o) \geq d_0, \end{cases} \tag{13.19}$$

d_0 can be interpreted as a threshold, above which the filter output tends to favor the considered object in the configuration.

Figure 13.1 shows the detection of disks on a synthetical image. Data are taken at the object level, inducing some robustness of the solution with respect to noise. When using the likelihood as a data term, we can notice than the band, with some radiometric properties similar to the disks, is paved with disks. The fully object approach, considering local filters to define the data term, avoids these false alarms.

13.2.4 Reference Measure

The considered reference measure is usually the Poisson process measure. However, some information can also be embedded into this measure. When computing a simulated annealing scheme, the process converges to the configuration of objects that maximizes the density (see (13.5)). The result is not affected by the reference measure. At low temperature, when simulating $(h(\mathbf{x}))^{\frac{1}{T}} d\pi(\mathbf{x})$, the density $h(\cdot)$ brings the main contribution on the different sampled configurations. On the contrary, at

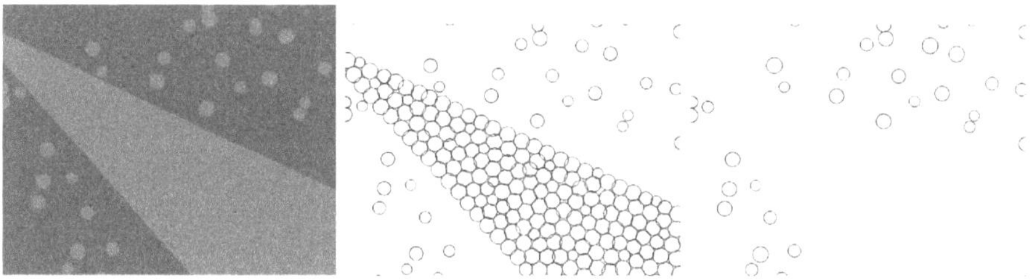

Fig. 13.1 Synthetical data (*left*), disks detection using the likelihood (*middle*) and the local filter approach (*right*)

high temperature, $(h(\mathbf{x}))^{\frac{1}{T}}$ tends to an uniform density. In this case, the main characteristics of the sampled configuration reflect the reference measure properties. The reference measure does not influence the final result but can be crucial for reducing the computation time. When modeling a given property, we then have the choice to include it in the density or in the reference measure. To make this decision, we have to answer to the following question: do I want to constraint the solution to match this property or do I only expect the solution to satisfy this property? In the former case, the property should be included in the density, but in the reference measure in the latter. A first example is to consider a Poisson measure as reference measure but with a non homogeneous intensity. The location of the objects in the final configuration will not be affected but, during the optimization, the objects will be searched preferably in areas with a high intensity. One such example is given in [335] for trees detection. The application concerns tree crown extraction in order to count individuals in a stand. A vegetation index is computed on the whole image. The intensity of the reference Poisson measure is proportional to this index. During the optimization process, new objects are then preferably proposed on pixels having a high value of vegetation, which saves time by avoiding to search trees in areas without vegetation. But the final result does not depend on this vegetation index, as it could introduce false alarms in meadows for instance.

A density, embedding some interaction, can also be included in the reference measure. Consider the example of tree crown extraction but in the case of a plantation. There is a periodicity in the tree location. We would like to favor the configurations which are consistent with this periodicity. This can be done by introducing some attractive property in the density by defining pairwise interaction between objects satisfying the periodicity constraint. On the other hand, if there is a lack in the plantation, we do not want to favor an object where the tree is missing. In this case, the periodicity interaction should not be included in the density $h(\cdot)$ but in the density of the reference measure $f(\cdot)$.

13.3 Optimization

Once the model is defined, the solution is obtained by optimizing the density. The partition function, or normalization constant, of the model, is unknown and cannot be estimated numerically. Therefore, we have to perform the optimization by an iterative algorithm converging toward the solution. This is done by using a simulated annealing scheme, which consists in sampling the model at a given temperature, as defined in (13.5). The different algorithms are characterized by the sampling scheme they employ. Basically, we can distinguish two main approaches. The first one is discrete and is based on a Metropolis-Hastings dynamics. It consists in randomly perturbing the current configuration. At each step, a new configuration is proposed and is accepted or not depending on a specific acceptation rate ensuring the convergence to the target distribution. The second class of algorithms is based on a continuous process which consists in randomly adding and removing objects in the current configuration. They are referred to as birth and death processes. Some hybrid approaches, combining both aspect, are called jump and diffusion processes. In this section, we present the classical RJMCMC scheme, belonging to the first category, and a recently proposed birth and death process, which allows adding several objects at the same time (after discretization).

13.3.1 Reversible Jump MCMC

One of the most common ways to simulate a point process is to define a Markov chain allowing two types of transitions: the addition of a point to the configuration (birth) and the deletion of a point from the configuration (death). The mixing properties of the chain can be increased by means of a Metropolis-Hastings (MH) dynamics [185]. This dynamics enables us to add moves that modify the characteristics of an object. Moves, such as translation, rotation, dilation, are important for speeding up the convergence in practice [190]. This class of algorithms is known under the name “Reversible Jump Monte Carlo Markov Chains” (RJMCMC), and has been widely applied to several image analysis problems in [269, 323, 365].

Suppose the Markov chain is in state $\mathbf{o}$. We randomly choose a move of type i, which transforms $\mathbf{o}$ to $\mathbf{o}'$ with probability $\gamma_i(\mathbf{o} \to d\mathbf{o}')$. The new state $\mathbf{o}'$ is then accepted with probability:

$$\alpha_i(\mathbf{o} \to \mathbf{o}') = \min\left\{1, \frac{h(\mathbf{o}')\gamma_i(\mathbf{o}' \to d\mathbf{o})d\pi(\mathbf{o})}{h(\mathbf{o})\gamma_i(\mathbf{o} \to d\mathbf{o}')d\pi(\mathbf{o}')}\right\}. \tag{13.20}$$

Green presented in [190] the RJMCMC method for jumping between states in spaces of different dimensions. It is necessary to match the dimension of state $\mathbf{o}$ to $\mathbf{o}'$. This is done by sampling a vector u of continuous random variables independently of $\mathbf{o}$. $\mathbf{o}'$ is obtained by using an invertible deterministic function $\varphi(\mathbf{o}, u)$. The acceptance probability is modified by the Jacobian of the transformation:

$$\alpha_i(\mathbf{o} \to \mathbf{o}') = \min\left\{1, \frac{h(\mathbf{o}')}{h(\mathbf{o})\gamma(u^{(i)})}\left|\frac{\partial\varphi}{\partial(\mathbf{o}, u)}\right|\frac{p(\text{proposing the reverse jump})}{p(\text{proposing the fowards jump})}\right\}, \tag{13.21}$$

where $\gamma(u^{(i)})$ is the density of the random vector u.

If the dimension of the configuration space is fixed, we obtain the classical Metropolis-Hastings dynamics.

13.3.2 Births and Deaths

We now describe an algorithm consisting of a multiple births and deaths process. The main point is that, at each iteration, a configuration of objects, and not only one object, is first added randomly, and then each object in the current configuration is killed with a given probability. Therefore, during the birth step several objects are involved. Besides, the birth of objects depends neither on the energy nor on the temperature. This algorithm has been recently proposed in [102] and appears to be an interesting alternative to RJMCMC algorithms. This process is a discretization of a continuous process defined by a stochastic differential equation. The convergence of the continuous process toward the stationary measure and the convergence of the discretization scheme toward the continuous process have been proved in [102, 349]. In this chapter, we only describe the discrete process, which provides the optimization algorithm, parameter δ being the time discretization step. At each iteration, a configuration $\mathbf{o}$ is transformed into a configuration $\mathbf{o}' = \mathbf{o}_1 \cup \mathbf{o}_2$, where $\mathbf{o}_1 \subseteq \mathbf{o}$, and, $\mathbf{o}_2$ is a configuration such that $\mathbf{o}_1 \cap \mathbf{o}_2 = \emptyset$ and is distributed following a Poisson law of intensity z.

This transformation thus contains a birth part, given by $\mathbf{o}_2$ and a death part given by $\mathbf{o} \setminus \mathbf{o}_1$.

The transition associated with the birth of an object $o = (x, m)$ in a small volume $\Delta x \subset K$ is given by:

$$q_\delta(o) = \begin{cases} z\Delta x\delta, & \text{if } \mathbf{o} \to \mathbf{o} \cup \{o\}, \\ 1 - z\Delta x\delta, & \text{if } \mathbf{o} \to \mathbf{o} \text{ (no birth)}, \end{cases} \tag{13.22}$$

where δ is the time discretization step of the corresponding stochastic differential equation.

The death probability of an object o from configuration $\mathbf{o}$ is given by:

$$p_\delta(o) = \begin{cases} \frac{\delta a(o)}{1+\delta a(o)}, & \text{if } \mathbf{o} \to \mathbf{o} \setminus u, \\ \frac{1}{1+\delta a(o)}, & \text{if } \mathbf{o} \to \mathbf{o} \text{ (object } o \text{ is not killed)} \end{cases} \tag{13.23}$$

with $a(o) = (\frac{h(\mathbf{o}/\{o\})}{h(\mathbf{o})})^\beta$, where β is interpreted as the inverse temperature.

Moreover, all new objects are added independently, and configurations $\mathbf{o}$ and $\mathbf{o}_2$ are independent.

The algorithm simulating the process is defined as follows:

- *Main program*: initialize the inverse temperature parameter $\beta = \beta_0$ and the discretization step $\delta = \delta_0$ and alternate birth and death steps

- *Birth step*: Select randomly a number of new objects following a Poisson law of average δ.
- *Death step*: for each object o in the current configuration $\mathbf{o}$, compute the death rate as follows:

$$d(o) = \frac{\delta a_\beta(o)}{1 + \delta a_\beta(o)}, \tag{13.24}$$

  where $a(o) = (\frac{h(\mathbf{o}/\{o\})}{h(\mathbf{o})})^\beta$, then the object o dies with probability $d(o)$.
- *Convergence test*: if the process has not converged, decrease the temperature and the discretization step by a given factor and go back to the birth step. In practice, the convergence is obtained when all the objects added during the birth step, and only these ones, have been killed during the death step.

13.4 Applications

In this section, we describe some real applications to show the relevance of marked point processes for analyzing high resolution images. We only describe simple prior models consisting only in local attractive or repulsive constraints between different objects. More sophisticated priors, such as alignment or paving, can be found to address more complex structural patterns such as road networks or building footprint detection [269, 323].

13.4.1 Segments

In this subsection we describe a first model based on segments (see [101] for full details), for detecting small lesions in the brain. Different types of small focal lesions are typically found in MRI scans of elderly subjects. Their neuropathological substrate and their influence on cognitive abilities is still under debate [220]. The neurobiological background of this work is to aid in discriminating healthy from pathological aging as revealed by MRI brain data sets. Estimating the lesion count and describing their position is tedious due to their multitude. So typically, these lesions are evaluated visually in the acquired data sets and rated by semiquantitative scales [373]. One such lesion type is called "enlarged Virchow-Robin space" (VRS) that corresponds to a small gap around a deep penetrating artery supplying the white matter. Such lesions appear as small tubular structures filled with cerebrospinal fluid (CSF) that are perpendicular to the brain surface. Typically, they are close to the spatial resolution limit of current MRI methods (1–3 mm in diameter and 3–15 mm long), and a single brain may contain hundreds of such lesions. We define a marked point process approach for detecting these small lesions, that are non-uniformly distributed in space, by modeling their relative position. Structure and distribution of VRS have motivated the following approach. A VRS can be modelled by a tube containing several voxels (Descombes et al. [101]). Thus, an approach based on a geometrical object appears suited for this problem. Secondly, VRS are not uniformly localized and some clustering property can be observed (Descombes et al. [101]). We define a prior which favors clustering of segments while penalizing overlapping between segments.

Prior The prior model takes the geometry of a detected object and its interactions into account. Each object is represented by a point with attributes (or marks) that define its geometry. The resulting configuration is a set of marked points $\mathbf{o} = \{o_1 = (x_1, m_1), \ldots, o_n = (x_n, m_n)\}$ where $x_i \in K \subset \mathbb{R}^3$ and $m_i = (\rho_i, \theta_i, \psi_i) \in [\rho_{\min}, \rho_{\max}] \times [0, 2\pi] \times [-\pi, \pi]$ (length and orientations of a segment). We restrict the volume K to the space of points that satisfy $\max(F_u(o_i), F_v(o_i), F_w(o_i)) \geq 0$, where F_u, F_v and F_w are the filters defining the data term (see below).

We define a density with respect to the Poisson measure of intensity λ. For a given configuration $\mathbf{o} = \{o_1 = (x_1, m_1), \ldots, o_n = (x_n, m_n)\}$, the density $h(\mathbf{o})$ is written as follows:

$$h(\mathbf{o}) \propto \prod_{k \in \langle 1,n \rangle} q_1(\rho_k) \prod_{k,k' \in \langle 1,n \rangle : k \sim k'} q_2(o_k, o_{k'}), \tag{13.25}$$

where $\sim$ defines a neighborhood relation. The term $q_1(\rho_k)$ represents a prior on the segment length and is defined as follows:

$$q_1(\rho_k) = \exp\left(-\left[A\left(\frac{\rho_{\max} - \rho_k}{\rho_{\max} - \rho_{\min}}\right)^2\right]\right). \tag{13.26}$$

In our experiments, we have chosen $\rho_{\min} = 2$ and $\rho_{\max} = 15$, according to the length of VRS observed in our datasets and reported in the literature [220]. The term $q_2(o_k, o_{k'})$ defines interactions between neighboring segments. We consider three kinds of interactions (see Fig. 13.2): (i) an explicit term which penalizes intersecting segments, (ii) a clustering term which favors neighboring segments with similar orientation and (iii) a repulsive term which penalizes neighboring segments with different orientations. To define the different interactions, we consider a discretization $S(o) = [o^{(1)}, o^{(2)}]$, of the segment o, where $o^{(1)} = (o_u^{(1)}, o_v^{(1)}, o_w^{(1)})$ and $o^{(2)} = (o_u^{(2)}, o_v^{(2)}, o_w^{(2)})$ are the extremity voxels of $S(o)$. We use the following explicit interaction to penalize intersecting segments:

$$q_2(o_k, o_{k'}) = \exp[-B] \quad \text{if } S(o_k) \cap S(o_{k'}) \neq \emptyset. \tag{13.27}$$

The vessels are distributed perpendiculary to the brain surface. Therefore, we define a local clustering interaction between segments of similar direction. Two segments o and o' have a similar u-direction if and only if:

$$\begin{cases} |o_u^{(2)} - o_u^{(1)}| > \max(|o_v^{(2)} - o_v^{(1)}|, |o_w^{(2)} - o_w^{(1)}|), \\ |o'^{(2)}_u - o'^{(1)}_u| > \max(|o'^{(2)}_v - o'^{(1)}_v|, |o'^{(2)}_w - o'^{(1)}_w|). \end{cases} \tag{13.28}$$

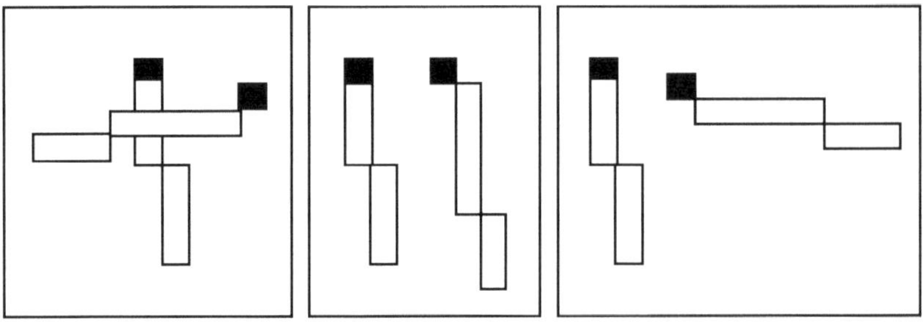

Fig. 13.2 Explicit (*left*), clustering (*middle*) and repulsive (*right*) interactions

If $o = (x, m)$ and $o' = (x', m')$ have a similar u-direction, we have a clustering interaction:

$$q_2^{clu}(o, o') = \exp\left[+C \min\left(|o_u^{(2)} - o_u^{(1)}|, |o'^{(2)}_u - o'^{(1)}_u|\right)\right]$$
$$\text{if } |x_u - x'_u| < c. \tag{13.29}$$

This interaction is easily extended to segments having similar v- or w-directions. Finally if two segments have no similar directions, we define a repulsive interaction as follows:

$$q_2^{rep}(o, o') = \exp[-D] \quad \text{if } \|x - x'\| < d. \tag{13.30}$$

Note that $q_2^{clu}(o, o')$ and $q_2^{rep}(o, o')$ are mutually excluding.

Data Term We first define a local filter which can be interpreted as a VRS indicator. This filter extracts the information provided by the radiometry (grey level of voxels). In a first approach, VRS can been modeled as small tubular structures, with a diameter close to the data resolution (typically 1–3 mm). Their length varies with an average of 3–4 mm but can reach up to 15 mm in exceptional cases. The data representation of these structures is discrete due to the discrete nature of the image lattice. The induced discretization is far from being isotropic. Therefore, we can only consider the three main directions (u, v, w) defined by the data without loosing significant information. For these three directions we design filters that take the diameter variability and the partial volume effect into account. A filter lies in the plane perpendicular to the considered direction, say u. It consists of a central voxel s, the eight surrounding neighbors $t_1 \in \mathcal{N}_u^1(s)$, and the next 16 neighbors $t_2 \in \mathcal{N}_u^2(s)$ (see Fig. 13.3). A VRS is characterized by three properties: (i) VRS contain CSF that should appear as low intensity voxels in T_1-weighted datasets. Surrounding tissue (i.e., white matter or grey matter voxels) may increase lesion intensities due to the partial volume effect. Thus, neighboring voxels will have higher intensity (ii) and are contrasted with the VRS voxels (iii). We define three filters corresponding to these three properties $F^{black}(i_s)$, $F^{white}(\min(i_t, t \in \mathcal{N}_u^2(s)))$ and $F^{contrast}(\sum_{t \in \mathcal{N}_u^2(s)} i_t/16)$, where i_s represents the grey level of voxel s in the data. Note that these three properties are not equivalent. Considering only one of them

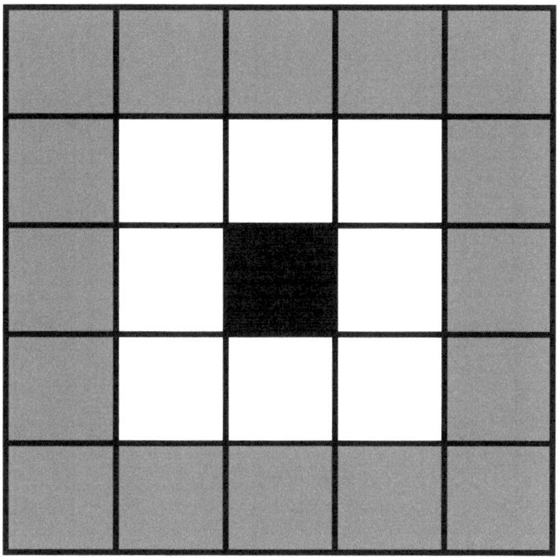

Fig. 13.3 VRS indicator filter: central voxel s *in black*, first neighborhood $\mathcal{N}_u^1(s)$ *in white* and second neighborhood $\mathcal{N}_u^2(s)$ *in grey*

leads to false alarms in the CSF compartment for property (i), in the white matter for property (ii) and in the thin grey matter structures for property (iii). The three defined functions depend on the data statistics and are defined on Fig. 13.4. We combine these three properties as follows:

$$f_u(s) = \min\left(F^{black}(i_s),\ F^{white}(\min(i_t, t \in \mathcal{N}_u^2(s))),\ F^{contrast}\left(\frac{\sum_{t\in\mathcal{N}_u^2(s)} i_t}{16}\right)\right). \tag{13.31}$$

To avoid multiple detections of the same VRS, the final filter is written as follows:

$$F_u(s) = \begin{cases} f_u(s) & \text{if } \forall t \in \mathcal{N}_u^1(s),\ i_s \le i_t, \\ \min(0, f_u(s)) & \text{otherwise.} \end{cases} \tag{13.32}$$

We now define the data term $U_d(\mathbf{o})$. Assuming that data are independent conditionally to the segments we can write:

$$U_d(\mathbf{o}) = \sum_{k\in\langle 1,n\rangle} u_d(o_k). \tag{13.33}$$

The data term corresponding to an object o consists of two parts. The first is proportional to the filter values along the segment. In the second part, we assume that voxels at the segment end have a high intensity as they do not belong to the VRS. The data term is then written as follows:

$$u_d(o) = -E\left(\sum_{t\in S(o)} F_u(t) + \frac{card(S(o))}{2} F_u^{end}(o)\right)$$

$$\text{if } |o_u^{(2)} - o_u^{(1)}| \ge \max\left(|o_v^{(2)} - o_v^{(1)}|, |o_w^{(2)} - o_w^{(1)}|\right),$$

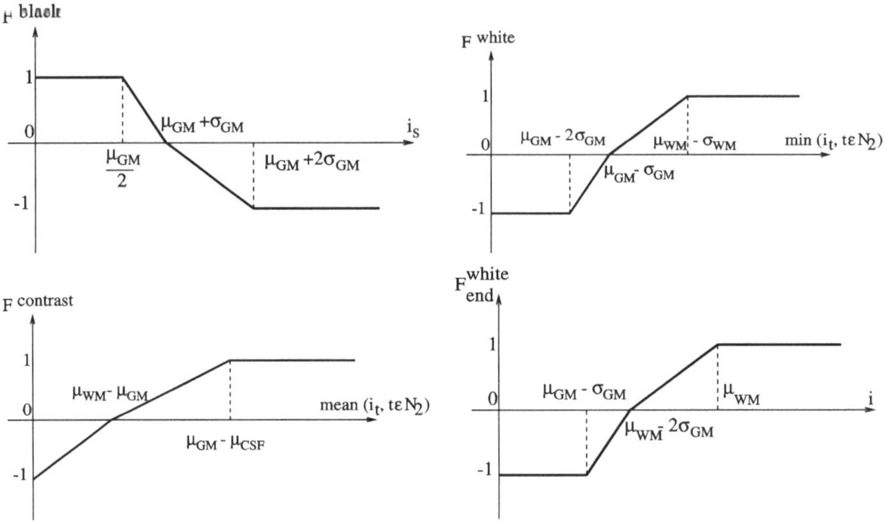

Fig. 13.4 *Top and bottom left*: The three filters characterizing VRS data, μ_{CFS}, μ_{GM} and μ_{WM} represent the mean of CSF, grey matter and white matter, σ_x being the associated standard deviations. *Bottom right*: The prolongating filter, μ_{CFS}, μ_{GM} and μ_{WM} represent the mean of CSF, grey matter and white matter, σ_x correspond to the associated standard deviations

$$
\begin{aligned}
u_d(o) &= -E\left(\sum_{t\in S(o)} F_v(t) + \frac{card(S(o))}{2} F_v^{end}(o)\right) \\
&\quad \text{if } |o_v^{(2)} - o_v^{(1)}| \geq \max\left(|o_u^{(2)} - o_u^{(1)}|, |o_w^{(2)} - o_w^{(1)}|\right), \\
u_d(o) &= -E\left(\sum_{t\in S(o)} F_w(t) + \frac{card(S(o))}{2} F_w^{end}(o)\right) \\
&\quad \text{if } |o_w^{(2)} - o_w^{(1)}| \geq \max\left(|o_v^{(2)} - o_v^{(1)}|, |o_u^{(2)} - o_u^{(1)}|\right),
\end{aligned}
\tag{13.34}
$$

where $card(S(o))$ is the number of voxel in $S(o)$ and:

$$
F_u^{end}(o) = \begin{cases}
F_{end}^{white}(\min(i_{o^{(2)}+(1,0,0)}, i_t, t \in \mathcal{N}_u^1(i_{o^{(2)}+(1,0,0)}))) \\
\quad + F_{end}^{white}(\min(i_{o^{(1)}+(-1,0,0)}, i_t, t \in \mathcal{N}_u^1(i_{o^{(1)}+(-1,0,0)}))) \\
\quad \text{if } o_u^{(2)} > o_u^{(1)}, \\
F_{end}^{white}(\min(i_{o^{(1)}+(1,0,0)}, i_t, t \in \mathcal{N}_u^1(i_{o^{(1)}+(1,0,0)}))) \\
\quad + F_{end}^{white}(\min(i_{o^{(2)}+(-1,0,0)}, i_t, t \in \mathcal{N}_u^1(i_{o^{(2)}+(-1,0,0)}))) \\
\quad \text{if } o_u^{(2)} \leq o_u^{(1)},
\end{cases}
\tag{13.35}
$$

where $F_{end}^{white}(\cdot)$ is defined on Fig. 13.4 bottom right. Similar definitions are used for $F_v^{end}(o)$ and $F_w^{end}(o)$.

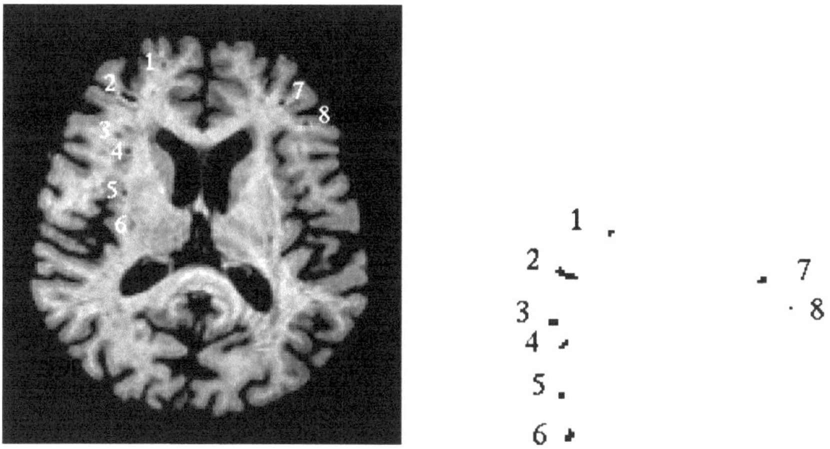

Fig. 13.5 An example of VRS detection: Original axial slice (*left*) and detected VRS (*right*)

Optimization We consider an RJMCMC algorithm embedded into a simulated annealing scheme. In this context, defining relevant moves to speed up the convergence is a crucial issue. We first consider a birth and death move. This move is essential to establish the real number of segments in the configuration. Besides, it guaranties the irreducibility property. This move is chosen with probability p_1. It consists in adding a new segment with probability P_b or removing a segment from the configuration with probability $P_d = 1 - P_b$. In the case of a death move, the segment is chosen uniformly among the segments of the current configuration. For the birth move, we propose a new segment uniformly in the parameter space. To speed up the convergence, we also consider some moves which modify one segment in the configuration. We choose a segment uniformly within the configuration and select a new location and new marks uniformly within the parameter space. The move is made with probability p_2. With this move, we can change the segments having a low probability. To allow a local improvement of the segments location with respect to the data we consider the move of end points. This move is made with probability p_3. We choose a new location of the end point uniformly in a neighborhood of its current position. This move may also extend segments in partially detected VRS. A VRS can be covered by several non-overlapping segments during optimization. Extending segments has a very low acceptance ratio in this configuration because it may induce overlapping, which is penalized by the explicit interaction term. A death move followed by an extension has also a low acceptance ratio because all the segments within this local configuration fit the data well. Therefore, we have introduced a move which merges neighboring segments. The reverse move corresponds to splitting a segment in two parts. This move is made with probability p_4. A merging is proposed with probability P_m and a splitting with probability $P_s = 1 - P_m$.

Results We present results on a particular dataset. This dataset has been selected among cases with a high prevalence of VRS. Figure 13.5 shows the result on a

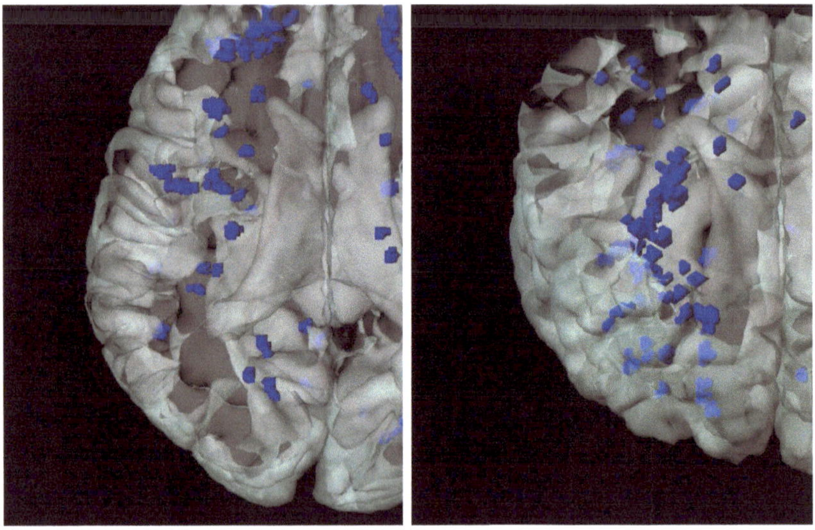

Fig. 13.6 Detected VRS in relation to the brain surface in the example dataset of Fig. 13.5, views from top and top-frontal

sample axial slice and the detected lesions on Fig. 13.5b. A 3D vizualization of the result is shown on Fig. 13.6. To validate the approach, we classify the datasets into three classes depending on the number of detected VRS. This classification has been compared with the one obtained by an expert. We obtained a correlation of 0.84 between the two classifications.

13.4.2 Ellipses

In this subsection, we consider an ellipses based on model for detecting and counting breeding Greater Flamingos (*Phoenicopterus Roseus*) on aerial photographs of their colonies [239].

The 2D model, used to extract flamingos, consists of a marked point process of ellipses. The associated state space is:

$$K \times M = [0, X_M] \times [0, Y_M] \times [a_m, a_M] \times [b_m, b_M] \times [0, \pi[,$$

where X_M and Y_M are respectively the width and the length of the image I, (a_m, a_M) and (b_m, b_M) respectively the minimum and the maximum semimajor axis and semiminor axis, and $\theta \in [0, \pi[$ the orientation of the objects.

Prior As we aim at detecting individuals in dense populations, we model flamingos as possibly slightly overlapping ellipses $o_i \sim_r o_j$. Then, the prior energy $U_p(\mathbf{o})$ that introduces interactions, penalizes configurations according to the overlapping

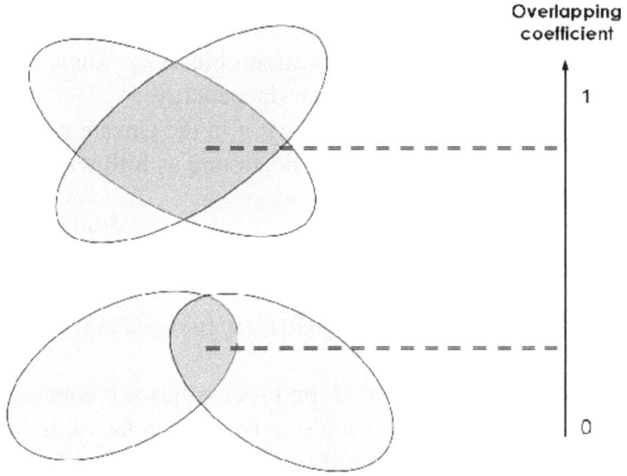

Fig. 13.7 Overlapping ellipses

objects area (see Fig. 13.7), see [100] for more details:

$$U_p(\mathbf{o}) = \gamma_p \sum_{o_i \in \mathbf{o}} \max_{o_j \sim_r o_i} \mathscr{A}(o_i, o_j), \tag{13.36}$$

where $\mathscr{A}(o_i, o_j) \in [0, 1]$ is an overlapping coefficient, as defined in (13.10) and γ_p is a weight which ponders the repulsion between the objects of the process. Each object is penalized depending on the maximal overlapping it exhibits with neighboring ellipses.

Data Term In flamingo populations, each flamingo can be modeled as a bright ellipse surrounded by a darker background. Thus, we define the boundary of an ellipse $\mathscr{F}(o)$ as the subset of K contained between the given ellipse $o = (x, m)$, where $m = (a, b, \theta)$ are the marks, and a concentric one $o' = (x, m')$, with $m' = (a + \rho, b + \rho, \theta)$. This boundary will stand for the background. To evaluate the contrast between the ellipses and the background, we calculate the distance $d(o, \mathscr{F}(o))$ given in (13.18) and consider the corresponding data term given in (13.19).

Optimization For optimizing the model, we consider a simulated annealing based on a multiple birth and death process as described in Sect. 13.3.2. An extension of this first scheme, speeding up the convergence, is employed:

- *Main program*: initialize the inverse temperature parameter $\beta = \beta_0 = 50$ and the discretization step $\delta = \delta_0 = 20000$ and alternate birth and death steps
  - *Birth step*: for each $s \in L$, if no object is already alive, we add an object in s with probability $\delta B(s)$ where $B(s)$ is proportional to the data term obtained with a disk of fixed radius:

$$\forall s \in I, \quad B(s) = \frac{zb(s)}{\sum_{t \in I} b(t)}, \tag{13.37}$$

where z is a parameter of the process.

- *Sorting step*: once the birth step is finished, we compute the data term $u_d(o_i)$ of the current configuration objects o_i. Then, we sort them, in the decreasing order, according to their data energy.
- *Death step*: for each object u in the current configuration $\mathbf{x}$, taken in the previous order, compute the death rate as follows:

$$d(u) = \frac{\delta a_\beta(u)}{1 + \delta a_\beta(u)}, \tag{13.38}$$

  where $a_\beta(u) = \exp(-[\beta(U(\mathbf{x}/\{u\}) - U(\mathbf{x}))])$, then the object u dies with probability $d(u)$.
- *Convergence test*: if the process has not converged, decrease the temperature and the discretization step by a given factor and go back to the birth step. The convergence is obtained when all the objects added during the birth step, and only these ones, have been killed during the death step.

Note that the birth consists in adding a sample of a Poisson process with an intensity proportional to a birth map, depending on the data term. Therefore, objects are added preferably in locations corresponding to a high response of the local filter obtained for a predefined object. Secondly, the death step is performed after having sorted the objects. We first propose to eliminate objects corresponding to a low response of the associated local filter, which helps to escape from local minima of the energy. These heuristics do not change the convergence properties but allow a smarter scanning of the configuration space.

Results We present in Descamps et al. [100] some result on a real aerial image. This image represents a colony in Camargue, the unique place in France, called Fangassier island, where flamingos are grouped for reproduction. The image, taken in 2002, is slightly blurred and flamingos are very close to each other. The obtained result, shown in Descamps et al. [100] is nevertheless satisfactory. We estimate the size of the colony as 10894 individuals. The computation time is 40 mn on a 2 GHz processor (image size: 5028 × 3408, resolution: 5 cm). The expert needs 5 hours and counts 10182 flamingos. The slight over-detections of the proposed method are due to either heads or wings of some flamingos.

To quantify the results, we consider samples extracted from different images, taken in different places (France, Turkey, Mauritania) and presenting different levels of difficulty. To evaluate the difficulty the counts have been performed by one expert and by five non specialists. Depending on the consistency between the counts of the expert and the non specialists, we have classified the samples in three levels: easy, middle and hard. In all cases, the automatic count obtained with the proposed approach is closer to the expert count than those obtained by the non specialists.

The Table 13.1 summarizes the obtained results in terms of good detections and false alarms. For studying the dynamic of the population, an error rate of 5% is acceptable. The obtained results globally reach this goal. However, higher resolution images should allow reducing the false alarms rate.

Table 13.1 Good detection rate and false alarms on several samples

Image	Difficulty	Good detection	False alarms
Fangassier 02	Hard, Middle, Easy	93%, 98%, 97%	7%, 3%, 2%
Fangassier 05	Middle, Middle, Middle	98%, 97%, 97%	9%, 8%, 16%
Kione 05	Hard, Hard, Hard	95%, 93%, 87%	12%, 12%, 15%
Tuz Lake 04	Easy, Easy, Easy	98%, 100%, 92%	4%, 0%, 0%
Tuz Lake 06	Easy, Middle, Middle	100%, 99%, 99%	1%, 2%, 0%

13.5 Conclusion

In this chapter, we have presented a general framework, based on marked point processes, to extract a collection of objects from images. We have shown how to model known information on the result either in a prior term or in the reference measure. The data term is based on the output of a locally adaptive filter. Two algorithms for optimizing the model, the RJMCMC scheme and the multiple birth and death process, have been detailed. Finally, two applications, concerning object counting, in MRI brain scan and in aerial images, have been detailed. This modeling, by considering an object level, appears to be well suited for high resolution images. It overcomes the limits of random fields when addressing geometrical information.

We do believe that this framework can find numerous applications in image analysis in different fields such as medical images, biological images or remote sensing. A last point that should be addressed concerns the parameters estimation. In the applications proposed in this chapter, the different parameters, involved in the model or in the optimization process, have been calibrated. Estimating them, with an acceptable algorithmic complexity, is still a challenge.

Chapter 14
Visual Data Recognition and Modeling Based on Local Markovian Models

Michal Haindl

Abstract An exceptional 3D wide-sense Markov model which can be completely solved analytically and easily synthesized is presented. The model can be modified to faithfully represent complex local data by adaptive numerically robust recursive estimators of all its statistics. Illumination invariants can be derived from some of its recursive statistics and exploited in content based image retrieval, supervised or unsupervised image recognition. Its modeling efficiency is demonstrated on several analytical and modeling image applications, in particular on unsupervised image or range data segmentation, bidirectional texture function (BTF) synthesis and compression, dynamic texture synthesis and adaptive multispectral and multichannel image and video restoration.

14.1 Introduction

Recognition and processing of multi-dimensional data (or set of spatially related objects) is more accurate and efficient if we take into account all interdependencies between single objects. Objects to be processed like for example multi-spectral pixels in a digitized image, are often mutually dependent (e.g., correlated) with a dependency degree related to a distance between two objects in their corresponding data space. These relations can be incorporated into a pattern recognition process through appropriate multi-dimensional data model. If such a model is probabilistic we can use consistent Bayesian framework for solving many pattern recognition tasks.

Features derived from multi-dimensional data models are information preserving in the sense that they can be used to synthesise data spaces closely resembling original measurement data space as can be illustrated on the recent best visual representation of real material surfaces in the form of bidirectional texture function [206]. Virtual or augmented reality systems require object surfaces covered with realistic nature-like color textures to enhance realism in virtual scenes. Similarly, realistic textures are used in computer games, CAD systems and some other computer

M. Haindl (✉)
Institute of Information Theory and Automation of the ASCR, Prague, Czech Republic
e-mail: haindl@utia.cz

L. Florack et al. (eds.), *Mathematical Methods for Signal and Image Analysis and Representation*, Computational Imaging and Vision 41,
DOI 10.1007/978-1-4471-2353-8_14,

graphics applications. Such textures can be either digitized natural textures or textures synthesized from an appropriate mathematical model. However digitized 3D multispectral or even 7D BTF textures are far less convenient alternatives, because of extreme virtual system memory demands, visible discontinuities and several other drawbacks [200].

Mathematical multi-dimensional data models, see also Chaps. 13 and 15, are useful for describing many of the multi-dimensional data types provided that we can assume some data homogeneity so some data characteristics are translation invariant. While the 1D models like time series are relatively well researched and they have rich application history in control theory, econometric, medicine and many other recognition applications, multi-dimensional models are much less known and their applications are still limited. The reason is not only unsolved theory difficulties but mainly their huge computing power demands which prevented their wider use until recently.

We introduced in [207] a fast multiresolution Markov random field (MRF) based model and the simultaneous causal autoregressive random field model [208], respectively. Although the former method avoids the time consuming Markov chain Monte Carlo simulation so typical for applications of Markov models, cf. Chaps. 13 and 15, it requires several approximations. The latter method is very efficient for multispectral image representation not only because it does not suffer from some problems of alternative options (see [200, 203] for details) but it is also easy to analyze as well as to synthesise and last but not least it is still flexible enough to imitate a large set of natural and artificial textures or other spatial data.

It is possible to divide data models applications into two broad categories: analysis (Fig. 14.1) and synthesis (Figs. 14.2, 14.3, 14.4 and 14.5). Analytical applications include data classifications or unsupervised segmentation (cf. Chaps. 13 and 15), data space directionality analysis, motion detection and some others. Frequent synthesis applications are missing data reconstruction, restoration (for deterministic alternatives see also Chaps. 2, 3, 4 and 16), image compression and static or dynamic texture synthesis.

In the application Sect. 14.4 and further we demonstrate advantages and weak points of the studied Markovian model on several multispectral image recognition and modeling examples.

14.2 3D Causal Simultaneous Autoregressive Model

Modeling visual data requires non-standard multi-dimensional (three-dimensional for static color textures, 4D for videos or even 7D for static BTFs) models. However if such a nD data space can be factorized then these data can be also approximated using a set of lower-dimensional probabilistic models. Although full nD models allow unrestricted spatial-spectral-temporal-angular correlation modeling their main drawback is large amount of parameters to be estimated, and in the case of some models (e.g. Markov models) also the necessity to estimate all these parameters simultaneously. The 3D causal simultaneous autoregressive model (3DCAR) is an exceptional model which can be utilized to build much more complex nD data models. For example, the 7D BTF models illustrated in Fig. 14.5 are composed from up to one hundred 3DCARs.

A digitized image Y is assumed to be defined on a finite rectangular $N \times M \times d$ lattice I, $r = \{r_1, r_2, r_3\} \in I$ denotes a pixel multiindex with the row, columns and spectral indices, respectively. The notation $\bullet$ has the meaning of all possible values of the corresponding index and $I_r^c \subset I$ is a causal or unilateral neighbourhood of pixel r, i.e.

$$I_r^c \subset I_r^C = \{s : 1 \leq s_1 \leq r_1, 1 \leq s_2 \leq r_2, s \neq r\}.$$

The 3D causal simultaneous autoregressive model (3DCAR) is the wide-sense Markov model which can be written in thc following regression equation form:

$$\tilde{Y}_r = \sum_{s \in I_r^c} A_s \tilde{Y}_{r-s} + e_r \quad \forall r \in I, \tag{14.1}$$

where A_s are matrices (14.2) and the zero mean white Gaussian noise vector e_r has uncorrelated components with data indexed from I_r^c but noise vector components can be mutually correlated.

$$A_{s_1,s_2} = \begin{pmatrix} a_{1,1}^{s_1,s_2} & \dots & a_{1,d}^{s_1,s_2} \\ \vdots & \ddots & \vdots \\ a_{d,1}^{s_1,s_2} & \dots & a_{d,d}^{s_1,s_2} \end{pmatrix} \tag{14.2}$$

are $d \times d$ parameter matrices. The model can be expressed in the matrix form (14.31) where

$$X_r = [\tilde{Y}_{r-s}^T : \forall s \in I_r^c], \tag{14.3}$$

X_r is a $d\eta \times 1$ vector, $\eta = card(I_r^c)$ and γ

$$\gamma = [A_1, \dots, A_\eta] \tag{14.4}$$

is a $d \times d\eta$ parameter matrix. To simplify notation the multiindexes $r, s, \dots$ have only two components further on in this section.

An optimal support can be selected as the most probable model given past data

$$Y^{(r-1)} = \{Y_{r-1}, Y_{r-2}, \dots, Y_1, X_r, X_{r-1}, \dots, X_1\},$$

i.e., $\max_j\{p(M_j|Y^{(r-1)})\}$.

$$p(Y^{(r-1)}|M_j) = \iint p(Y^{(r-1)}|\gamma, \Sigma^{-1}) p(\gamma, \Sigma^{-1}|M_j)\, d\gamma\, d\Sigma^{-1} \tag{14.5}$$

and for implemented uniform priors start we get a decision rule [213]:

Theorem 14.1 *The most probable AR model given past data $Y^{(r-1)}$, the normal-Wishart parameter prior and the uniform model prior is the model M_i for which*

$$i = \arg\max_j\{D_j\},$$

$$
\begin{aligned}
D_j = &-\frac{d}{2}\ln|V_{x(r-1)}| - \frac{\beta(r)-d\eta+d+1}{2}\ln|\lambda_{(r-1)}| + \frac{d^2\eta}{2}\ln\pi \\
&+\sum_{i=1}^{d}\left[\ln\Gamma\left(\frac{\beta(r)-d\eta+d+2-i}{2}\right)\right. \\
&\left.-\ln\Gamma\left(\frac{\beta(0)-d\eta+d+2-i}{2}\right)\right],
\end{aligned}
\tag{14.6}
$$

where $V_{x(r-1)} = \tilde{V}_{x(r-1)} + V_{x(0)}$ *with* $\tilde{V}_{x(r-1)}$ *defined in* (14.12), $V_{x(0)}$ *is an appropriate part of* V_0 (14.13), $\beta(r)$ *is defined in* (14.7), (14.8) *and* $\lambda_{(r-1)}$ *is* (14.9).

Proof [199]

$$\beta(r) = \beta(0) + r - 1 = \beta(r-1) + 1, \tag{14.7}$$

$$\beta(0) > \eta - 2, \tag{14.8}$$

and

$$\lambda_{(r)} = V_{y(r)} - V_{xy(r)}^T V_{x(r)}^{-1} V_{xy(r)}, \tag{14.9}$$

$$V_{r-1} = \tilde{V}_{r-1} + V_0, \tag{14.10}$$

$$\tilde{V}_{r-1} = \begin{pmatrix} \tilde{V}_{y(r-1)} & \tilde{V}_{xy(r-1)}^T \\ r\,\tilde{V}_{xy(r-1)} & \tilde{V}_{x(r-1)} \end{pmatrix}, \tag{14.11}$$

$$\tilde{V}_{y(r-1)} = \sum_{k=1}^{r-1} Y_k Y_k^T, \tag{14.12}$$

$$\tilde{V}_{xy(r-1)} = \sum_{k=1}^{r-1} X_k Y_k^T, \tag{14.13}$$

$$\tilde{V}_{x(r-1)} = \sum_{k=1}^{r-1} X_k X_k^T. \tag{14.14}$$

Marginal densities $p(\gamma|Y^{(r-1)})$ and $p(\Sigma^{-1}|Y^{(r-1)})$ can be evaluated from (14.15), (14.16), respectively.

$$p(\gamma|Y^{(r-1)}) = \int p(\gamma, \Sigma^{-1}|Y^{(r-1)})\, d\Sigma^{-1}, \tag{14.15}$$

$$p(\Sigma^{-1}|Y^{(r-1)}) = \int p(\gamma, \Sigma^{-1}|Y^{(r-1)})\, d\gamma. \tag{14.16}$$

The marginal density $p(\Sigma^{-1}|Y^{(r-1)})$ is the Wishart distribution density [199]

$$p(\Sigma^{-1}\,|\,Y^{(r-1)}) = \frac{\pi^{\frac{d(1-d)}{4}}\,|\Sigma^{-1}|^{\frac{\beta(r)-d\eta}{2}}}{2^{\frac{d(\beta(r)-d\eta+d+1)}{2}}\prod_{i=1}^{d}\Gamma(\frac{\beta(r)-d\eta+2+d-i}{2})}|\lambda_{(r-1)}|^{\frac{\beta(r)-d\eta+d+1}{2}} \times \exp\left\{-\frac{1}{2}\,\mathrm{tr}\{\Sigma^{-1}\lambda_{(r-1)}\}\right\} \tag{14.17}$$

with

$$E\{\Sigma^{-1}\,|\,Y^{(r-1)}\} = (\beta(r) - d\eta + d + 1)\lambda_{(r-1)}^{-1}, \tag{14.18}$$

$$E\big\{(\Sigma^{-1} - E\{\Sigma^{-1}\,|\,Y^{(r-1)}\})^T(\Sigma^{-1} - E\{\Sigma^{-1}\,|\,Y^{(r-1)}\})\,|\,Y^{(r-1)}\big\} = \frac{2(\beta(r) - d\eta + 1)}{\lambda_{(r-1)}\lambda_{(r-1)}^T}. \tag{14.19}$$

The marginal density $p(\gamma\,|\,Y^{(r-1)})$ is matrix t distribution density [199]:

$$p(\gamma\,|\,Y^{(r-1)}) = \frac{\prod_{i=1}^{d}\Gamma(\frac{\beta(r)+d+2-i}{2})}{\prod_{i=1}^{d}\Gamma(\frac{\beta(r)-d\eta+d+2-i}{2})}\pi^{-\frac{d^2\eta}{2}}|\lambda_{(r-1)}|^{-\frac{d\eta}{2}}|V_{x(r-1)}|^{\frac{d}{2}} \times \left|I + \lambda_{(r-1)}^{-1}(\gamma - \hat{\gamma}_{r-1})V_{x(r-1)}(\gamma - \hat{\gamma}_{r-1})^T\right|^{-\frac{\beta(r)+d+1}{2}} \tag{14.20}$$

with the mean value

$$E\{\gamma\,|\,Y^{(r-1)}\} = \hat{\gamma}_{r-1} \tag{14.21}$$

and covariance matrix

$$E\big\{(\gamma - \hat{\gamma}_{r-1})^T(\gamma - \hat{\gamma}_{r-1})\,|\,Y^{(r-1)}\big\} = \frac{V_{x(r-1)}^{-1}\lambda_{(r-1)}}{\beta(r) - d\eta}. \tag{14.22}$$

Similar statistics can be easily derived [199] for the alternative Jeffreys non-informative parameter prior. □

Theorem 14.2 *The one-step-ahead predictive posterior density for the normal-Wishart parameter prior has the form of d-dimensional Student's probability density* (14.23)

$$p(Y_r\,|\,Y^{(r-1)}) = \frac{\Gamma(\frac{\beta(r)-d\eta+d+2}{2})}{\Gamma(\frac{\beta(r)-d\eta+2}{2})\,\pi^{\frac{d}{2}}(1 + X_r^T V_{x(r-1)}^{-1}X_r)^{\frac{d}{2}}|\lambda_{(r-1)}|^{\frac{1}{2}}} \times \left(1 + \frac{(Y_r - \hat{\gamma}_{r-1}X_r)^T\lambda_{(r-1)}^{-1}(Y_r - \hat{\gamma}_{r-1}X_r)}{1 + X_r^T V_{x(r-1)}^{-1}X_r}\right)^{-\frac{\beta(r)-d\eta+d+2}{2}}, \tag{14.23}$$

with $\beta(r) - d\eta + 2$ degrees of freedom, if $\beta(r) > d\eta$ then the conditional mean value is

$$E\{Y_r \mid Y^{(r-1)}\} = \hat{\gamma}_{r-1} X_r, \tag{14.24}$$

and

$$E\{(Y_r - \hat{\gamma}_{r-1} X_r)(Y_r - \hat{\gamma}_{r-1} X_r)^T \mid Y^{(r-1)}\} = \frac{1 + X_r V_{x(r-1)}^{-1} X_r^T}{(\beta(r) - d\eta)} \lambda_{(r-1)}. \tag{14.25}$$

Proof [199]. □

14.2.1 Adaptivity

The 3DCAR model can be made adaptive if we modify its recursive statistics using exponential forgetting factor, i.e. a constant $\varphi \approx 0.99$. This forgetting factor smaller than 1 is used to weigh the influence of older data:

$$\hat{\gamma}_r^T = \hat{\gamma}_{r-1}^T + (\varphi^2 + X_r^T V_{x(r-1)}^{-1} X_r)^{-1} V_{x(r-1)}^{-1} X_r (Y_r - \hat{\gamma}_{r-1} X_r)^T,$$
$$|V_{x(t)}| = |V_{x(t-1)}| \varphi^{2\eta} (1 + X_t^T V_{x(t-1)}^{-1} X_t),$$
$$\lambda_t = \lambda_{t-1} (1 + (Y_t - \hat{\gamma}_{t-1}^T X_t)^T \lambda_{t-1}^{-1} (Y_t - \hat{\gamma}_{t-1}^T X_t)(\varphi^2 + X_t^T V_{x(t-1)}^{-1} X_t)^{-1}).$$

14.2.2 Numerical Stability

The numerical stability of 3DCAR can be quaranteed if all its recursive statistics use the square-root factor updating applying either the Cholesky or LDL^T decomposition [202], respectively. Let us denote a lower triangular matrix L_t and a matrix B_t as

$$B_t = V_{i,t}^{-1} = L_t L_t^T, \tag{14.26}$$
$$B_t = (B_{t-1}^{-1} \pm d_t d_t^T)^{-1}, \tag{14.27}$$
$$B_t = (\tilde{B}_{t-1}^{-1} \varphi^{-2} \pm \alpha^2 \tilde{d}_t d_t^T)^{-1}, \tag{14.28}$$

where d_t is an updating vector with exponential forgetting φ and data normalization α. The square-root updating factor of the inversion data gathering matrix B_t can be computed recursively:

$$L_{t,i,j} = \frac{\tilde{\zeta}_{t,j+1}}{\varphi \tilde{\zeta}_{t,j}} \left[L_{t-1,i,j} \mp \frac{\tilde{f}_{t,j}}{\tilde{\zeta}_{t,j+1}^2} \tilde{g}_{t,j+1}^{(i)} \right], \tag{14.29}$$

where $\tilde{f}_t = L_{t-1}^T d_t$ and

$$\tilde{g}_{t,j+1}^{(i)} = \sum_{k=j+1}^{i} L_{t-1,i,k} \tilde{f}_{t,k}$$

and

$$\zeta_{t,i} = \frac{\alpha}{\varphi} \sqrt{\frac{\varphi^2}{\alpha^2} \pm \sum_{j=i}^{n} \tilde{f}_{t,j}^2} = \frac{\alpha}{\varphi} \tilde{\zeta}_{t,i}.$$

14.2.3 3DCAR Model Properties

The 3DCAR (analogously also the 2DCAR model) model has advantages in analytical solutions (Bayes, ML, or LS estimates) for I_r, $\hat{\gamma}$, $\hat{\sigma}^2$, $\hat{Y}_r$ statistics. It allows extremely simple fast synthesis, adaptivity and building efficient recursive application algorithms. Its major drawback in some image representation applications might be its mostly artificial causality which can introduce a directional bias into modelled image data.

14.3 Illumination Invariants

Textures are important clues to specify objects present in a visual scene. However, the appearance of natural textures is highly illumination and view angle dependent. As a consequence, most recent realistic texture based classification or segmentation methods require multiple training images [425] captured under all possible illumination and viewing conditions for each class. Such learning is obviously clumsy, probably expensive and very often even impossible if required measurements are not available.

If we assume fixed positions of viewpoint and illumination sources, uniform illumination sources and Lambertian surface reflectance, than two images $\tilde{Y}, Y$ acquired with different illumination spectra can be linearly transformed to each other:

$$\tilde{Y}_r = B Y_r \quad \forall r. \tag{14.30}$$

It is possible to show that assuming (14.30) the following features are illumination invariant:

1. trace: *trace* A_m, $m = 1, \ldots, \eta K$,
2. eigenvalues: $\nu_{m,j}$ of A_m, $m = 1, \ldots, \eta K$, $j = 1, \ldots, C$,
3. $1 + X_r^T V_x^{-1} X_r$,
4. $\sqrt{\sum_r (Y_r - \hat{\gamma} X_r)^T \lambda^{-1} (Y_r - \hat{\gamma} X_r)}$,
5. $\sqrt{\sum_r (Y_r - \mu)^T \lambda^{-1} (Y_r - \mu)}$, μ is the mean value of vector Y_r.

Above textural features derived from the 3DCAR model are robust to illumination direction changes, invariant to illumination brightness and spectrum changes, and simultaneously also robust to Gaussian noise degradation. This property was extensively verified on University of Bonn BTF texture measurements [310], where illumination sources are spanned over 75% of possible illumination half-sphere.

14.4 Unsupervised Image Recognition

Unsupervised or supervised texture segmentation is the prerequisite for successful content-based image retrieval, scene analysis, automatic acquisition of virtual models, quality control, security, medical applications and many others. Although more than 1000 different methods were already published [461], this problem is still far from being solved. This is among others due to missing reliable performance comparison between different techniques because very limited effort was spent [212] to develop suitable quantitative measures of segmentation quality that can be used to evaluate and compare segmentation algorithms. Spatial interaction models and especially Markov random field-based models are increasingly popular for texture representation [200, 248, 355], etc. Several researchers dealt with the difficult problem of unsupervised segmentation using these models, see for example [9, 201, 209–211, 299, 328], or Chap. 15.

Our unsupervised segmenter is illustrated on a multiscale unsupervised automatic detection of potentially cancerous regions of interest containing fibroglandular tissue in digital screening mammography. The mammogram tissue textures are locally represented by four causal multispectral random field models recursively evaluated for each pixel and several scales. The segmentation part of the algorithm is based on the underlying Gaussian mixture model and starts with an over segmented initial estimation which is adaptively modified until the optimal number of homogeneous mammogram segments is reached.

Our method segments pseudo-color multiresolution mammograms each created from the original grey scale mammogram and its two nonlinear gamma transformations. We assume to down-sample input image Y into $M = 3$ different resolutions $Y^{(m)} = \downarrow^{\iota_m} Y$ with sampling factors ι_m, $m = 1, \dots, M$ identical for both directions and $Y^{(1)} = Y$. Local texture for each pixel $Y_r^{(m)}$ is represented using the 3D CAR model parameter space $\Theta_r^{(m)}$. The concept of decision fusion for high-performance pattern recognition is well known and widely accepted in the area of supervised classification where (often very diverse) classification technologies, each providing complementary sources of information about class membership, can be integrated to provide more accurate, robust and reliable classification decisions than the single classifier applications. The proposed method circumvents the problem of multiple unsupervised segmenters combination [211] by fusing multiple-processed measurements into a single segmenter feature vector.

Smooth pseudo-color mammogram textures require three dimensional models for adequate representation. We assume that single multi spectral texture can be

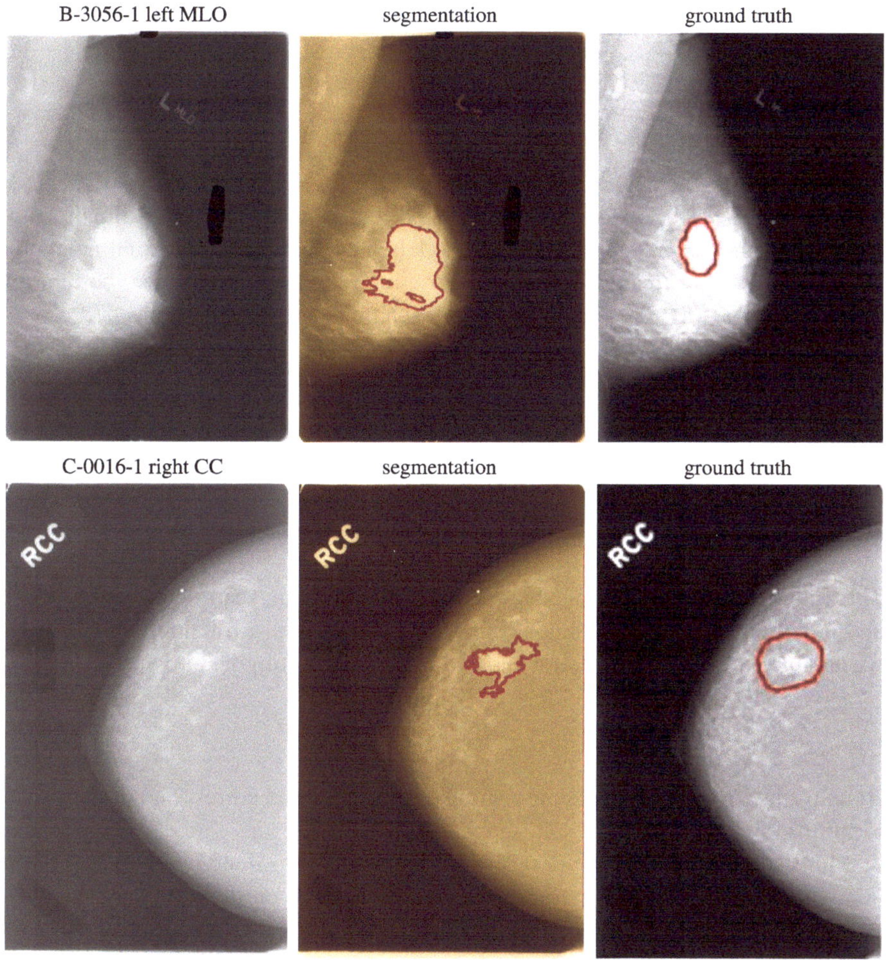

Fig. 14.1 Cancerous mammograms (patients age 58 (*top*) and 80 (*bottom*)), radiologist associated ground truth and detected regions of interest using the multiple segmenter approach, respectively

locally modelled using a 3D simultaneous causal autoregressive random field model (3DCAR). This model can be expressed as a stationary causal uncorrelated noise driven 3D autoregressive process [213]:

$$Y_r = \gamma X_r + e_r, \tag{14.31}$$

where $\gamma = [A_1, \ldots, A_\eta]$ is the $3 \times 3\eta$ parameter matrix, e_r is a white Gaussian noise vector with zero mean and a constant but unknown variance, X_r is a corresponding vector of the contextual neighbors Y_{r-s} and $r, r-1, \ldots$ is a chosen direction of movement on the image index lattice I. The optimal neighborhood (I_r^c) as well as the Bayesian parameters estimation of a 3DCAR model can be found analytically under few additional and acceptable assumptions using the Bayesian approach

(14.6). The recursive Bayesian parameter estimation of the 3DCAR model is [213]:

$$\hat{\gamma}_{r-1}^T = \hat{\gamma}_{r-2}^T + \frac{V_{x(r-2)}^{-1} X_{r-1} (Y_{r-1} - \hat{\gamma}_{r-2} X_{r-1})^T}{(1 + X_{r-1}^T V_{x(r-2)}^{-1} X_{r-1})}, \tag{14.32}$$

where $V_{x(r-1)} = \sum_{k=1}^{r-1} X_k X_k^T + V_{x(0)}$. Each matrix contains local estimations of the 3DCAR model parameters. These models have identical contextual neighborhood I_r^c but they differ in their major movement direction (top-down, bottom-up, rightward, leftward). The local texture for each pixel and M resolutions $\alpha_1, \ldots, \alpha_M$ is represented by four parametric matrices t, b, r, l e.g. $\hat{\gamma}_r^{i,\alpha_j}$ for $i \in \{t, b, r, l\}$, $j = 1, \ldots, M$ which are subsequently compressed using the local PCA (for computational efficiency) into $\tilde{\gamma}_r^{i,\alpha_j}$. Single resolution compressed parameters are composed into M parametric matrices:

$$\tilde{\gamma}_r^{\alpha_j T} = \{\tilde{\gamma}_r^{t,\alpha_j}, \tilde{\gamma}_r^{b,\alpha_j}, \tilde{\gamma}_r^{r,\alpha_j}, \tilde{\gamma}_r^{l,\alpha_j}\}^T, \quad j = 1, \ldots, M.$$

The parametric space $\tilde{\gamma}^{\alpha_j}$ is subsequently smoothed out, rearranged into a vector and its dimensionality is reduced using the PCA feature extraction ($\bar{\gamma}^{\alpha_j}$). Finally we add the average local spectral values $\zeta_r^{\alpha_j}$ to the resulting feature vector:

$$\Theta_r = [\bar{\gamma}_r^{\alpha_1}, \zeta_r^{\alpha_1}, \ldots, \bar{\gamma}_r^{\alpha_M}, \zeta_r^{\alpha_M}]^T. \tag{14.33}$$

Rough scale pixels parameters are simply mapped to the corresponding fine scale locations.

Multi-spectral, multiresolution texture segmentation is done by clustering in the combined 3DCAR models parameter space Θ defined on the lattice I where Θ_r is the modified parameter vector (14.33) computed for the lattice location r. We assume that this parametric space can be represented using the Gaussian mixture model (GM) with diagonal covariance matrices due to the previous 3DCAR parametric space decorrelation. The Gaussian mixture model for 3DCAR parametric representation is as follows:

$$p(\Theta_r) = \sum_{i=1}^{K} p_i \, p(\Theta_r \mid \nu_i, \Sigma_i), \tag{14.34}$$

$$p(\Theta_r \mid \nu_i, \Sigma_i) = \frac{|\Sigma_i|^{-\frac{1}{2}}}{(2\pi)^{\frac{d}{2}}} e^{-\frac{(\Theta_r - \nu_i)^T \Sigma_i^{-1} (\Theta_r - \nu_i)}{2}}. \tag{14.35}$$

The mixture model equations (14.34), (14.35) are solved using a modified EM algorithm. The algorithm is initialized using ν_i, Σ_i statistics estimated from the corresponding regions obtained by regular division of the input detected breast area. An alternative initialisation can be random choice of these statistics. For each possible couple of regions the Kullback Leibler divergence

$$D(p(\Theta_r \mid \nu_i, \Sigma_i) \parallel p(\Theta_r \mid \nu_j, \Sigma_j)) = \int_{\Omega} p(\Theta_r \mid \nu_i, \Sigma_i) \log\left(\frac{p(\Theta_r \mid \nu_i, \Sigma_i)}{p(\Theta_r \mid \nu_j, \Sigma_j)} \right) d\Theta_r \tag{14.36}$$

is evaluated and the most similar regions, i.e.,

$$\{i, j\} = \arg\min_{k,l} D(p(\Theta_r \mid \nu_l, \Sigma_l) \parallel p(\Theta_r \mid \nu_k, \Sigma_k))$$

are merged together in each step. This initialisation results in K_{ini} subimages and recomputed statistics ν_i, Σ_i. $K_{ini} > K$ where K is the optimal number of textured segments to be found by the algorithm. Two steps of the EM algorithm are repeating after the initialisation. The components with smaller weights than a fixed threshold ($p_j < \frac{0.01}{K_{ini}}$) are eliminated. For every pair of components we estimate their Kullback Leibler divergence (14.36). From the most similar couple, the component with the weight smaller than the threshold is merged to its stronger partner and all statistics are actualized using the EM algorithm. The algorithm stops when either the likelihood function has negligible increase ($\mathcal{L}_t - \mathcal{L}_{t-1} < 0.01$) or the maximum iteration number threshold is reached.

The parametric vectors representing texture mosaic pixels are assigned to the clusters according to the highest component probabilities, i.e., Y_r is assigned to the cluster ω_{j^*} if

$$\pi_{r,j^*} = \max_j \sum_{s \in I_r} w_s \, p(\Theta_{r-s} \mid \nu_j, \Sigma_j),$$

where w_s are fixed distance-based weights, I_r is a rectangular neighbourhood and $\pi_{r,j^*} > \pi_{thre}$ (otherwise the pixel is unclassified). The area of single cluster blobs is evaluated in the post-processing thematic map filtration step. Regions with similar statistics are merged. Thematic map blobs with area smaller than a given threshold are attached to its neighbour with the highest similarity value. Finally, regions which have grey level mean value difference from the median mean value (over the same type of digitized mammograms) of cancerous ground truth regions larger than a specified threshold are eliminated.

14.5 Multispectral Image Restoration

Physical imaging systems and a recording medium are imperfect and thus a recorded image represents a degraded version of the original scene. Similarly an image is usually further corrupted during its processing, transmission or storage. The image restoration task is to recover an unobservable image given the observed corrupted image with respect to some statistical criterion. Image restoration is the busy research area for already several decades and many restoration algorithms have been proposed [3, 8, 183], see also Chaps. 2, 3, 4 and 16.

Fig. 14.2 Original, corrupted, and reconstructed Cymbidium image

The image degradation is supposed to be approximated by the linear degradation model:

$$X_r = \sum_{s \in I_r} f_s Y_{r-s} + e_r, \tag{14.37}$$

where f is a discrete representation of the unknown point-spread function. The point-spread function can be non-homogeneous but we assume that it changes slowly relative to the size of an image. I_r is some contextual support set, and the degradation noise e is uncorrelated with the unobservable image, i.e.,

$$E\{Ye\} = 0. \tag{14.38}$$

The point-spread function is unknown but such that we can assume the unobservable image Y to be reasonably well approximated by the expectation of the corrupted image

$$\hat{Y} = E\{X\} \tag{14.39}$$

in regions with gradual pixel value changes. Pixels with steep step discontinuities are left unrestored to avoid excessive blurring, i.e.,

$$\hat{Y}_r = \begin{cases} E\{X_r\} & \text{if } |E\{X_r\} - X_r| < \frac{1}{n_s} \sum_s |E\{X_{r-s}\} - X_{r-s}|, \\ X_r & \text{otherwise.} \end{cases} \tag{14.40}$$

The expectation (14.39) can be expressed as follows:

$$\begin{aligned} E\{X\} &= \int X p(X)\, dX \\ &= \int \begin{pmatrix} X_1 & X_2 & \dots & X_M \\ X_{M+1} & X_{M+2} & \dots & X_{2M} \\ \vdots & \vdots & \ddots & \vdots \\ X_{NM-M+1} & X_{NM-M+2} & \dots & X_{NM} \end{pmatrix} \prod_{r=1}^{NM} p(X_r \mid X^{(r-1)})\, dX_1 \cdots dX_{NM}, \end{aligned} \tag{14.41}$$

where

$$X^{(r-1)} = \{X_{r-1}, \ldots, X_1\} \tag{14.42}$$

is a set of noisy pixels in some chosen but fixed ordering. For single matrix elements in (14.41) it holds

$$\begin{aligned}
E\{X_j\} &= \int X_j \prod_{r=1}^{NM} p(X_r \,|\, X^{(r-1)})\, dX_1 \cdots dX_{NM} \\
&= \int X_j \prod_{r=1}^{j} p(X_r \,|\, X^{(r-1)})\, dX_1 \cdots dX_j \\
&= \int E\{X_j \,|\, X^{(j-1)}\} \prod_{r=1}^{j-1} p(X_r \,|\, X^{(r-1)})\, dX_1 \cdots dX_{j-1} \\
&= E\{E\{X_j \,|\, X^{(j-1)}\}\}.
\end{aligned} \tag{14.43}$$

Let us approximate after having observed $X^{(j-1)}$ the mean value $\hat{Y}_j = E\{X_j\}$ by the $E\{X_j \,|\, X^{(j-1)} = x^{(j-1)})$ where $x^{(j-1)}$ are known past realisation for j. Thus we suppose that all other possible realisations $x^{(j-1)}$ than the true past pixel values have negligible probabilities. This assumption implies conditional expectations approximately equal to unconditional ones, i.e.,

$$E\{X_j\} \approx E\{X_j \,|\, X^{(j-1)}\}, \tag{14.44}$$

and

$$\hat{Y} = E\{X\} \approx \begin{pmatrix} E\{X_1 \,|\, X^{(0)}\} & E\{X_2 \,|\, X^{(1)}\} & \ldots & E\{X_M \,|\, X^{(M-1)}\} \\ E\{X_{M+1} \,|\, X^{(M)}\} & E\{X_{M+2} \,|\, X^{(M+1)}\} & \ldots & E\{X_{2M} \,|\, X^{(2M-1)}\} \\ \vdots & \vdots & \ddots & \vdots \\ E\{X_{NM-M+1} \,|\, X^{(NM-M)}\} & E\{X_{NM-M+2} \,|\, X^{(NM-M+1)}\} & \ldots & E\{X_{NM} \,|\, X^{(NM-1)}\} \end{pmatrix}.$$

Suppose also that the noisy image X can be represented by a causal simultaneous autoregressive model (14.1), then the conditional mean (14.44) values needed for the estimation $\hat{Y}$ is (14.52) if we replace in above equations $X \rightarrow Y, Z \rightarrow X$. The estimator (14.44) can be efficiently computed using the following recursion

$$\hat{\gamma}_r^T = \hat{\gamma}_{r-1}^T + (1 + Z_r^T V_{z(r-1)}^{-1} Z_r)^{-1} V_{z(r-1)}^{-1} Z_r (X_r - \hat{\gamma}_{r-1} Z_r)^T. \tag{14.45}$$

The selection of an appropriate model support (I_r^c) is important to obtain good restoration results. The optimal Bayesian decision rule for this selection is either (14.6) or analogous statistics depending on the parameter prior.

14.5.1 Local Estimation of the Point-Spread Function

If we assume a non-homogeneous slowly changing point-spread function, we can estimate its local value using the local least square estimate

$$\hat{\psi}_r = \min_{\psi_r}\left\{\sum_{\forall r\in J_r} (X_r - \psi_r \hat{W}_r)^2\right\}. \qquad (14.46)$$

The locally optimal estimate is

$$\hat{\psi}_r^T = \tilde{V}_{\hat{W}(r)}^{-1}\, \tilde{V}_{\hat{W}\,X(r)}, \qquad (14.47)$$

where $\tilde{V}_{\hat{W}(r)}$, $\tilde{V}_{\hat{W}\,X(r)}$ are corresponding local data gathering matrices analogous to (14.14), (14.12), but using only data from local sub-lattice $J_r \subset I$, $r \in J_r$. This estimator can be efficiently evaluated using the fast recursive square-root filter introduced in Sect. 14.2.2. If the point-spread function is constant for all lattice positions both PSF estimators (local and global) are equivalent.

14.6 Multichannel Image Restoration

The major degradation of a ground-based telescope is caused by random fluctuations originating mostly in the Earth's atmosphere (*seeing*) along the optical path between the object space and the image formation device. The image degradation by seeing is a very complicated process due to blurring, motion, and distortion. The image degradation is described by the changing complex point-spread-function (PSF) of the telescope, which embodies all the important behaviour of the optical image formation system. For the restoration we assume one unknown degradation function involving all degradation aspects.

Suppose Y represents a true but unobservable monospectral image defined on the finite rectangular $N \times M$ underlying lattice I. Suppose further that we have a set of d observable images $\mathcal{X}$ where each $X_{\bullet,i} \in \mathcal{X}$ is the ith version of Y distorted by the unknown PSF and noise independent of the signal. The notation $\bullet$ designates of all possible values of the corresponding multiindex (e.g. the multiindex $r = \{r_1, r_2\}$ which has the row and columns indices, respectively). We assume knowledge of all pixels from the reconstructed scene. For the treatment of the more difficult problem when some data are missing see [214, 215]. The image degradation is supposed to be approximated by the linear discrete spatial domain degradation model

$$X_{r,\bullet} = \sum_{s \in I_r} H_s Y_{r-s} + \varepsilon_{r,\bullet}, \tag{14.48}$$

where H is a discrete representation of the unknown point-spread function, $X_{r,\bullet}$ is the $d \times 1$ vector of the rth pixel in different distortions and Y_{r-s} are ideal (unobservable) image pixels. The point-spread function is assumed to be either homogeneous or it can be non-homogeneous but in this case we assume it slowly changes relative to the size of an image. I_r is some contextual support set, and a noise vector ε is uncorrelated with the true image, i.e., $E\{Y\varepsilon_{\bullet,i}\} = 0$. The point-spread function is unknown but such that we can assume the unobservable image Y to be reasonably well approximated by the expectation of the corrupted image

$$\hat{Y} = E\{X_{\bullet,i}\} \tag{14.49}$$

in regions with gradual pixel value changes, and the ith degraded image $X_{\bullet,i} \in \mathscr{X}$ is the least degraded image from the set $\mathscr{X}$. The index i of the least degraded image is excluded from the following equations (14.50)–(14.52) to simplify the corresponding notation. The above method (14.49) changes all pixels in the restored image and thus blurs discontinuities present in the scene although to much less extent than the classical restoration methods due to our restoration model (14.31) adaptivity. This excessive blurring can be avoided if pixels with steep step discontinuities are left unrestored, i.e.,

$$\hat{Y}_r = \begin{cases} E\{X_r\} & \text{if } p(X_r \mid X^{(r-1)}) > \kappa, \\ X_r & \text{otherwise,} \end{cases} \tag{14.50}$$

where κ is a probabilistic threshold based on the prediction density. Single matrix elements in the expectation $E\{X\}$ are approximated [216] by the conditional expectation $E\{X_j \mid X^{(j-1)} = x^{(j-1)}\}$ where $x^{(j-1)}$ are known past realisation for j. Thus we suppose that all other possible realisation $x^{(j-1)}$ than the true past pixel values have negligible probabilities. This assumption implies conditional expectations approximately equal to unconditional ones, i.e., then the expectation is $E\{X_j\} \approx E\{X_j \mid X^{(j-1)}\}$.

Suppose further that a noisy image can be represented by an adaptive 2.5D causal simultaneous autoregressive model

$$X_{r,i} = \gamma Z_r + \varepsilon_r, \tag{14.51}$$

where $\gamma = [A_1, \ldots, A_\eta]$, $\eta = card(I_r^c)$ is a $1 \times d\eta$ parameter matrix, Z_r is a corresponding vector of X_{r-s}, ε_r is a white Gaussian noise vector with zero mean, and a constant but unknown covariance matrix Σ. The noise vector is uncorrelated with data from a causal neighbourhood I_r^c. $A_s = [a_{s,1}, \ldots, a_{s,d}]$ $\forall s$ are parameter vectors. The model adaptivity is introduced using the exponential forgetting factor technique in parameter learning part of the algorithm. The conditional mean value

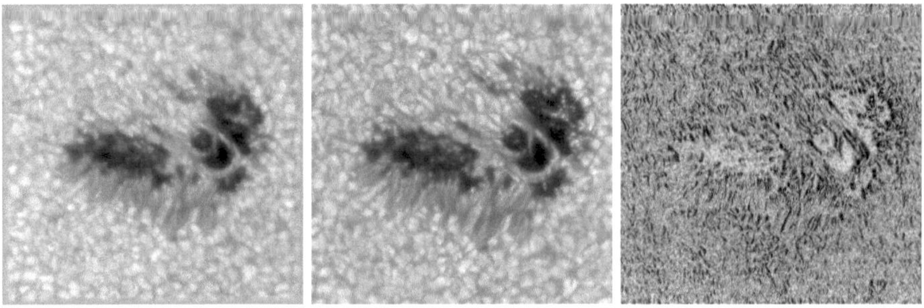

Fig. 14.3 The measured degraded (*left*), reconstructed sunspot image using our method, and its corresponding prediction probability image

can be derived under few acceptable conditions [216] in the following form:

$$E\{X_r|X^{(r-1)}\} = V_{zx(r-1)}^T V_{zz(r-1)}^{-T} Z_r, \tag{14.52}$$

where $V_{r-1} = \tilde{V}_{r-1} + I$ and

$$\tilde{V}_{r-1} = \begin{pmatrix} \sum_{k=1}^{r-1} X_k X_k^T & \sum_{k=1}^{r-1} X_k Z_k^T \\ \sum_{k=1}^{r-1} Z_k X_k^T & \sum_{k=1}^{r-1} Z_k Z_k^T \end{pmatrix} = \begin{pmatrix} \tilde{V}_{xx(r-1)} & \tilde{V}_{zx(r-1)}^T \\ \tilde{V}_{zx(r-1)} & \tilde{V}_{zz(r-1)} \end{pmatrix}.$$

An appropriate model support (I_r^c) can be found using the Bayesian decision rule (cf. [216]).

The proposed recursive multitemporal blur minimizing reconstruction method is very fast (approximately five times faster than the median filter) robust and its reconstruction results surpass some standard reconstruction methods, which we were able to implement for the verification. Our causal model has the advantage to have the analytical solution for all needed model statistics. Possible artifacts introduced by this type of models are diminished by introducing *adaptivity* into the model. This novel formulation allows us to obtain extremely fast adaptive multichannel/multitemporal restoration and it can be easily parallellized as well as generalized for multispectral (e.g. color, multispectral satellite images) or registered images, which is seldom the case for alternative methods.

14.7 Video Restoration

Every movie deteriorates with usage and time irrespective of any care it gets. Movies (on both optical and magnetic materials) suffer with blotches, dirt, sparkles and noise, scratches, missing or heavily corrupted frames, mold, flickering, jittering, image vibrations and some other problems. For each kind of the defect usually a different kind of restoration algorithm is needed. The scratch notion in this section means every coherent region with missing data (simultaneously in all spectral

Fig. 14.4 A car frame restoration (original, scratch, quadratic interpolation method, 3D CAR, 3.5D CAR)

bands) in a color movie frame. Our method [204] reconstructs missing multispectral (e.g., color) pixels from available data in neighboring frames and pixels from the corrupted frame as well. A digitized color movie is supposed to be represented with the 3.5D causal AR model (see Sect. 14.2):

$$Y_{r_1,r_2,\bullet,r_4} = \gamma\, X_{r_1,r_2,\bullet,r_4} + e_{r_1,r_2,\bullet,r_4} \quad \forall r \in I. \tag{14.53}$$

The missing scratch data are reconstructed from the topologically nearest known data in the lattice I using temporal and spatial correlation in the neighbourhood. Scratch pixels are computed from the set of one-step-ahead predictions using the conditional mean predictor

$$\tilde{Y}_r = E\{Y_{r_1,r_2,\bullet,r_4} \mid Y^{(r-1)}\} = \hat{\gamma}_{r-1} X_{r_1,r_2,\bullet,r_4}, \tag{14.54}$$

where

$$Y^{(r-1)} = \{Y_{r-1}, Y_{r-2}, \ldots, Y_1\}$$

is the known process history and $\hat{\gamma}_{r-1}$ is the estimator of unknown model parameter matrix γ (see Sect. 14.2).

A model movement towards the scratch is assumed. When the model reaches the scratch, the corrupted pixel prediction is evaluated. This is performed for each line in the scratch from top and bottom edge of the scratch using two symmetrical downwards and upwards moving models and their results are averaged. This helps to counterbalance artificial restriction on the contextual neighbourhood which has to be causal. Similarly another couple of models is moving in the opposite direction. Two computed predictions for each missing pixel have to be combined. Simple averaging is not appropriate, because each of both predictors has different distance from the last known original data and consequently it has also a different precision. Hence the exponential interleaving was used to weight the data influence from each side of the scratch as a function of the horizontal position of the predicted pixel on a scratch line to be reconstructed.

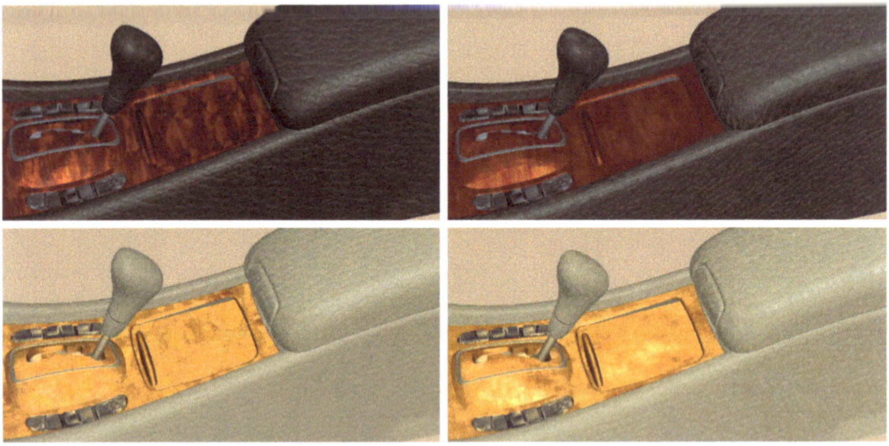

Fig. 14.5 BTF measurements mapped on part of a car gearbox. Original BTF data (enlarged using image tiling) (*left*) compared with synthesized BTF (*right*) for four distinct materials: *wood01*, *foil01*, *wood02*, *foil02* (3D model courtesy of DaimlerChrysler)

14.8 Texture Synthesis and Compression

Texture synthesis methods may be divided primarily into intelligent sampling and model-based methods. Sampling approaches [94, 107, 124, 125, 219, 456, 459] rely on sophisticated sampling from real texture measurements while the model-based techniques [31, 32, 186, 200, 207, 208, 326, 466] describe texture data using multidimensional mathematical models and their synthesis is based on the estimated model parameters only.

There are several texture modeling approaches published [207, 208, 247] and some survey articles are also available [200, 203]. Most published texture models are restricted only to monospectral textures for few models developed for multispectral (mostly color) textures refer [31, 32, 207, 208].

The Bidirectional Texture Function (BTF) [143, 205, 206, 217] is the most advanced representation of visual properties for realistic real-world materials. BTF describes rough texture appearance for varying illumination and viewing conditions. Such a function can be represented by thousands of measurements (images) per material sample. The resulting BTF size excludes its direct rendering in graphical applications and some compression of these huge BTF data spaces is obviously inevitable. The BTF modeling ultimate aim is to create a visual impression of the same material without a pixel-wise correspondence to the original measurements. The cornerstone of our BTF compression and modeling method is the replacement of a huge number of original BTF measurements by their efficient parametric estimates derived from an underlying set of 2DCAR or 3DCAR spatial probabilistic models.

The off-line part of the algorithm [206] starts with the BTF illumination/view $(\theta_i, \phi_i/\theta_v, \phi_v)$ space segmentation into several subspace images using the K-means algorithm on color cumulative histograms features. Thus we trade off between an extreme compression ratio and the visual quality by using several probabilistic BTF subspace dedicated models. The overall roughness of a textured surface significantly influences the BTF texture appearance. Such a surface can be specified using its range map, which is estimated by the photometric stereo approach. The subspace 2DCAR texture model starts with a spectral PCA-based decorrelation of subspace image (result of BTF segmentation) into mono-spectral factors. The 3DCAR model does not need this decorrelation step. Each of these factors is subsequently decomposed into sub-band components using the multi-resolution Gaussian-Laplacian pyramid. This allows to use simpler 2D/3D CAR models to model wide range of textures. Each such sub-band component is analyzed by a dedicated CAR factor model to obtain a compact set of model parameters.

The 3DCAR model [143, 206] offers a huge BTF compression ratio unattainable by any alternative sampling-based BTF synthesis method. Simultaneously this model can be used to reconstruct missing parts of the BTF measurement space. Dynamic textures [144] can be modelled using this model as well. The method [144] is based on eigen-analysis of dynamic texture images and subsequent preprocessing and modelling of temporal interpolation eigen-coefficients using a 3DCAR model. This method compresses significantly the original data and enables extremely fast synthesis of artificial sequences, which can be easily performed by means of contemporary graphics hardware.

14.9 Conclusion

The 3DCAR models are among rare exceptions in the Markovian model family that allow to derive extremely efficient and fast data processing algorithms. All their statistics can be evaluated recursively and they do not need any Monte Carlo sampling typical for other Markovian models. The 3DCAR models have the advantage over non causal (3DAR) ones that they can be treated analytically. It is possible to find analytical solution of model parameters, optimal model support, model predictor, etc. Similarly the 3DCAR model synthesis is very simple and a causal SAR RF can be directly generated from model equation. The disadvantage on the other hand is the causality which is usually rather artificially imposed on image data for algorithmic reasons and it is seldom supported by real image data. A causal model introduces an arbitrary directional bias, which depends on the orientation of a causal neighbourhood used. The CAR model can represent only wide-sense stationary data and data which have linear mutual relationships.

Acknowledgements This research was supported by the project GAČR 102/08/0593 and partially by the projects MŠMT 1M0572 DAR, GAČR 103/11/0335, CESNET 387/2010.

Chapter 15
Locally Specified Polygonal Markov Fields for Image Segmentation

Michal Matuszak and Tomasz Schreiber

Abstract We introduce a class of polygonal Markov fields driven by local activity functions. Whereas the local rather than global nature of the field specification ensures substantial additional flexibility for statistical applications in comparison to classical polygonal fields, we show that a number of simulation algorithms and graphical constructions, as developed in our previous joint work with M.N.M. van Lieshout and R. Kluszczynski, carry over to this more general framework. Moreover, we provide explicit formulae for the partition function of the model, which directly implies the availability of closed form expressions for the corresponding likelihood functions. Within the framework of this theory we develop an image segmentation algorithm based on Markovian optimization dynamics combining the simulated annealing ideas with those of Chen-style stochastic optimization, in which successive segmentation updates are carried out simultaneously with adaptive optimization of the local activity functions.

15.1 Introduction

The polygonal Markov fields, originally introduced by Arak and Surgailis [10–12] and then studied by a number of authors [13, 319, 377–379, 395], arise as continuum ensembles of non-intersecting polygonal contours in the plane. One of the sources of theoretical interest in these processes lies in that they share a number of salient features with the two-dimensional Ising model, including the geometry of phase transitions and phase separation phenomenon [319, 377, 378] as well as the availability of explicit formulae for important numerical characteristics [11, 12, 379] yielding in particular closed form expressions for the likelihood functions. The idea that the polygonal Markov fields can carry out image processing tasks traditionally reserved for lattice-indexed Markov fields (see [448] for a comprehensive survey, cf.

M. Matuszak (✉) · T. Schreiber
Faculty of Mathematics & Computer Science, Nicolaus Copernicus University, ul. Chopina 12/18, 87-100 Toruń, Poland
e-mail: gruby@mat.uni.torun.pl

T. Schreiber
e-mail: tomeks@mat.uni.torun.pl

L. Florack et al. (eds.), *Mathematical Methods for Signal and Image Analysis and Representation*, Computational Imaging and Vision 41,
DOI 10.1007/978-1-4471-2353-8_15,

also Chap. 14 and the references therein for new developments) has emerged quite early and originates from Clifford, Middleton and Nicholls [79] who formulated it in a Bayesian setting. The obvious crucial advantage of polygonal fields in this context is their continuum nature which makes them completely free of lattice artifacts in image processing applications. The significant problem which slowed down the progress of this early work was the lack of efficient samplers and simulation algorithms for polygonal fields. These were introduced a decade later in a series of our joint papers with M.N.M. van Lieshout and R. Kluszczynski [256, 257, 377, 380, 424] where a polygonal field optimization approach for image segmentation was advocated. Although these methods were quite succesful in global shape recognition, the problem we faced in that work was related to the lack of local parametrization tools designed to deal with intermediate scale image characteristics—even though the applied simulated annealing algorithm would eventually converge to the target polygonal segmentation, we were looking for a more efficient explicit mechanism to drive the local search. Introducing such mechanisms and applying them to image segmentation is the principal purpose of the present paper. We construct a class of polygonal Markov fields with local activity functions (Sect. 15.2) and discuss their properties and graphical representations (Sect. 15.3). Next, in Sects. 15.4 and 15.5 we develop a Markovian optimization dynamics for image segmentation, under which both the polygonal configuration and the underlying local activity function are subject to optimization—whereas the polygonal configuration evolves according to a simulated annealing scheme in the spirit of [256, 257], the local activity function is initially chosen to reflect the image gradient information, whereupon it undergoes adaptive updates in the spirit of the celebrated Chen algorithm, see [73] and 10.2.4.c. in [333], with the activity profile reinforced along polygonal paths contributing to the improvement of the overall segmentation quality and faded along paths which deteriorate the segmentation quality. The sample results of our software are presented in the final Sect. 15.6.

15.2 Locally Specified Polygonal Markov Fields

Fix an open bounded convex set D in the plane $\mathbb{R}^2$, referred to as the field domain in the sequel, and define the family Γ_D of admissible polygonal configurations in D, by taking all the finite planar graphs γ in $D \cup \partial D$, with straight-line segments as edges, such that

- The edges of γ do not intersect,
- All the interior vertices of γ (lying in D) are of degree 2,
- All the boundary vertices of γ (lying in ∂D) are of degree 1,
- No two edges of γ are colinear.

In other words, γ consists of a finite number of disjoint polygons, possibly nested and chopped off by the boundary. We shall write $\Gamma_D[k] \subset \Gamma_D$ for the set of all admissible polygonal configurations in D with precisely k edges.

For a Borel subset of $A \subseteq \mathbb{R}^2$ by $[\![A]\!]$ we shall denote the family of all straight lines hitting A so that in particular $[\![\mathbb{R}^2]\!]$ stands for the collection of all straight lines in $\mathbb{R}^2$. Further, we let μ be the standard isometry-invariant Haar-Lebesgue measure on the space $[\![\mathbb{R}^2]\!]$ of straight lines in $\mathbb{R}^2$. Recall that one possible construction of μ goes by identifying a straight line l with the pair $(\phi, \rho) \in [0, \pi) \times \mathbb{R}$, where $(\rho \sin(\phi), \rho \cos(\phi))$ is the vector orthogonal to l, and joining it to the origin, and then by endowing the parameter space $[0, \pi) \times \mathbb{R}$ with the usual Lebesgue measure. Note that the above parametrisation of $[\![\mathbb{R}^2]\!]$ with $[0, \pi) \times \mathbb{R}$ endows $[\![\mathbb{R}^2]\!]$ with a natural metric, topology and Borel σ-field which will be used in this paper.

On $[\![D]\!] \times D$ we consider a non-negative bounded *local activity function* $\mathcal{M}(\cdot\,;\cdot)$ which will determine the local activity structure of the polygonal field. Define the formal Hamiltonian $L^{\mathcal{M}} : \Gamma_D \to \mathbb{R}_+$ given by

$$L^{\mathcal{M}}(\gamma) := \sum_{e \in \mathrm{Edges}(\gamma)} \int_{l \in [\![e]\!]} \mathcal{M}(l; l \cap e)\mu(dl), \quad \gamma \in \Gamma_D. \tag{15.1}$$

We note that the energy function $L^{\mathcal{M}}$ should be regarded as an anisotropic environment-specific version of the length functional. Indeed, for a line l hitting a graph edge $e \in \mathrm{Edges}(\gamma)$ at their intersection point $x = l \cap e$, the local activity $\mathcal{M}(l; l \cap e)$ shall be interpreted as the likelihood of a new edge being created along l intersecting and hence fracturing at x the edge e in γ. Under this interpretation we see that, roughly speaking, the value of $\int_{l \in [\![e]\!]} \mathcal{M}(l; l \cap e)\mu(dl)$ determines how likely the edge e is to be fractured by another edge present in the environment. In other words, $L^{\mathcal{M}}(\gamma)$ determines *how difficult it is to maintain* the whole graph $\gamma \in \Gamma_D$ without fractures in the environment whose local activity profile is characterised by $\mathcal{M}(\cdot\,;\cdot)$—note that due to the anisotropy of the environment there may be graphs of a higher (lower) total edge length than γ and yet of lower (higher) energy and thus easier (more difficult) to maintain and to keep unfractured due to the lack (presence) of high local activity lines likely to fracture their edges. In the particular case where $\mathcal{M}$ is constant, $L^{\mathcal{M}}$ is readily verified to be a multiple of the usual length functional, see e.g. p. 554 in [11].

We assume that a measurable *anchor mapping* $\mathbb{A} : [\![D]\!] \to D$ is given on the set of lines crossing D, assigning to each of them its *anchor point*, also interpreted as the *initial point* of the line. This allows us to define for each bounded linear segment/graph edge e in D its initial point $\iota[e]$ which is the point of e closest to the anchor $\mathbb{A}(l[e])$, where $l[e]$ is the straight line extending e. In particular, if $\mathbb{A}(l[e]) \in e$ then $\iota[e] = \mathbb{A}(l[e])$, otherwise $\iota[e]$ is the endpoint of e closest to $\mathbb{A}(l[e])$.

The polygonal Markov field $\mathcal{A}_D^{\mathcal{M}}$ with local activity function $\mathcal{M}$ in D is defined by

$$\begin{aligned} &\mathbb{P}(\mathcal{A}_D^{\mathcal{M}} \in d\gamma) \\ &\quad \propto \exp(-L^{\mathcal{M}}(\gamma)) \prod_{e \in \mathrm{Edges}(\gamma)} [\mathcal{M}(l[e]; \iota[e])\mu(dl[e])], \quad \gamma \in \Gamma_D. \end{aligned} \tag{15.2}$$

In other words, the probability of having $\mathcal{A}_D^{\mathcal{M}} \in d\gamma$ is proportional to the Boltzmann factor $\exp(-L^{\mathcal{M}}(\gamma))$ times the product of local edge activities $\mathcal{M}(l[e];$

$\iota[e])\mu(dl[e])$, $e \in \mathrm{Edges}(\gamma)$. Observe that this construction should be regarded as a specific version of the general polygonal model given by Arak and Surgailis [11, 2.11] and an extension of the non-homogeneous polygonal fields considered in Schreiber [379] at their consistent regime (inverse temperature parameter fixed to 1). It should be also noted at this point that if the typical edge length for $\mathscr{A}_D^{\mathscr{M}}$ is much smaller than the characteristic scale for oscillations of $\mathscr{M}$, which is often the case in our applications below, then $\mathscr{M}(l[e]; \cdot)$ is usually approximately constant along the corresponding edge e and the formal dependency of the factor $\mathscr{M}(l[e]; \iota[e])\mu(dl[e])$ on the choice of initial segment for e becomes negligible in large systems. The finiteness of the partition function

$$\mathscr{Z}_D^{\mathscr{M}} := \sum_{k=0}^{\infty} \frac{1}{k!} \int_{\Gamma_D[k]} \exp(-L^{\mathscr{M}}(\gamma)) \prod_{e \in \mathrm{Edges}(\gamma)} [\mathscr{M}(l[e]; \iota[e])\mu(dl[e])] \tag{15.3}$$

is not difficult to verify, see [379], and in fact it will be explicitly calculated in the sequel.

The so-defined locally specified polygonal fields enjoy a number of striking features inherited from the previously developed polygonal models, see [11, 379]. One of these is the two-dimensional germ-Markov property stating that the conditional behaviour of the field $\mathscr{A}_D^{\mathscr{M}}$ inside a smooth closed curve θ depends on the outside field configuration only through the trace it leaves on θ, consisting of intersection points and the respective line directions, see [11] for details. This is where the term *polygonal Markov field* comes from. Further properties of the locally defined polygonal fields are going to be discussed in the next section, where their algorithmic construction is provided.

15.3 Dynamic Representation for Locally Specified Polygonal Fields

The present section is meant to extend the so-called *generalised dynamic representation* for consistent polygonal fields as developed in Schreiber [379] to cover the more general class of locally specified polygonal fields defined in Sect. 15.2 above. The name *generalised representation* comes from the fact that it generalises the original construction of homogeneous polygonal fields introduced by Arak and Surgailis [11]. In the sequel we will often omit the qualifier *generalised* for the sake of terminological brevity. To describe the generalised representation, fix the convex field domain D and let $(D_t)_{t\in[0,1]}$ be a time-indexed increasing family of compact convex subsets of $\bar{D}$, eventually covering the entire $\bar{D}$ and interpreted as a *growing window* gradually revealing increasing portions of the polygonal field under construction in the course of the time flow. In other words, under this interpretation, the portion of a polygonal field in a bounded open convex domain D *uncovered* by time t is precisely its intersection with D_t. To put it in formal terms, consider $(D_t)_{t\in[0,1]}$ satisfying

(D1) $(D_t)_{t \in [0,1]}$ is a strictly increasing family of compact convex subsets of $\bar{D} = D \cup \partial D$.
(D2) D_0 is a single point x in $\bar{D} = D \cup \partial D$.
(D3) D_1 coincides with $\bar{D}$.
(D4) D_t is continuous in the usual Hausdorff metric on compacts.

Clearly, under these conditions, for μ-almost each $l \in [\![D]\!]$ the intersection $l \cap D_{\tau_l}$ consists of precisely one point $\mathbb{A}(l)$, where $\tau_l = \inf\{t \in [0,1],\ D_t \cap l \neq \emptyset\}$. The point $\mathbb{A}(l)$ is chosen to be the *anchor point* for l, which induces the *anchor* mapping $\mathbb{A} : [\![D]\!] \to D$ as required for our construction in Sect. 15.2. Note that this choice of the anchor mapping implies that at each point of a line l the direction away from its anchor point $\mathbb{A}(l)$ coincides with the outwards direction with respect to the *growing window* (D_t). Consider now the following dynamics in time $t \in [0,1]$, with all updates, given by the rules below, performed independently of each other, see Fig. 15.1.

(*GE:Initialise*) Begin with empty field at the time 0.
(*GE:Unfold*) Between critical moments listed below, during the time interval $[t, t+dt]$ the unfolding field edges in D_t reaching ∂D_t extend straight to $D_{t+dt} \setminus D_t$.
(*GE:BoundaryHit*) When a field edge hits the boundary ∂D, it stops growing in this direction (note that μ-almost everywhere the intersection of a line with ∂D consists of at most two points).
(*GE:Collision*) When two unfolding field edges intersect in $D_{t+dt} \setminus D_t$, they are not extended any further beyond the intersection point (stop growing in the direction marked by the intersection point).
(*GE:DirectionalUpdate*) A field edge extending along $l \in [\![D_t]\!]$ updates its direction during $[t, t+dt]$ and starts unfolding along $l' \in [\![l^{[t,t+dt]}]\!]$, extending away from the anchor point $\mathbb{A}(l')$, with probability $\mathscr{M}(l'; l \cap l')\mu(dl')$, where $l^{[t,t+dt]} := l \cap (D_{t+dt} \setminus D_t)$. Directional updates of this type are all performed independently.
(*GE:LineBirth*) Whenever the anchor point $\mathbb{A}(l)$ of a line l falls into $D_{t+dt} \setminus D_t$, the line l is born at the time t at its anchor point with probability $\mathscr{M}(l; \mathbb{A}(l))\mu(dl)$, whereupon it begins extending in both directions with the growth of D_t (recall that l is μ-almost always tangential to ∂D_t here).
(*GE:VertexBirth*) For each intersection point of lines l_1 and l_2 falling into $D_{t+dt} \setminus D_t$, the pair of field lines l_1 and l_2 is born at $l_1 \cap l_2$ with probability $\mathscr{M}(l_1; l_1 \cap l_2)\mathscr{M}(l_2; l_1 \cap l_2)\mu(dl_1)\mu(dl_2)$, whereupon both lines begin unfolding in the directions away from their respective anchor points $\mathbb{A}(l_1)$ and $\mathbb{A}(l_2)$.

Observe that the evolution rule (*GE:VertexBirth*) means that pairs of lines are born at birth sites distributed according to a Poisson point process in D with intensity measure given by the *intersection measure* $\langle\!\langle \mathscr{M} \rangle\!\rangle$ of $\mathscr{M}$:

$$\langle\!\langle \mathscr{M} \rangle\!\rangle(A) := \frac{1}{2} \int_{\{(l_1,l_2),\, l_1 \cap l_2 \subset A\}} \mathscr{M}(l_1; l_1 \cap l_2)\mathscr{M}(l_2; l_1 \cap l_2)\mu(dl_1)\mu(dl_2). \quad (15.4)$$

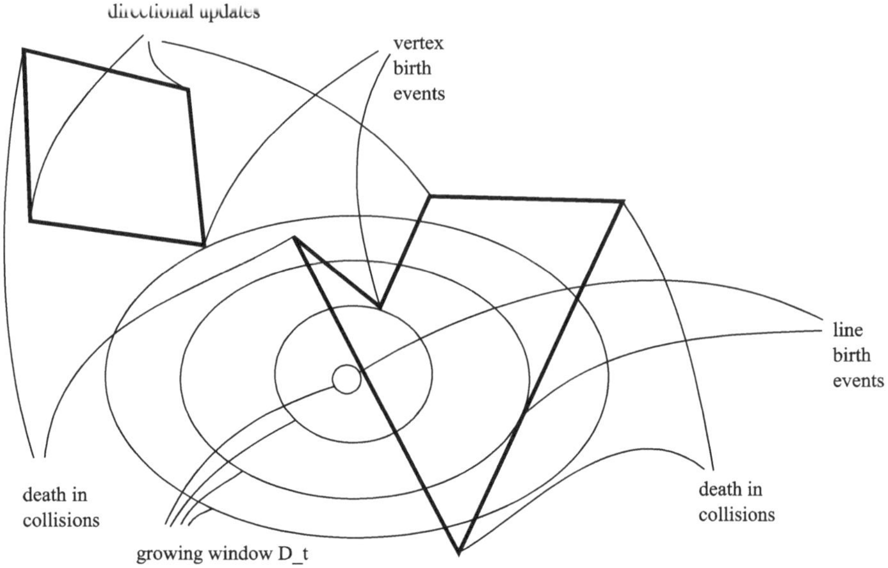

Fig. 15.1 Dynamic representation

Likewise, the evolution rule (*GE:LineBirth*) implies that individual lines are born at their anchor points according to a Poisson point process in $[\![D]\!]$ with intensity measure given by the *anchor measure* $\langle \mathscr{M} \rangle$ of $\mathscr{M}$:

$$\langle \mathscr{M} \rangle (B) := \int_{\{l,\ \mathbb{A}(l) \in B\}} \mathscr{M}(l, \mathbb{A}(l))\mu(dl). \tag{15.5}$$

The main theorem of this section is the following extension of Theorem 3 in [379].

Theorem 15.1 *The random contour ensemble resulting from the above construction (GE) coincides in law with* $\mathscr{A}_D^{\mathscr{M}}$*. Moreover, we have*

$$\log \mathscr{Z}_D^{\mathscr{M}} = \langle\!\langle \mathscr{M} \rangle\!\rangle (D) + \langle \mathscr{M} \rangle ([\![D]\!]). \tag{15.6}$$

Proof We pick some $\gamma \in \Gamma_D$ and calculate the probability that the outcome of the above dynamic construction falls into $d\gamma$. To this end, we note that:

- Each edge $e \in \text{Edges}(\gamma)$ containing the anchor point $\mathbb{A}(l[e])$ and hence resulting from a line birth event due to the rule (*GE:LineBirth*), contributes to the considered probability the factor $\mathscr{M}(l[e]; \mathbb{A}(l[e]))\mu(dl[e])$ (line birth probability for $l[e]$) times $\exp[-\int_{[\![e]\!]} \mathscr{M}(l; l \cap e)\mu(dl)]$ (no directional updates along e).
- Each of the two edges $e_1, e_2 \in \text{Edges}(\gamma)$ stemming from a common interior birth vertex $l[e_1] \cap l[e_2] = \iota[e_1] = \iota[e_2]$ yields the factor $\mathscr{M}(l[e_i]; \iota[e_i])\mu(dl[e_i]),\ i = 1, 2$, (coming from the vertex birth probability due to the rule (*GE:VertexBirth*)) times $\exp[-\int_{[\![e_i]\!]} \mathscr{M}(l; l \cap e_i)\mu(dl)]$ (no directional updates along e_i).

- Each of the edges $e \in \text{Edges}(\gamma)$ arising in (*GE:DirectionalUpdate*) yields the factor $\mathcal{M}(l[e]; \iota[e])$ (directional update probability) times $\exp[-\int_{[\![e]\!]} \mathcal{M}(l; l \cap e)\mu(dl)]$ (no directional updates along e).
- The absence of interior birth sites in $D \setminus \gamma$ yields the factor $\exp[-\langle\!\langle \mathcal{M} \rangle\!\rangle(D)]$.
- Finally, the absence of line birth events for all lines in $[\![D]\!]$ except for the finite collection $\{l[e],\ e \in \text{Edges}(\gamma),\ \mathbb{A}(l[e]) \in e\}$ yields the additional factor $\exp[\langle -\mathcal{M} \rangle([\![D]\!])]$.

Putting these observations together we conclude that the probability element of γ resulting from the generalized construction above is

$$\frac{\exp(-L^{\mathcal{M}}(\gamma)) \prod_{e \in \text{Edges}(\gamma)} [\mathcal{M}(l[e]; \iota[e])\mu(dl[e])]}{\exp[\langle\!\langle \mathcal{M} \rangle\!\rangle(D)] \exp[\langle \mathcal{M} \rangle([\![D]\!])]}$$

and thus, upon comparing with (15.2) and (15.3), the field obtained by this construction coincides in law with $\mathcal{A}_D^{\mathcal{M}}$ as required and (15.6) follows as well. This completes the proof of the theorem. □

15.4 Disagreement Loop Dynamics

In this section we discuss a random dynamics on the space Γ_D of admissible polygonal configurations which leaves the law of the field $\mathcal{A}_D^{\mathcal{M}}$ invariant and reversible. This dynamics will be used in the sequel as a mechanism for update proposal generation in stochastic optimization schemes for image segmentation. We build upon [377, 379] in our presentation of the dynamics based on an important concept of a *disagreement loop*.

To proceed we place ourselves within the context of the dynamic representation discussed in Sect. 15.3 above and suppose that we observe a particular realisation $\gamma \in \Gamma_D$ of the polygonal field $\mathcal{A}_D^{\mathcal{M}}$ and that we modify the configuration by adding an extra (*GE:VertexBirth*) vertex birth site at $x_0 \in D$ to the existing collection of vertex births for γ, while keeping unchanged the remaining evolution rules (*GE*) for all the edges, including the two newly added ones. Denote the resulting new (random) polygonal configuration by $\gamma \oplus x_0$. A simple yet crucial observation is that for $x_0 \in D$ the symmetric difference $\gamma \triangle [\gamma \oplus x_0]$ is almost surely a single loop (a closed polygonal curve), possibly self-intersecting and possibly chopped off by the boundary (becoming a path then). Indeed, this is seen as follows. Each point in $x \in D$ can be attributed its *time coordinate* which is just the time moment at which x is first hit by ∂D_t. Then the chronologically initial point of the loop $\gamma \triangle [\gamma \oplus x_0]$ is of course x_0. Each of the two *new* polygonal curves p_1, p_2 initiated by edges e_1, e_2 emitted from x_0 unfold independently, according to (*GE*), each giving rise to a *disagreement path*. The initial segments of such a disagreement path correspond to the growth of the curve, say p_1, before its annihilation in the first collision. If this is a collision with the boundary, the disagreement path gets chopped off and terminates there. If this is a collision with a segment of the original configuration γ corresponding to a certain *old* polygonal curve p_3 emitted from a prior vertex birth site, the *new* curve p_1 dies but the disagreement path continues along the part of the trajectory of p_3 which is contained in γ but not in $\gamma \oplus x_0$. At some further moment p_3 dies itself in γ, touching the boundary or killing another polygonal curve p_4 in γ. In the second case, however, this collision only happens for γ and not for $\gamma \oplus x_0$ so the polygonal curve p_4 survives (for some time) in $\gamma \oplus x_0$ yielding a further connected portion of the disagreement path initiated by p_1, which is contained in $\gamma \oplus x_0$ but not in γ etc. A recursive continuation of this construction shows that the disagreement path initiated by p_1 at x_0 consists alternately of connected polygonal sub-paths contained in $[\gamma \oplus x_0] \setminus \gamma$ (call these *creation phase* sub-paths) and in $\gamma \setminus [\gamma \oplus x_0]$ (call these *annihilation phase* sub-paths). Note that this disagreement path is self-avoiding and, in fact, it can be represented as the graph of some piecewise linear function $t \mapsto x(t) \in \partial D_t$. Clearly, the same applies for the disagreement path initiated by p_2 at x_0. An important observation is that whenever two *creation phase* or two *annihilation phase* sub-paths of the two disagreement paths hit each other, both disagreement paths die at this point and the disagreement loop closes (as opposed to intersections of segments of different phases which do not have this effect). Obviously, if the disagreement loop does not close in the above way, it gets eventually chopped off by the boundary. We shall write $\Delta^{\oplus}[x_0; \gamma] = \gamma \triangle [\gamma \oplus x_0]$ to denote the (random) disagreement loop constructed above. A similar argument shows that an extra (*GE:LineBirth*) line birth event added for $l \in [\![D]\!]$ at its anchor point $\mathbb{A}(l)$, while keeping the remaining evolution rules unchanged, also gives rise to a disagreement loop $\Delta^{\oplus}[l; \gamma]$ which coincides with the symmetric difference $\gamma \triangle [\gamma \oplus l]$, where $\gamma \oplus l$ is the polygonal configuration resulting from γ upon adding the line birth site at $\mathbb{A}(l)$.

Likewise, a disagreement loop arises if we *remove* one vertex birth site $x_0 \in D$ from the collection of vertex birth sites of an admissible polygonal configuration $\gamma \in \Gamma_D$, while keeping the remaining evolution rules. We write $\gamma \ominus x_0$ for the configuration obtained from γ by removing x_0 from the list of vertex birth sites, while the resulting random disagreement loop is denoted by $\Delta^{\ominus}[x_0; \gamma]$ so that $\Delta^{\ominus}[x_0; \gamma] = \gamma \triangle [\gamma \ominus x_0]$. In full analogy, we define $\gamma \ominus l$ and $\Delta^{\ominus}[l; \gamma]$ where $l = l[e]$ is the field line extending an edge $e \in \text{Edges}(\gamma)$ with $\mathbb{A}(l) \in e$ and $\gamma \ominus l$ is the configuration obtained from γ upon killing the line l at its anchor $\mathbb{A}(l)$ whereas $\Delta^{\ominus}[x_0; \gamma]$ is the resulting disagreement loop. We refer the reader to Sect. 2.1 in [377] for further discussion.

With the above terminology we are in a position to describe a random dynamics on the configuration space Γ_D, which leaves invariant the law of the polygonal process $\mathscr{A}_D^{\mathscr{M}}$. Particular care is needed, however, to distinguish between the notion of time considered in the dynamic representation of the field as well as throughout the construction of the disagreement loops above, and the notion of time to be introduced for the random dynamics on Γ_D constructed below. To make this distinction clear we shall refer to the former as to the *representation time* (r-time for short) and shall reserve for it the notation t, while the latter will be called the *simulation time* (s-time for short) and will be consequently denoted by s in the sequel.

Consider the following pure jump birth and death type Markovian dynamics on Γ_D, with $\gamma_s = \gamma_s^D$ standing for the current configuration

(*DL:Birth*) With intensity $\langle\langle \mathcal{M} \rangle\rangle(dx)ds$ for $x \in D$ and with intensity $\langle \mathcal{M} \rangle(dl)ds$ for $l \in [\![D]\!]$ set $\gamma_{s+ds} := \gamma_s \oplus x$ and $\gamma_{s+ds} := \gamma_s \oplus l$ respectively.
(*DL:Death*) For each vertex birth site x in γ_s with intensity ds set $\gamma_{s+ds} := \gamma_s \ominus x$. For each line birth site $\mathbb{A}(l[e]) \in e$, $e \in \text{Edges}(\gamma)$ with intensity ds set $\gamma_{s+ds} := \gamma_s \ominus l[e]$.

If none of the above updates occurs we keep $\gamma_{s+ds} = \gamma_s$. It is convenient to perceive the above dynamics in terms of generating random disagreement loops λ and setting $\gamma_{s+ds} := \gamma_s \Delta \lambda$, with the loops of the type $\Delta^{\oplus}[\cdot,\cdot]$ corresponding to the rule (*DL:Birth*) and $\Delta^{\ominus}[\cdot,\cdot]$ to the rule (*DL:Death*).

As a direct consequence of the dynamic representation of the field $\mathscr{A}_D^{\mathcal{M}}$ as developed in Sect. 15.3, we obtain

Theorem 15.2 *The distribution of the polygonal field $\mathscr{A}_D^{\mathcal{M}}$ is the unique invariant law of the dynamics given by* (DL:Birth) *and* (DL:Death). *The resulting s-time stationary process is reversible. Moreover, for any initial distribution of γ_0 the laws of the polygonal fields γ_s converge in variational distance to the law of $\mathscr{A}_D^{\mathcal{M}}$ as $s \to \infty$.*

The uniqueness and convergence statements in the above theorem require a short justification. They both follow by the observation that, in finite volume, regardless of the initial state, the process γ_s spends a non-null fraction of time in the empty state (no polygonal contours). Indeed, this observation allows us to conclude the required uniqueness and convergence by a standard coupling argument, e.g. along the lines of the proof of Theorem 1.2 in [289].

15.5 Adaptive Optimization Scheme for Image Processing

To provide a formal description of our image segmentation procedure we represent the image processed by a continuously differentiable function $\phi : D \to [-1,1]$ defined on an open bounded convex image domain D. By segmentations of ϕ we shall understand admissible polygonal configurations $\gamma \in \Gamma_D$. Interpreting the contours of γ as curves separating regions of different signs in D we associate with γ two natural sign-functions $s_\gamma^+ : D \to \{-1,+1\}$ and $s_\gamma^- = -s_\gamma^+$. The quality of a segmentation is quantified in terms of an *energy function* $\mathscr{H}[\gamma] := \mathscr{H}[\gamma|\phi]$ which in our case is a positive linear combination of a L_1-type distance (multiple of pixel misclassification ratio), the length element and the number of edges, that is to say

$$\mathscr{H}[\gamma] := \alpha_2 \min\left(\int_D |\phi(x) - s_\gamma^+(x)|dx, \int_D |\phi(s) - s_\gamma^-(x)|dx\right)$$
$$+ \alpha_1 \,\text{length}(\gamma) + \alpha_0 \,\text{card}(\text{Edges}(\gamma)), \quad \alpha_i > 0,\ i = 0,1,2, \quad (15.7)$$

although clearly many other natural options are also possible, such as L_p-type metrics or various weighed versions thereof. Our optimization scheme (*OPT*) presented below is based on the (*DL*) evolution as described in Sect. 15.4 above, combined with the following ideas.

- The initial local activity function encodes the gradient information for ϕ.
- In the course of the dynamics, the local activity function undergoes adaptive updates in the spirit of the celebrated Chen algorithm, see [73] and 10.2.4.c in [333].
- The segmentation update proposals are accepted or rejected depending on the energy changes they induce, conforming to the simulated annealing scheme, see [1] for a general reference.

At each time moment $s \geq 0$ in the course of the (OPT) dynamics the local activity function $\mathscr{M}(\cdot\,;\cdot)$ is given by

$$\mathscr{M}_s(l; x) := |\mathbf{e}[l] \times \mathbf{G}_s(x)|, \tag{15.8}$$

where $\mathbf{e}[l]$ is a unit vector along l and $\times$ stands for the usual vector cross product. The vector field $\mathbf{G}_s$ evolves in (OPT) time together with the polygonal configuration γ_s as specified below, with the initial condition

$$\mathbf{G}_0(x) := \nabla\phi(x), \tag{15.9}$$

for practical reasons possibly modified by convolving ϕ with a small variance Gaussian kernel at the pre-processing stage. When combined, the relations (15.8) and (15.9) mean that we promote edges in directions perpendicular to local gradients and proportionally to the gradient lengths. In precise terms, our algorithm admits a description in terms of the following (non-homogeneous) pure-jump Markovian dynamics (OPT) unfolding in time $s \geq 0$.

$(OPT{:}Initialise)$ At time 0 set the initial activity function $\mathscr{M}_0(\cdot\,;\cdot)$ as specified by (15.8) and (15.9) and generate γ_0 according to $\mathscr{A}_D^{\mathscr{M}_0}$.

$(OPT{:}Birth)$ For $\mathscr{M}_s$ given as in (15.8), with intensity $\langle\langle \mathscr{M}_s \rangle\rangle(dx)ds$ for $x \in D$ and with intensity $\langle \mathscr{M}_s \rangle(dl)$ for $l \in [\![D]\!]$ do

[*GenerateDisagreementLoop*] Set $\delta := \gamma_s \oplus x$ and $\delta := \gamma_s \oplus l$ respectively, with λ standing for the respective disagreement loop $\Delta^{\oplus}[x;\gamma]$ or $\Delta^{\oplus}[l;\gamma]$ and with λ^+ and λ^- denoting its respective creation and annihilation phase sub-paths. Note that the disagreement loop is generated according to the current activity measure $\mathscr{M}_s$. Let $\Delta := \mathscr{H}(\delta) - \mathscr{H}(\gamma_s)$ be the energy difference between the current configuration γ_s and its update proposal δ.

[*ActivityUpdate*] Put

$$\begin{aligned}\mathbf{G}_{s+ds}(x) := \mathbf{G}_s(x) & \\ &+ \frac{\exp(-K_s\Delta) - 1}{2\pi\sigma_s^2}\int_{\lambda^+} \mathbf{n}[y]\langle \mathbf{n}[y], \mathbf{G}_s(x)\rangle \exp\left(-\frac{\operatorname{dist}^2(x,y)}{2\sigma_s^2}\right)dy \\ &+ \frac{\exp(K_s\Delta) - 1}{2\pi\sigma_s^2}\int_{\lambda^-} \mathbf{n}[y]\langle \mathbf{n}[y], \mathbf{G}_s(x)\rangle \exp\left(-\frac{\operatorname{dist}^2(x,y)}{2\sigma_s^2}\right)dy,\end{aligned}$$

where $\mathbf{n}[y]$ stands for the unit normal to λ at $y \in \lambda$, defined almost everywhere; whereas K_s and σ_s are positive deterministic parameter functions discussed in more detail below.

[*ConfigurationUpdate*] If $\Delta < 0$ then set $\gamma_{s+ds} := \delta$. Otherwise set $\gamma_{s+ds} := \delta$ with probability $\exp(-\beta_s \Delta)$ (*accept update*) and keep $\gamma_{s+ds} = \gamma_s$ with the complementary probability (*reject update*). The parameter function β_s, referred to as the inverse temperature according to the usual terminology, increases in time following the cooling protocol of our simulated annealing.

(*OPT:Death*) With $\mathscr{M}_s$ as given by (15.8), for each vertex birth site x in γ_s with activity ds, and for each line birth site $\mathbb{A}(l[e]) \in e$, $e \in \text{Edges}(\gamma)$ with intensity ds, do

[*GenerateDisagreementLoop*] Set $\delta := \gamma_s \ominus x$ and $\delta := \gamma_s \ominus l[e]$ respectively, with λ standing for the respective disagreement loop $\Delta^{\ominus}[x;\gamma]$ or $\Delta^{\ominus}[l[e];\gamma]$ and with λ^+ and λ^- denoting its respective creation and annihilation phase sub-paths. Note that the disagreement loop is generated according to the current activity measure $\mathscr{M}_s$. Let $\Delta := \mathscr{H}(\delta) - \mathscr{H}(\gamma_s)$ be the energy difference between the current configuration γ_s and its update proposal δ.

[*ActivityUpdate*] Put

$$\begin{aligned}\mathbf{G}_{s+ds}(x) &:= \mathbf{G}_s(x)\\ &+ \frac{\exp(-K_s\Delta)-1}{2\pi\sigma_s^2}\int_{\lambda^-}\mathbf{n}[y]\langle\mathbf{n}[y],\mathbf{G}_s(x)\rangle\exp\left(-\frac{\text{dist}^2(x,y)}{2\sigma_s^2}\right)dy\\ &+ \frac{\exp(K_s\Delta)-1}{2\pi\sigma_s^2}\int_{\lambda^+}\mathbf{n}[y]\langle\mathbf{n}[y],\mathbf{G}_s(x)\rangle\exp\left(-\frac{\text{dist}^2(x,y)}{2\sigma_s^2}\right)dy.\end{aligned}$$

[*ConfigurationUpdate*] If $\Delta < 0$ then set $\gamma_{s+ds} := \delta$. Otherwise set $\gamma_{s+ds} := \delta$ with probability $\exp(-\beta_s \Delta)$ (*accept update*) and keep $\gamma_{s+ds} = \gamma_s$ with the complementary probability (*reject update*).

Roughly speaking, our optimization dynamics (*OPT*) generates successive updates according to the disagreement loop dynamics (*DL*) driven by the current local activity function $\mathscr{M}_s$, whereupon it updates the activity function in the spirit of the Chen algorithm in [*ActivityUpdate*] phase, and then accepts or rejects the configuration update proposal for γ_s in [*ConfigurationUpdate*] conforming to the simulated annealing paradigm. Note that the activity update is carried out regardless of whether the configuration update proposal has been accepted or not. This is natural because in the activity update step the original and new configuration are compared for quality and then, along the disagreement segments present in the better of the two configurations, the normal component of the local gradient field is reinforced and, likewise, the normal component is subject to fading along the disagreement segments present in the worse configuration. The strength of this reinforcement/fading depends exponentially on the energy difference between the original configuration and its update, with rate controlled by time-dependent parameter K_s, which should increase over time starting from a low level to avoid erratic reinforcements induced by the initially chaotic nature of the early stage polygonal configurations γ_s. To keep the local activity function smooth we smear the activity updates over the domain by convolving them with a Gaussian kernel of time-dependent standard deviation parameter σ_s, as made precise in the [*ActivityUpdate*] formulae above. The

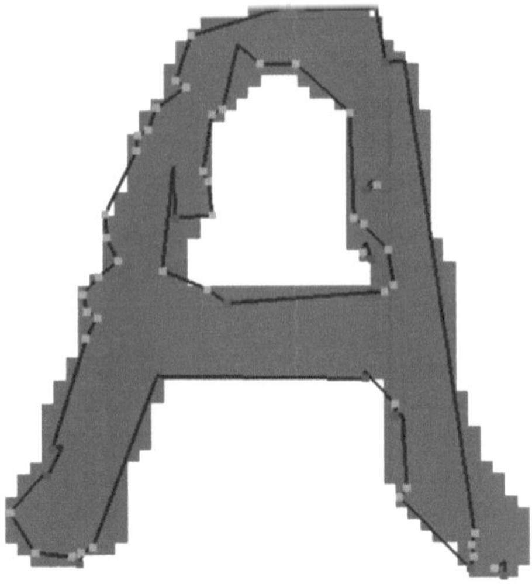

Fig. 15.2 Segmented handwritten A (30000 updates)

parameter σ_s should decrease over time to pass from global shape approximation to fine detail tuning at the later stages of the (OPT) dynamics. The update proposals for the polygonal configurations are accepted or rejected according to the standard simulated annealing scheme with time-dependent inverse temperature parameter β_s which increases over time—to be precise, our software employs a linear cooling schedule $\beta_s = \beta s$ for some constant $\beta > 0$.

15.6 Results and Discussion

In this final section we present applications of our algorithm on sample images. The software, implemented in D programming language, is in a rather early stage of development and will be further optimised. The segmentations shown in Figs. 15.2, 15.3 and 15.4 have been obtained after about 30000 (accepted) updates under a linear cooling schedule, with mean execution time 0.05 sec per single update on Intel Pentium M 2 GHz CPU and 2 GB RAM memory.

A large number of segmentation techniques are available in the literature. But, there does not exist a general algorithm that can perform the segmentation task for all images. Classification of image segmentation methods can be divided into several categories. Starting with the simplest one, the thresholding method, that uses a global property of the image, usually intensity, to classify individual pixels from the image as object pixels, if the value of the pixels property exceeds threshold value, or as background pixels otherwise. The main disadvantage of the method is a narrow range of application, because it works only for a subclass of images in which objects are distinct from background in intensity. The adjustment of the threshold parameter is also a nontrivial task and often requires human interaction. Another well known method is the K-means algorithm [293]. It is an unsupervised clustering algorithm that classifies the pixels from the image into multiple classes based on their inherent

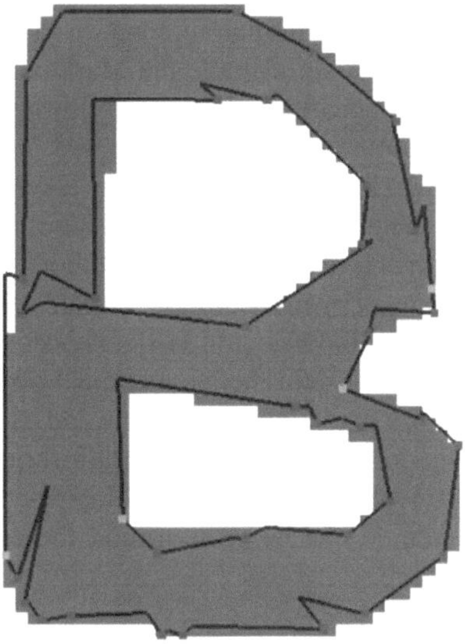

Fig. 15.3 Segmented handwritten B (30000 updates)

Fig. 15.4 Segmented gingerbread-man (30000 updates)

distance from each other. For small values of k the algorithm gives good results, but for larger values of k, the segmentation is very coarse, many clusters appear in the images at discrete places. Selection of parameter k is crucial in that algorithm and inappropriate choice may yield wrong results.

One of the most popular methods in segmentation uses morphological approach: the watershed transformation. That approach was introduced in [34] and consists of placing a spring of water in each selected region, the water will relief from sources, and construct barriers when water from different sources meet. The resulting barriers are the segmentation of the image. The main disadvantage of the algorithm is the need of human interaction to locate points where flooding should start. Another group of algorithms: the Markov random fields (MRF)-based methods are of great importance, for their ability to model a prior belief about the continuity of image features such as textures, edges or region labels [457], but obtain unsatisfied results when the prior knowledge is taken seriously.

The approach described here uses models that operate on the pixel level. Alternative intermediate level methods focus on the partition of the image that is the outcome of a segmentation. Green [190] and Møller and Skare [313] propose Voronoi-based models, and [318] suggests triangulations. One of the main advantage of our method is a higher conceptual level than most of listed algorithms i.e. the real world is not a collection of pixels and as we do not know what is in the image we cannot model the objects. The algorithm achieves reasonable global behaviour. Another benefit from our algorithm is easy and fast implementation. The drawbacks of our method can be seen around the edges, which will require more finetuning in the future.

Acknowledgements We gratefully acknowledge the support from the Polish Minister of Science and Higher Education grant N N201 385234 (2008–2010).

Chapter 16
Regularization with Approximated L^2 Maximum Entropy Method

Jean-Michel Loubes and Paul Rochet

Abstract We tackle the inverse problem of reconstructing an unknown finite measure μ from a noisy observation of a generalized moment of μ defined as the integral of a continuous and bounded operator Φ with respect to μ. When only a quadratic approximation Φ_m of the operator is known, we introduce the L^2 approximate maximum entropy solution as a minimizer of a convex functional subject to a sequence of convex constraints. Under several assumptions on the convex functional, the convergence of the approximate solution is established and rates of convergence are provided.

16.1 Introduction

A number of inverse problems may be stated in the form of reconstructing an unknown measure μ from observations of generalized moments of μ, i.e., moments y of the form

$$y = \int_{\mathscr{X}} \Phi(x) d\mu(x),$$

where $\Phi : \mathscr{X} \to \mathbb{R}^k$ is a given map. Such problems are encountered in various fields of sciences, time-series analysis, speech processing, spectroscopy, geophysical sciences, image reconstruction, crystallography, see for example [95, 197, 225, 387]. This approach also enables to deal with image denoising issues, which are treated in Chaps. 1 and 4, as well as tomography problems discussed in Chaps. 5 and 11. Recovering the unknown measure μ is generally an ill-posed problem, which turns out to be difficult to solve in the presence of noise, i.e., one observes y^{obs} given by

$$y^{obs} = \int_{\mathscr{X}} \Phi(x) d\mu(x) + \varepsilon. \tag{16.1}$$

J.-M. Loubes (✉) · P. Rochet
Institut de Mathématiques de Toulouse, UMR 5219, Université Toulouse 3,
118 Route de Narbonne, 31062 Toulouse cedex 9, France
e-mail: loubes@math.univ-toulouse.fr

P. Rochet
e-mail: rochet@math.univ-toulouse.fr

L. Florack et al. (eds.), *Mathematical Methods for Signal and Image Analysis and Representation*, Computational Imaging and Vision 41,
DOI 10.1007/978-1-4471-2353-8_16,

For inverse problems with known operator Φ, regularization techniques allow the solution to be stabilized by giving favor to those solutions which minimize a regularizing functional J, i.e., one minimizes $J(\mu)$ over μ subject to the constraint that $\int_{\mathcal{X}} \Phi(x)d\mu(x) = y$ when y is observed, or $\int_{\mathcal{X}} \Phi(x)d\mu(x) \in K_Y$ in the presence of noise, for some convex set K_Y containing y^{obs}. Several types of regularizing functionals have been introduced in the literature. In this general setting, the inversion procedure is deterministic, i.e., the noise distribution is not used in the definition of the regularized solution. Bayesian approaches to inverse problems allow one to handle the noise distribution, provided it is known, yet in general, a distribution like the normal distribution is postulated (see [129] for a survey). However in many real-world inverse problems, the noise distribution is unknown, and only the output y is easily observable, contrary to the input to the operator. Consequently very few paired data is available to reliably estimate the noise distribution, thereby causing robustness deficiencies on the retrieved parameters. Nonetheless, even if the noise distribution is unavailable to the practitioner, she often knows the *noise level*, i.e., the maximal magnitude of the disturbance term, say $\rho > 0$, and this information may be reflected by taking a constraint set K_Y of diameter 2ρ.

As an alternative to standard regularizations such as Tikhonov or Galerkin, see for instance [127], we focus on a regularization functional with grounding in information theory, generally expressed as a negative entropy, leading to *maximum entropy* solutions to the inverse problem. In a deterministic framework, maximum entropy solutions have been studied in [45] and [46], while some others studies exist in a Bayesian setting [178, 179], in seismic tomography [141], in image analysis [197, 387]. Regularization with maximum entropy also provides one with a very simple and natural manner to incorporate constraints on the support and the range of the solution (see e.g. the discussion in [179]).

In many actual situations, however, the map Φ is unknown and only an approximation to it is available, say Φ_m, which converges in quadratic norm to Φ as m goes to infinity. In this paper, following lines devised in [178, 179, 294], we introduce an approximate maximum entropy on the mean (AMEM) estimate $\hat{\mu}_{m,n}$ of the measure μ_X to be reconstructed. This estimate is expressed in the form of a discrete measure concentrated on n points of $\mathcal{X}$. In our main result, we prove that $\hat{\mu}_{m,n}$ converges to the solution of the initial inverse problem as $m \to \infty$ and $n \to \infty$ and provide a rate of convergence for this estimate.

The paper is organized as follows. Section 16.2 introduces some notation and the definition of the AMEM estimate. In Sect. 16.3, we state our main result (Theorem 16.2). Section 16.4 is devoted to the proofs of our results.

16.2 Notation and Definitions

16.2.1 Problem Position

Let Φ be a continuous and bounded map defined on a subset $\mathcal{X}$ of $\mathbb{R}^d$ and taking values in $\mathbb{R}^k$. The set of finite measures on $(\mathcal{X}, \mathcal{B}(\mathcal{X}))$ will be denoted by $\mathcal{M}(\mathcal{X})$, where $\mathscr{B}(\mathscr{X})$ denotes the Borel σ-field of $\mathscr{X}$. Let $\mu_X \in \mathscr{M}(\mathscr{X})$ be an unknown finite measure on $\mathscr{X}$ and consider the following equation:

$$y = \int_{\mathscr{X}} \Phi(x) d\mu_X(x). \tag{16.2}$$

Suppose that we observe a perturbed version y^{obs} of the response y:

$$y^{obs} = \int_{\mathscr{X}} \Phi(x) d\mu_X(x) + \varepsilon,$$

where ε is an error term supposed bounded in norm from above by some positive constant η, representing the maximal noise level. Based on the data y^{obs}, we aim at reconstructing the measure μ_X with a maximum entropy procedure. As explained in the introduction, the true map Φ is *unknown* and we assume knowledge of an *approximating sequence* Φ_m to the map Φ, such that

$$\|\Phi_m - \Phi\|_{\mathbb{L}^2(P_X)} = \sqrt{\mathbb{E}(\|\Phi_m(X) - \Phi(X)\|^2)} \to 0,$$

at a rate φ_m.

Let us first introduce some notation. For all probability measures ν on $\mathbb{R}^n$, we shall denote by $\mathscr{L}_\nu$, Λ_ν, and Λ_ν^* the Laplace, log-Laplace, and Cramer transforms of ν, respectively defined for all $s \in \mathbb{R}^n$ by:

$$\begin{aligned}
\mathscr{L}_\nu(s) &= \int_{\mathbb{R}^n} \exp\langle s, x\rangle d\nu(x), \\
\Lambda_\nu(s) &= \log \mathscr{L}_\nu(s), \\
\Lambda_\nu^*(s) &= \sup_{u \in \mathbb{R}^n} \{\langle s, u\rangle - \Lambda_\nu(u)\}.
\end{aligned}$$

Define the set

$$K_Y = \{y \in \mathbb{R}^k : \|y - y^{obs}\| \leq \eta\},$$

i.e., K_Y is the closed ball centered at the observation y^{obs} and of radius η.

Let $\mathscr{X}$ be a set, and let $\mathscr{P}(\mathscr{X})$ be the set of probability measures on $\mathscr{X}$. For $\nu, \mu \in \mathscr{P}(\mathscr{X})$, the relative entropy of ν with respect to μ is defined by

$$H(\nu|\mu) = \begin{cases} \int_{\mathscr{X}} \log(\frac{d\nu}{d\mu}) d\nu & \text{if } \nu \ll \mu, \\ +\infty & \text{otherwise.} \end{cases}$$

Given a set $\mathscr{C} \in \mathscr{P}(\mathscr{X})$ and a probability measure $\mu \in \mathscr{P}(\mathscr{X})$, an element $\mu^\star$ of $\mathscr{C}$ is called an *I-projection* of μ on $\mathscr{C}$ if

$$H(\mu^\star|\mu) = \inf_{\nu \in \mathscr{C}} H(\nu|\mu).$$

Now we let $\mathscr{X}$ be a locally convex topological vector space of finite dimension. The dual of $\mathscr{X}$ will be denoted by $\mathscr{X}'$. The following two Theorems, due to [87], characterize the entropic projection of a given probability measure on a convex set. For their proofs, see Theorem 3 and Lemma 3.3 in [87], respectively.

Theorem 16.1 *Let μ be a probability measure on $\mathscr{X}$. Let $\mathscr{C}$ be a convex subset of $\mathscr{X}$ whose interior has a non-empty intersection with the convex hull of the support of μ. Let*

$$\Pi(\mathscr{X}) = \left\{P \in \mathscr{P}(\mathscr{X}) : \int_{\mathscr{X}} x dP(x) \in \mathscr{C}\right\}.$$

Then the I-projection $\mu^\star$ of μ on $\Pi(\mathscr{C})$ is given by the relation

$$d\mu^\star(x) = \frac{\exp \lambda^\star(x)}{\int_{\mathscr{X}} \exp \lambda^\star(u) d\mu(u)} d\mu(x),$$

where $\lambda^\star \in \mathscr{X}'$ is given by

$$\lambda^\star = \arg\max_{\lambda \in \mathscr{X}'} \left[\inf_{x \in \mathscr{C}} \lambda(x) - \log \int_{\mathscr{X}} \exp \lambda(x) d\mu(x)\right].$$

Now let ν_Z be a probability measure on $\mathbb{R}_+$. Let P_X be a probability measure on $\mathscr{X}$ having full support, and define the convex functional $I_{\nu_Z}(\mu|P_X)$ by:

$$I_{\nu_Z}(\mu|P_X) = \begin{cases} \int_{\mathscr{X}} \Lambda^*_{\nu_Z}(\frac{d\mu}{dP_X}) dP_X & \text{if } \mu \ll P_X, \\ +\infty & \text{otherwise.} \end{cases}$$

Within this framework, we consider as a solution of the inverse problem (16.2) a minimizer of the functional $I_{\nu_Z}(\mu|P_X)$ subject to the constraint

$$\mu \in S(K_Y) = \left\{\mu \in \mathscr{M}(\mathscr{X}) : \int_{\mathscr{X}} \Phi(x) d\mu(x) \in K_Y\right\}.$$

16.2.1.1 The AMEM Estimate

We introduce the approximate maximum entropy on the mean (AMEM) estimate as a sequence $\hat{\mu}_{m,n}$ of discrete measures on $\mathscr{X}$. In all of the following, the integer m indexes the approximating sequence Φ_m to Φ, while the integer n indexes a random discretization of the space $\mathscr{X}$. For the construction of the AMEM estimate, we proceed as follows.

Let $(X_1, \ldots, X_n)$ be an i.i.d sample drawn from P_X. Thus the empirical measure $\frac{1}{n}\sum_{i=1}^n \delta_{X_i}$ converges weakly to P_X.

Let L_n be the discrete measure with random weights defined by

$$L_n = \frac{1}{n}\sum_{i=1}^n Z_i \delta_{X_i},$$

where $(Z_i)_i$ is a sequence of i.i.d. random variables on $\mathbb{R}$.

For $\mathscr{S}$ a set we denote by $\operatorname{co}\mathscr{S}$ its convex hull. Let $\Omega_{m,n}$ be the probability event defined by

$$\Omega_{m,n} = [K_Y \cap \operatorname{co}\operatorname{Supp} F_* \nu_Z^{\otimes n} \neq \emptyset], \tag{16.3}$$

where $F : \mathbb{R}^n \to \mathbb{R}^k$ is the linear operator associated with the matrix $\mathbf{A}_{m,n} = \frac{1}{n}(\Phi_m^i(X_j))_{(i,j)\in[1,k]\times[1,n]}$ and where $F_*\nu_Z^{\otimes n}$ denotes the image measure of $\nu_Z^{\otimes n}$ by F. For ease of notation, the dependence of F on m and n will not be explicitly written throughout.

Denote by $\mathscr{P}(\mathbb{R}^n)$ the set of probability measures on $\mathbb{R}^n$. For any map $\Psi : \mathscr{X} \to \mathbb{R}^k$ define the set

$$\Pi_n(\Psi, K_Y) = \left\{ \nu \in \mathscr{P}(\mathbb{R}^n) : \mathbb{E}_\nu \left[\int_{\mathscr{X}} \Psi(x) dL_n(x) \right] \in K_Y \right\}.$$

Let $\nu_{m,n}^\star$ be the I-projection of $\nu_Z^{\otimes n}$ on $\Pi_n(\Phi_m, K_Y)$.

Then, on the event $\Omega_{m,n}$, we define the AMEM estimate $\hat{\mu}_{m,n}$ by

$$\hat{\mu}_{m,n} = \mathbb{E}_{\nu_{m,n}^\star}[L_n], \tag{16.4}$$

and we extend the definition of $\hat{\mu}_{m,n}$ to the whole probability space by setting it to the null measure on the complement $\Omega_{m,n}^c$ of $\Omega_{m,n}$. In other words, letting $(z_1, \ldots, z_n)$ be the expectation of the measure $\nu_{m,n}^\star$, the AMEM estimate may be rewritten more conveniently as

$$\hat{\mu}_{m,n} = \frac{1}{n} \sum_{i=1}^{n} z_i \delta_{X_i} \tag{16.5}$$

with $z_i = \mathbb{E}_{\nu_{m,n}^\star}(Z_i)$ on $\Omega_{m,n}$, and as $\hat{\mu}_{m,n} \equiv 0$ on $\Omega_{m,n}^c$. It is shown in [294] that $\mathbb{P}(\Omega_{m,n}) \to 1$ as $m \to \infty$ and $n \to \infty$. Hence for m and n large enough, the AMEM estimate $\hat{\mu}_{m,n}$ may be expressed as in (16.5) with high probability, and asymptotically with probability 1.

Remark 16.1 The construction of the AMEM estimate relies on a discretization of the space $\mathscr{X}$ according to the probability P_X. Therefore by varying the support of P_X, the practitioner may easily incorporate some a-priori knowledge concerning the support of the solution. Similarly, the AMEM estimate also depends on the measure ν_Z, which determines the domain of $\Lambda_{\nu_Z}^*$, and so the range of the solution.

16.3 Convergence of the AMEM Estimate

16.3.1 Main Result

Assumption 16.1 The minimization problem admits at least one solution, i.e., there exists a continuous function $g_0 : \mathscr{X} \to \mathrm{co}\,\mathrm{Supp}\,\nu_Z$ such that

$$\int_{\mathscr{X}} \Phi(x) g_0(x) dP_X(x) \in K_Y.$$

Assumption 16.2

(i) $\mathrm{dom}\,\Lambda_{\nu_Z} := \{s : |\Lambda_{\nu_Z}(s)| < \infty\} = \mathbb{R}$;
(ii) Λ'_{ν_Z} is bounded.

Assumption 16.3 The approximating sequence Φ_m converges to Φ in $L^2(\mathscr{X}, P_X)$. Its rate of convergence is given by

$$\|\Phi_m - \Phi\|_{\mathbb{L}^2} = O(\varphi_m^{-1}).$$

Assumption 16.4 Λ_{ν_Z} is a convex function.

Assumption 16.5 For all m, the components of Φ_m are linearly independent.

Assumption 16.6 Λ'_{ν_Z} and Λ''_{ν_Z} are continuous functions.

We define $\mathscr{C}_b = \{g : \mathscr{X} \to \mathbb{R}, g \textit{ continuous and bounded}\}$.

We are now in a position to state our main result.

Theorem 16.2 (Convergence of the AMEM estimate) *Suppose that Assumptions* 16.1 *and* 16.2 *hold. Let* μ^* *be the minimizer of the functional*

$$I_{\nu_Z}(\mu|P_X) = \int_{\mathscr{X}} \Lambda^*_{\nu_Z}\left(\frac{d\mu}{dP_X}\right) dP_X$$

subject to the constraint $\mu \in S(K_Y) = \{\mu \in \mathscr{M}(\mathscr{X}) : \int_{\mathscr{X}} \Phi(x) d\mu(x) \in K_Y\}$.

- *Then the AMEM estimate* $\hat{\mu}_{m,n}$ *is given by*

$$\hat{\mu}_{m,n} = \frac{1}{n} \sum_{i=1}^{n} \Lambda'_{\nu_Z}(\langle \hat{v}_{m,n}, \Phi_m(X_i)\rangle) \delta_{X_i},$$

where $\hat{v}_{m,n}$ *minimizes on* $\mathbb{R}^k$

$$H_n(\Phi_m, v) = \frac{1}{n} \sum_{i=1}^{n} \Lambda_{\nu_Z}(\langle v, \Phi_m(X_i)\rangle) - \inf_{y \in K_Y} \langle v, y\rangle.$$

- *Moreover, under Assumptions* 16.4, 16.2, *and* 16.3, *it converges weakly to* μ^* *as* $m \to \infty$ *and* $n \to \infty$. *Its rate of convergence is given by*

$$\forall g \in \mathscr{C}_b, \quad \left| \int_{\mathscr{X}} g(d\hat{\mu}_{m,n} - d\mu^*) \right| = \kappa_{m,n} + O_P\left(\frac{1}{\sqrt{n}}\right),$$

where $\kappa_{m,n} = O_P(\varphi_m^{-1})$, *uniformly for all* $n \in \mathbb{N}$.

Remark 16.2 Assumption 16.2(i) ensures that the function $H(\Phi, v)$ in Theorem 16.2 attains its minimum at a unique point $v^\star$ belonging to the interior of its domain. If this assumption is not met, [45, 179] have shown that the minimizers of $I_{\nu_Z}(\mu|P_X)$ over $S(K_Y)$ may have a singular part with respect to P_X.

Proof The first part of the theorem is proved in Theorem 3.1 in [294] using Theorem 16.1. We here focus on the proof of the second part. The rate of convergence of the AMEM estimate depends both on the discretization n and the convergence of the approximated operator m. Hence we consider

$$\begin{aligned}
\hat{v}_{m,\infty} &= \underset{v\in\mathbb{R}^k}{\operatorname{argmin}}\, H(\Phi_m, v) = \underset{v\in\mathbb{R}^k}{\operatorname{argmin}} \left\{ \int_{\mathscr{X}} \Lambda_{\nu_Z}(\langle \Phi_m(x), v\rangle) dP_X - \inf_{y\in K_Y} \langle v, y\rangle \right\}, \\
\hat{\mu}_{m,n} &= \Lambda'_{\nu_Z}(\langle \hat{v}_{m,n}, \Phi_m(\cdot)\rangle) P_n, \\
\hat{\mu}_{m,\infty} &= \Lambda'_{\nu_Z}(\langle \Phi_m(\cdot), \hat{v}_{m,\infty}\rangle) P_X.
\end{aligned}$$

For a given $g \in \mathscr{C}_b$, we have the following upper bound

$$\left| \int_{\mathscr{X}} g(d\hat{\mu}_{m,n} - d\mu^*) \right| \le \left| \int_{\mathscr{X}} g(d\hat{\mu}_{m,n} - d\hat{\mu}_{m,\infty}) \right| + \left| \int_{\mathscr{X}} g(d\hat{\mu}_{m,\infty} - d\mu^*) \right|,$$

where each term must be tackled separately.

First, let us consider $|\int_{\mathscr{X}} g(d\hat{\mu}_{m,n} - d\hat{\mu}_{m,\infty})|$.

$$\begin{aligned}
&\left| \int_{\mathscr{X}} g(d\hat{\mu}_{m,n} - d\hat{\mu}_{m,\infty}) \right| \\
&\quad = \left| \int_{\mathscr{X}} g(\Lambda'_{\nu_Z}(\langle \Phi_m, \hat{v}_{m,n}\rangle) dP_n - \Lambda'_{\nu_Z}(\langle \Phi_m, \hat{v}_{m,\infty}\rangle) dP_X) \right| \\
&\quad \le \left| \int_{\mathscr{X}} g(\Lambda'_{\nu_Z}(\langle \Phi_m, \hat{v}_{m,n}\rangle) - \Lambda'_{\nu_Z}(\langle \Phi_m, \hat{v}_{m,\infty}\rangle)) dP_n \right| \\
&\qquad + \left| \int_{\mathscr{X}} g(\Lambda'_{\nu_Z}(\langle \Phi_m, \hat{v}_{m,\infty}\rangle) dP_n - \Lambda'_{\nu_Z}(\langle \Phi_m, \hat{v}_{m,\infty}\rangle) dP_X) \right|.
\end{aligned}$$

To bound the first term $|\int_{\mathscr{X}} g(\Lambda'_{\nu_Z}(\langle \Phi_m, \hat{v}_{m,n}\rangle) - \Lambda'_{\nu_Z}(\langle \Phi_m, \hat{v}_{m,\infty}\rangle)) dP_n|$, write

$$\left|\int_{\mathscr{X}} g(\Lambda'_{\nu_Z}(\langle \Phi_m, \hat{v}_{m,n}\rangle) - \Lambda'_{\nu_Z}(\langle \Phi_m, \hat{v}_{m,\infty}\rangle))dP_n\right|$$
$$\leq \|g\|_\infty \|\Lambda''_{\nu_Z}\|_\infty \frac{1}{n}\sum_{i=1}^{n}\langle \Phi_m(X_i), \hat{v}_{m,n} - \hat{v}_{m,\infty}\rangle$$
$$\leq \|g\|_\infty \|\Lambda''_{\nu_Z}\|_\infty \|\hat{v}_{m,n} - \hat{v}_{m,\infty}\| \frac{1}{n}\sum_{i=1}^{n}\|\Phi_m(X_i)\|,$$

where we have used Cauchy-Schwarz inequality. Since $(\Phi_m)_m$ converges in $\mathbb{L}^2(P_X)$, it is bounded in $\mathbb{L}^2(P_X)$-norm, yielding that $\frac{1}{n}\sum_{i=1}^{n}\|\Phi_m(X_i)\|$ converges almost surely to $\mathbb{E}\|\Phi_m(X)\| < \infty$. Hence, there exists $K_1 > 0$ such that

$$\left|\int_{\mathscr{X}} g(\Lambda'_{\nu_Z}(\langle \Phi_m, \hat{v}_{m,n}\rangle) - \Lambda'_{\nu_Z}(\langle \Phi_m, \hat{v}_{m,\infty}\rangle))dP_n\right| \leq K_1\|\hat{v}_{m,n} - \hat{v}_{m,\infty}\|.$$

For the second term, we obtain by Assumption 16.2

$$\left|\int_{\mathscr{X}} g(d\hat{\mu}_{m,n} - d\hat{\mu}_{m,\infty})\right| = O_P\left(\frac{1}{\sqrt{n}}\right),$$

uniformly for all $n \in \mathbb{N}$. Hence we get

$$\left|\int_{\mathscr{X}} g(d\hat{\mu}_{m,n} - d\mu^*)\right| \leq K_1\|\hat{v}_{m,n} - \hat{v}_{m,\infty}\| + O_P\left(\frac{1}{\sqrt{n}}\right).$$

The second step is to consider $|\int_{\mathscr{X}} g(d\hat{\mu}_{m,\infty} - d\mu^*)|$ and to follow the same guidelines. So, we get

$$\left|\int_{\mathscr{X}} g(d\hat{\mu}_{m,\infty} - d\mu^*)\right| = \left|\int_{\mathscr{X}} g(\Lambda'_{\nu_Z}(\langle \Phi_m, \hat{v}_{m,\infty}\rangle) - \Lambda'_{\nu_Z}(\langle \Phi, v^*\rangle))dP_X\right|$$
$$\leq \left|\int_{\mathscr{X}} g(\Lambda'_{\nu_Z}(\langle \Phi_m, \hat{v}_{m,\infty}\rangle) - \Lambda'_{\nu_Z}(\langle \Phi_m, v^*\rangle))dP_X\right|$$
$$+ \left|\int_{\mathscr{X}} g(\Lambda'_{\nu_Z}(\langle \Phi_m, v^*\rangle) - \Lambda'_{\nu_Z}(\langle \Phi, v^*\rangle))dP_X\right|.$$

We can write still using Cauchy-Schwarz inequality that

$$\int_{\mathscr{X}} g(x)(\Lambda'_{\nu_Z}(\langle \Phi_m(x), \hat{v}_{m,\infty}\rangle) - \Lambda'_{\nu_Z}(\langle \Phi_m(x), v^*\rangle))dP_X(x)$$
$$\leq \int_{\mathscr{X}} g(x)\Lambda''_{\nu_Z}(\xi)\langle \Phi_m(x), \hat{v}_{m,\infty} - v^*\rangle dP_X(x)$$
$$\leq \|\Lambda''_{\nu_Z}\|_\infty \sqrt{\mathbb{E}(g(X))^2}\sqrt{\mathbb{E}(\|\Phi_m(X)\|^2)}\;\|\hat{v}_{m,\infty} - v^*\|.$$

Hence there exists $K_2 > 0$ such that

$$\left|\int_{\mathscr{X}} g(\Lambda'_{\nu_Z}(\langle \Phi_m, \hat{v}_{m,\infty}\rangle) - \Lambda'_{\nu_Z}(\langle \Phi_m, v^*\rangle))dP_X\right| \leq K_2\|\hat{v}_{m,\infty} - v^*\|.$$

Finally, the last term $|\int_{\mathscr{X}} g(\Lambda'_{\nu_Z}(\langle \Phi_m, v^*\rangle) - \Lambda'_{\nu_Z}(\langle \Phi, v^*\rangle))dP_X|$ can be bounded. Indeed,

$$\begin{aligned}
&\left|\int_{\mathscr{X}} g(x)(\Lambda'_{\nu_Z}(\langle \Phi_m(x), v^*\rangle) - \Lambda'_{\nu_Z}(\langle \Phi(x), v^*\rangle))dP_X(x)\right| \\
&= \left|\int_{\mathscr{X}} g(x)\Lambda''_{\nu_Z}(\xi_x)\langle \Phi_m(x) - \Phi(x), v^*\rangle dP_X(x)\right| \\
&\leq \left|\int_{\mathscr{X}} g(x)\Lambda''_{\nu_Z}(\xi_x)\|\Phi_m(x) - \Phi(x)\|\|v^*\|dP_X(x)\right| \\
&\leq \|v^*\|\|\Lambda''_{\nu_Z}\|_\infty\sqrt{\mathbb{E}(g(X))^2}\sqrt{\mathbb{E}(\|\Phi_m(X) - \Phi(X)\|^2)}.
\end{aligned}$$

Hence there exists $K_3 > 0$ such that

$$\left|\int_{\mathscr{X}} g(\Lambda'_{\nu_Z}(\langle \Phi_m, v^*\rangle) - \Lambda'_{\nu_Z}(\langle \Phi, v^*\rangle))dP_X\right| \leq K_3\|\Phi_m - \Phi\|_{\mathbb{L}^2}.$$

We finally obtain the following bound

$$\begin{aligned}
\left|\int_{\mathscr{X}} g(d\hat{\mu}_{m,n} - d\mu^*)\right| &\leq K_1\|\hat{v}_{m,n} - \hat{v}_{m,\infty}\| + K_2\|\hat{v}_{m,\infty} - v^*\| + K_3\|\Phi_m - \Phi\|_{\mathbb{L}^2} \\
&\quad + O_P\left(\frac{1}{\sqrt{n}}\right).
\end{aligned}$$

Using Lemmas 16.1 and 16.2, we obtain that

$$\begin{aligned}
\|\hat{v}_{m,n} - \hat{v}_{m,\infty}\| &= O_P\left(\frac{1}{\sqrt{n}}\right), \\
\|\hat{v}_{m,\infty} - v^*\| &= O_P(\varphi_m^{-1}),
\end{aligned}$$

uniformly for all $n \in \mathbb{N}$. Finally, we get

$$\left|\int_{\mathscr{X}} g(d\hat{\mu}_{m,n} - d\mu^*)\right| = \kappa_{m,n} + O_P\left(\frac{1}{\sqrt{n}}\right),$$

where $\kappa_{m,n} = O_P(\varphi_m^{-1})$ uniformly for all $n \in \mathbb{N}$, which proves the result. □

16.3.2 Application to Remote Sensing

In remote sensing of aerosol vertical profiles, one wishes to recover the concentration of aerosol particles from noisy observations of the radiance field (i.e., a radiometric quantity), in several spectral bands (see e.g. [175, 176]). More specifically, at a given level of modeling, the noisy observation y^{obs} may be expressed as

$$y^{obs} = \int_{\mathscr{X}} \Phi(x; t^{obs}) d\mu_X(x) + \varepsilon, \tag{16.6}$$

where $\Phi : \mathscr{X} \times \mathscr{T} \to \mathbb{R}^k$ is a given operator, and where t^{obs} is a vector of angular parameters observed simultaneously with y^{obs}. The aerosol vertical profile is a function of the altitude x and is associated with the measure μ_X to be recovered, i.e., the aerosol vertical profile is the Radon-Nikodym derivative of μ_X with respect to a given reference measure (e.g., the Lebesgue measure on $\mathbb{R}$). The analytical expression of Φ is fairly complex as it sums up several models at the microphysical scale, so that basically Φ is available in the form of a computer code. So this problem motivates the introduction of an efficient numerical procedure for recovering the unknown μ_X from y^{obs} and arbitrary t^{obs}.

More generally, the remote sensing of the aerosol vertical profile is in the form of an inverse problem where some of the inputs (namely t^{obs}) are observed simultaneously with the noisy output y^{obs}. Suppose that random points $X_1, \dots, X_n$ of $\mathscr{X}$ have been generated. Then, applying the maximum entropy approach would require the evaluations of $\Phi(X_i, t^{obs})$ each time t^{obs} is observed. If one wishes to process a large number of observations, say (y_i^{obs}, t_i^{obs}), for different values t_i^{obs}, the computational cost may become prohibitive. So we propose to replace Φ by an approximation Φ_m, the evaluation of which is faster in execution. To this aim, suppose first that $\mathscr{T}$ is a subset of $\mathbb{R}^p$. Let $T_1, \dots, T_m$ be random points of $\mathscr{T}$, independent of $X_1, \dots, X_n$, and drawn from some probability measure μ_T on $\mathscr{T}$ admitting a density f_T with respect to the Lebesgue measure on $\mathbb{R}^p$ such that $f_T(t) > 0$ for all $t \in \mathscr{T}$. Next, consider the operator

$$\Phi_m(x, t) = \frac{1}{f_T(t)} \frac{1}{m} \sum_{i=1}^{m} K_{h_m}(t - T_i) \Phi(x, T_i),$$

where $K_{h_m}(\cdot)$ is a symmetric kernel on $\mathscr{T}$ of smoothing sequence h_n. It is a classical result to prove that Φ_m converges to Φ in quadratic norm provided h_m tends to 0 at a suitable rate, which ensures that Assumption 16.3 of Theorem 16.2 is satisfied. Since the T_i's are independent from the X_i, one may see that Theorem 16.2 applies, and so the solution to the approximate inverse problem

$$y^{obs} = \int_{\mathscr{X}} \Phi_m(x; t^{obs}) d\mu_X(x) + \varepsilon,$$

will converge to the solution to the original inverse problem in Eq. (16.6). In terms of computational complexity, the advantage of this approach is that the construction of the AMEM estimate requires, for each new observation (y^{obs}, t^{obs}), the evaluation of the m kernels at t^{obs}, i.e., $K_{h_m}(t^{obs} - T_i)$, the $m \times n$ outputs $\Phi(X_i, T_j)$ for $i = 1, \dots, n$ and $j = 1, \dots, m$ having evaluated once and for all.

16.3.3 Application to Deconvolution Type Problems in Optical Nanoscopy

The Maximum Entropy provides also a natural frame for image denoising for inverse problems, when little is known about the operator. For instance, in optical nanoscopy, the number of photons counted can be expressed using a convolution of $p(x - y, y)$ the probability of recording a photon emission at point y when illuminating point x, with $d\mu(y) = f(y)dy$ the measure of the fluorescent markers.

$$g(x) = \int p(x - y, x) f(y) dy.$$

Here $p(x - y, y) = p(x, y, \phi(x))$. Reconstruction of μ can be achieved using L^2-AMEM technics.

16.4 Technical Lemmas

Recall the following definitions

$$\hat{v}_{m,\infty} = \underset{v \in \mathbb{R}^k}{\operatorname{argmin}}\, H(\Phi_m, v) = \underset{v \in \mathbb{R}^k}{\operatorname{argmin}} \left\{ \int_{\mathscr{X}} \Lambda_{\nu_Z}(\langle \Phi_m(x), v \rangle) d P_X - \inf_{y \in K_Y} \langle v, y \rangle \right\},$$

$$\hat{v}_{m,n} = \underset{v \in \mathbb{R}^k}{\operatorname{argmin}}\, H_n(\Phi_m, v) = \underset{v \in \mathbb{R}^k}{\operatorname{argmin}} \left\{ \frac{1}{n} \sum_{i=1}^{n} \Lambda_{\nu_Z}(\langle v, \Phi_m(X_i) \rangle) - \inf_{y \in K_Y} \langle v, y \rangle \right\},$$

$$v^* = \underset{v \in \mathbb{R}^k}{\operatorname{argmin}}\, H(\Phi, v) = \underset{v \in \mathbb{R}^k}{\operatorname{argmin}} \left\{ \int_{\mathscr{X}} \Lambda_{\nu_Z}(\langle \Phi(x), v \rangle) d P_X(x) - \inf_{y \in K_Y} \langle v, y \rangle \right\}.$$

Lemma 16.1 (Uniform Convergence at a Given Approximation Level m) *For all m, we get*

$$\| \hat{v}_{m,n} - \hat{v}_{m,\infty} \| = O_P\left(\frac{1}{\sqrt{n}} \right).$$

Proof $\hat{v}_{m,n}$ is defined as the minimizer of an empirical contrast function $H_n(\Phi_m, .)$. Indeed, set

$$h_m(v, x) = \Lambda_{\nu_Z}(\langle \Phi_m(x), v \rangle) - \inf_{y \in K_Y} \langle v, y \rangle,$$

hence

$$H(\Phi_m, v) = P_X h_m(v, \cdot).$$

Using a classical theorem from the theory of M-estimation, we get the convergence in probability of $\hat{v}_{m,n}$ towards $\hat{v}_{m,\infty}$ provided that the contrast converges uniformly over every compact set of $\mathbb{R}^k$ towards $H(\Phi_m, \cdot)$ when $n \to \infty$. More precisely Corollary 5.53 in [421] states that if we consider $x \mapsto h_m(v, x)$ a measurable function and $\dot{h}_m$ a function in $L^2(P)$, such that for all v_1 and v_2 in a neighbourhood of v^*

$$|h_m(v_1, x) - h_m(v_2, x)| \leq \dot{h}_m(x)\|v_1 - v_2\|.$$

Moreover if $v \mapsto P h_m(v, \cdot)$ has a Taylor expansion of order at least 2 around its unique minimum v^* and if the Hessian matrix at this point is positive, hence provided $\mathbb{P}_n h_m(\hat{v}_n, \cdot) \leq \mathbb{P}_n h_m(v^*, \cdot) + O_P(n^{-1})$ then

$$\sqrt{n}(\hat{v}_n - v^*) = O_P(1).$$

We want to apply this result to our problem. Let η be an un upper bound for $\|\varepsilon\|$, we set $h_m(v, x) = \Lambda_{\nu_Z}(\langle \Phi_m(x), v\rangle) - \langle v, y^{obs}\rangle - \inf_{\|y - y^{obs}\| \leq \eta}\langle v, y - y_{obs}\rangle$. Now note that $z \mapsto \langle v, z\rangle$ reaches its minimum on $\mathscr{B}(0, \eta)$ at the point $-\eta\frac{v}{\|v\|}$, so

$$h_m(v, x) = \Lambda_{\nu_Z}(\langle \Phi_m(x), v\rangle) - \langle v, y^{obs}\rangle + \eta\|v\|.$$

For all $v_1, v_2 \in \mathbb{R}^k$, we have

$$\begin{aligned}
&|h_m(v_1, x) - h_m(v_2, x)| \\
&= \left|\Lambda_{\nu_Z}(\langle \Phi_m(x), v_1\rangle) - \inf_{y \in K_Y}\langle v_1, y\rangle - \Lambda_{\nu_Z}(\langle \Phi_m(x), v_2\rangle) + \inf_{y \in K_Y}\langle v_2, y\rangle\right| \\
&\leq |\Lambda_{\nu_Z}(\langle \Phi_m(x), v_1\rangle) - \Lambda_{\nu_Z}(\langle \Phi_m(x), v_2\rangle)| + \left|\inf_{y \in K_Y}\langle v_2, y\rangle - \inf_{y \in K_Y}\langle v_1, y\rangle\right| \\
&\leq |\Lambda_{\nu_Z}(\langle \Phi_m(x), v_1\rangle) - \Lambda_{\nu_Z}(\langle \Phi_m(x), v_2\rangle)| \\
&\quad + |\langle v_2 - v_1, y^{obs}\rangle - \eta(\|v_2\| - \|v_1\|)| \\
&\leq \left(\|\Lambda'_{\nu_Z}\|_\infty \|\Phi_m(x)\| + \|y^{obs}\| + \eta\right)\|v_1 - v_2\|.
\end{aligned}$$

Define $\dot{h}_m : x \mapsto \|\Lambda'_{\nu_Z}\|_\infty \|\Phi_m(x)\| + \|y^{obs}\| + \eta$. Since $(\Phi_m)_m$ is bounded in $\mathbb{L}^2(P_X)$, $(\dot{h}_m)_m$ is in $L^2(P_X)$ uniformly with respect to m, which entails that

$$\exists K, \forall m, \quad \int_{\mathscr{X}} \dot{h}_m^2 \, dP_X < K. \tag{16.7}$$

Hence the function $\dot{h}_m$ satisfies the first condition

$$|h_m(v_1, x) - h_m(v_2, x)| \leq \dot{h}_m(x)\|v_1 - v_2\|.$$

Now, consider $H(\Phi_m,\cdot)$ Let $V_{m,v}$ be the Hessian matrix of $H(\Phi_m,\cdot)$ at point v. We need to prove that $V_{m,\hat{v}_{m,\infty}}$ is non negative. Let ∂_i be the derivative with respect to the ith component. Set $v \neq 0$, we have

$$\begin{aligned}
{V_{m,v}}^{ij}(v) &= \partial_i\partial_j H(\Phi_m, v) = \int_{\mathscr{X}} \partial_i\partial_j h_m(v,x)dP_X \\
&= \int_{\mathscr{X}} \Phi_m^i(x)\Phi_m^j(x)\Lambda''_{\nu Z}(\langle\Phi_m(x), v\rangle)dP_X + \eta\,\partial_i\partial_j N(v),
\end{aligned}$$

where let N be $N : v \mapsto \|v\|$.

Hence the Hessian matrix $V_{m,\hat{v}_{m,\infty}}$ of $H(\Phi_m,\cdot)$ at point $\hat{v}_{m,\infty}$ can be split into the sum of the following matrices

$$\begin{aligned}
(M_1)_{ij} &= \int_{\mathscr{X}} \Phi_m^i(x)\Phi_m^j(x)\Lambda''_{\nu Z}(\langle\Phi_m(x), \hat{v}_{m,\infty}\rangle)dP_X, \\
(M_2)_{ij} &= \partial_i\partial_j N(\hat{v}_{m,\infty}).
\end{aligned}$$

Under Assumptions (A3) and (A5), $\Lambda''_{\nu Z}$ is positive and belongs to $L_1(P_X)$ since it is bounded. So we can define $\int_{\mathscr{X}} \Phi_m^i(x)\Phi_m^j(x)\Lambda''_{\nu Z}(\langle\Phi_m(x), \hat{v}_{m,\infty}\rangle)dP_X$ as the scalar product of Φ_m^i and Φ_m^j in the space $\mathbb{L}^2(\Lambda''_{\nu Z}(\langle\Phi_m(\cdot), \hat{v}_{m,\infty}\rangle P_X)$.

M_1 is a Gram matrix, hence using (A6) it is a non negative matrix.

M_2 can be computed as follows. For all $v \in \mathbb{R}^k/\{0\}$, we have

$$\begin{aligned}
N(v) &= \sqrt{\sum_{i=1}^k v_i^2}, \\
\partial_i N(v) &= \frac{v_i}{\|v\|}, \\
\partial_i\partial_j N(v) &= \begin{cases} -\frac{v_i v_j}{\|v\|^3} & \text{if } i \neq j, \\ \frac{\|v\|^2 - v_i^2}{\|v\|^3} & \text{if } i = j. \end{cases}
\end{aligned}$$

Hence for all $a \in \mathbb{R}^k$, we can write

$$\begin{aligned}
a^T M_2 a &= \sum_{1\le i,j\le k} \partial_i\partial_j N(\hat{v}_{m,\infty})a_i a_j \\
&= \sum_{i=1}^k \frac{\|\hat{v}_{m,\infty}\|^2 - \hat{v}_{m,\infty,i}^2}{\|\hat{v}_{m,\infty}\|^3} a_i^2 - \sum_{i\neq j} \frac{\hat{v}_{m,\infty,i}\hat{v}_{m,\infty,j}}{\|\hat{v}_{m,\infty}\|^3} a_i a_j \\
&= \frac{1}{\|\hat{v}_{m,\infty}\|^3}\left(\|\hat{v}_{m,\infty}\|^2 \sum_{i=1}^k a_i^2 - \sum_{i=1}^k a_i^2 \hat{v}_{m,\infty,i}^2\right.
\end{aligned}$$

$$- \sum_{1\le i,j\le k} a_i \hat{v}_{m,\infty,i} a_j \hat{v}_{m,\infty,j} + \sum_{i=1}^{k} a_i^2 \hat{v}_{m,\infty,i}^2 \Bigg)$$

$$= \frac{1}{\|\hat{v}_{m,\infty}\|^3} \Bigg(\|\hat{v}_{m,\infty}\|^2 \|a\|^2 - \sum_{1\le i,j\le k} a_i \hat{v}_{m,\infty,i} a_j \hat{v}_{m,\infty,j} \Bigg)$$

$$= \frac{1}{\|\hat{v}_{m,\infty}\|^3} (\|\hat{v}_{m,\infty}\|^2 \|a\|^2 - \langle a, \hat{v}_{m,\infty}\rangle^2)$$

$$\ge 0 \quad \text{using Cauchy-Schwarz's inequality.}$$

So M_2 is clearly non negative, hence $V_{m,\hat{v}_{m,\infty}} = M_1 + \eta M_2$ is also non negative. Finally we conclude that $H(\Phi_m, \cdot)$ undergoes the assumptions of Corollary 5.53 in [421]. □

Lemma 16.2

$$\|\hat{v}_{m,\infty} - v^*\| = O_P(\varphi_m^{-1}).$$

Proof First write,

$$|H(\Phi_m, v) - H(\Phi, v)| = \left| \int_{\mathscr{X}} \Lambda_{\nu_Z}(\langle \Phi_m(x), v\rangle) - \Lambda_{\nu_Z}(\langle \Phi(x), v\rangle) dP_X(x) \right|$$
$$\le \|\Lambda'_{\nu_Z}\|_\infty \|v\| \|\Phi_m - \Phi\|_{\mathbb{L}^2},$$

which implies uniform convergence over every compact set of $H(\Phi_m, \cdot)$ towards $H(\Phi, \cdot)$ when $m \to \infty$, yielding that $\hat{v}_{m,\infty} \to v^*$ in probability. To compute the rate of convergence, we use Lemma 16.3. As previously we can show that the Hessian matrix of $H(\phi, \cdot)$ at point v^* is positive. We need to prove uniform convergence of $\nabla H(\phi_m, \cdot)$ towards $\nabla H(\phi, \cdot)$. For this, write

$$\partial_i [H(\phi_m, \cdot) - H(\phi, \cdot)](v)$$
$$= \int_{\mathscr{X}} \Phi_m^i(x) \Lambda'_{\nu_Z}(\langle \Phi_m(x), v\rangle) - \Phi^i(x) \Lambda'_{\nu_Z}(\langle \Phi(x), v\rangle) dP_X(x)$$
$$= \int_{\mathscr{X}} (\Phi_m^i - \Phi^i)(x) \Lambda'_{\nu_Z}(\langle \Phi_m(x), v\rangle)$$
$$- \Phi^i(x) \Lambda''_{\nu_Z}(\xi) \langle (\Phi - \Phi_m)(x), v\rangle dP_X(x)$$
$$\le \|\Phi^i - \Phi_m^i\|_{\mathbb{L}^2} \|\Lambda'_{\nu_Z}\|_\infty + \|\Phi^i\|_{\mathbb{L}^2} \|\Lambda''_{\nu_Z}\|_\infty \|\Phi - \Phi_m\|_{\mathbb{L}^2} \|v\|$$

using again Cauchy-Schwarz's inequality. Finally we obtain

$$\|\nabla(H(\phi_m, \cdot) - H(\Phi, \cdot))(v)\| \le (C_1 + C_2\|v\|)\ \|\Phi - \Phi_m\|_{\mathbb{L}^2}$$

for positive constants C_1 and C_2. For any compact neighbourhood of v^*, $\mathscr{S}$, the function $v \mapsto \|\nabla(H(\phi_m, \cdot) - H(\Phi, \cdot))(v)\|$ converges uniformly to 0. But for m

large enough, $\hat{v}_{m,\infty} \in \mathscr{S}$ almost surely. Using 2. in Lemma 16.3 with the function $v \mapsto \|\nabla(H(\phi_m, \cdot) - H(\Phi, \cdot))(v)\| \mathbf{1}_{\mathscr{S}}(v)$ converging uniformly to 0, implies that

$$\|\hat{v}_{m,\infty} - v^*\| = O_P(\varphi_m^{-1}).$$

□

Lemma 16.3 *Let f be defined on $\mathscr{S} \subset \mathbb{R}^d \to \mathbb{R}$, which reaches a unique minimum at point θ_0. Let $(f_n)_n$ be a sequence of continuous functions which converges uniformly towards f. Let $\hat{\theta}_n = \operatorname{argmin} f_n$. If f is twice differentiable on a neighbourhood of θ_0 and provided its Hessian matrix V_{θ_0} is non negative, we get*

1. *There exists a positive constant C such that*

$$\|\hat{\theta}_n - \theta_0\| \le C\sqrt{\|f - f_n\|_\infty}.$$

2. *Moreover if $\theta \mapsto V_\theta$ is continuous in a neighbourhood of θ_0 and $\|\nabla f_n(\cdot)\|$ uniformly converges towards $\|\nabla f(\cdot)\|$, there exists a constant C' such that*

$$\|\hat{\theta}_n - \theta_0\| \le C' \|\nabla(f - f_n)\|_\infty$$

with $\|g\|_\infty = \sup_{x \in \mathscr{S}} \|g(x)\|$.

Proof The proof of this classical result in optimization relies on easy convex analysis tricks. For sake of completeness, we recall here the main guidelines.

1. There are non negative constants C_1 and δ_0 such that

$$\forall 0 < \delta \le \delta_0, \qquad \inf_{d(\theta,\theta_0) > \delta} f(\theta) - f(\theta_0) > C_1 \delta^2.$$

   Set $\|f_n - f\|_\infty = \varepsilon_n$. For $0 < \delta_1 < \delta_0$, let n be chosen such that $2\varepsilon_n \le C_1 \delta_1^2$. Hence

$$\inf_{d(\theta,\theta_0) > \delta_1} f_n(\theta) \ge \inf_{d(\theta,\theta_0) > \delta_1} f(\theta) - \varepsilon_n > f(\theta_0) + \varepsilon_n \ge f_n(\theta_0).$$

   Finally $f_n(\theta_0) < \inf_{d(\theta,\theta_0) > \delta_1} f_n(\theta) \Longrightarrow \hat{\theta}_n \in \{\theta : d(\theta, \theta_0) \le \delta_1\}$, which enables to conclude setting $C = \sqrt{\frac{2}{C_1}}$.
2. We prove the result for $d = 1$, which can be easily extended for all d. Using Taylor-Lagrange expansion, there exists $\tilde{\theta}_n \in]\hat{\theta}_n, \theta_0[$ such that

$$f'(\theta_0) = 0 = f'(\hat{\theta}_n) + (\theta_0 - \hat{\theta}_n) f''(\tilde{\theta}_n).$$

   Remind that $f''(\tilde{\theta}_n) \underset{n \to \infty}{\longrightarrow} f''(\theta_0) > 0$. So, for n large enough there exits $C' > 0$ such that

$$|\theta_0 - \hat{\theta}_n| = \frac{|f'(\hat{\theta}_n) - f'(\theta_0)|}{|f''(\tilde{\theta}_n)|} \le C' \|f' - f_n'\|_\infty,$$

   which ends the proof.

□

16.5 Conclusion

Statistical models involving moment conditions are of fundamental importance in inverse problems. While the literature in this domain has focused on inverse problems with complete knowledge of the operator, it appears that many actual situations do not allow the operator to be exactly known by the statistician, whether because of noise in the observations or for computational feasibility reasons. By considering a situation with approximate operator, we extend moment condition models to a more general framework that provides a more realistic formulation of actual problems. This approach enables to have a better idea of the accuracy of regularization methods developed for such problems, and provides an insight on regularity conditions that are necessary for asymptotic efficiency when dealing with an approximate operator.

References

1. Aarts, E., Korst, J.: Simulated Annealing and Boltzmann Machines: A Stochastic Approach to Combinatorial Optimization and Neural Computing. Wiley, New York (1989)
2. Acton, S.T.: Multigrid anisotropic diffusion. IEEE Trans. Image Process. **7**(3), 280–291 (1998)
3. Acton, S.T., Bovik, A.C.: Piecewise and local image models for regularized image restoration using cross-validation. IEEE Trans. Image Process. **8**(5), 652–665 (1999)
4. Adler, R.J.: On excursion sets, tube formulae, and maxima of random fields. Ann. Appl. Probab. **10**(1), 1–74 (2000)
5. Adler, D., Nenadić, O., Zucchini, W.: RGL: a R-library for 3D visualization with OpenGL. In: Braverman, A. (ed.) Proceedings of the 35th Symposium on the Interface: Computing Science and Statistics 2003 Security and Infrastructure Protection, Interface 2003, Salt Lake City, USA, March 12–15, 2003. Computing Science and Statistics, vol. 35. Curran Associates, Rostrevor (2003)
6. Akian, M., Quadrat, J., Viot, M.: Bellman Processes. Lecture Notes in Control and Information Science, vol. 199, pp. 302–311. Springer, Berlin (1994)
7. Altunbasak, Y., Tekalp, M.: Occlusion-adaptive content-based mesh design and forward tracking. IEEE Trans. Image Process. **6**(9), 1270–1280 (1997)
8. Andrews, H.C., Hunt, B.R.: Digital Image Restoration. Prentice-Hall, Englewood Cliffs (1977)
9. Andrey, P., Tarroux, P.: Unsupervised segmentation of Markov random field modeled textured images using selectionist relaxation. IEEE Trans. Pattern Anal. Mach. Intell. **20**(3), 252–262 (1998)
10. Arak, T.: On Markovian random fields with finite number of values. In: Abstracts of Communications of the 4th USSR-Japan Symposium on Probability Theory and Mathematical Statistics, Tbilisi, p. 1 (1982)
11. Arak, T., Surgailis, D.: Markov fields with polygonal realizations. Probab. Theory Relat. Fields **80**(4), 543–579 (1989)
12. Arak, T., Surgailis, D.: Consistent polygonal fields. Probab. Theory Relat. Fields **89**(3), 319–346 (1991)
13. Arak, T., Clifford, P., Surgailis, D.: Point-based polygonal models for random graphs. Adv. Appl. Probab. **25**(2), 348–372 (1993)
14. Arsigny, V., Fillard, P., Pennec, X., Ayache, N.: Log-Euclidean metrics for fast and simple calculus on diffusion tensors. Magn. Reson. Med. **56**(2), 411–421 (2006)
15. Auger, F., Flandrin, P.: Improving the readability of time-frequency and time-scale representations by the reassignment method. IEEE Trans. Signal Process. **43**(5), 1068–1089 (1995)
16. August, J.: The curve indicator random field. PhD thesis, Yale (2001)
17. Aurich, V., Weule, J.: Non-linear Gaussian filters performing edge preserving diffusion. In: Sagerer, G., Posch, S., Kummert, F. (eds.) Mustererkennung 1995. Proceedings of the

L. Florack et al. (eds.), *Mathematical Methods for Signal and Image Analysis and Representation*, Computational Imaging and Vision 41,
DOI 10.1007/978-1-4471-2353-8, © Springer-Verlag London Limited 2012

17th DAGM-Symposium, Bielefeld, Germany, September 13–15, 1995. Informatik Aktuell, pp. 538–545. Springer, Berlin (1995)
18. Baddeley, A., van Lieshout, M.N.M.: Stochastic geometry models in high-level vision. Stat. Images **1**, 231–254 (1993)
19. Bänsch, E., Mikula, K.: A coarsening finite element strategy in image selective smoothing. Comput. Vis. Sci. **1**(1), 53–61 (1997)
20. Barash, D.: A fundamental relationship between bilateral filtering, adaptive smoothing and the nonlinear diffusion equation. IEEE Trans. Pattern Anal. Mach. Intell. **24**(6), 844–847 (2002)
21. Barash, D., Comaniciu, D.: A common framework for nonlinear diffusion, adaptive smoothing, bilateral filtering and mean shift. Image Vis. Comput. **22**(1), 73–81 (2004)
22. Baraud, Y., Huet, S., Laurent, B.: Adaptive tests of linear hypotheses by model selection. Ann. Stat. **31**(1), 225–251 (2003)
23. Barron, J.L., Fleet, D.J., Beauchemin, S.S.: Performance of optical flow techniques. Int. J. Comput. Vis. **12**(1), 43–77 (1994)
24. Basser, P.J., Pajevic, S.: Statistical artefacts in diffusion tensor MRI (DT-MRI) caused by background noise. Magn. Reson. Med. **44**(1), 41–50 (2000)
25. Basser, P.J., Pierpaoli, C.: Microstructural and physiological features of tissues elucidated by quantitative-diffusion-tensor MRI. J. Magn. Reson. **111**(3), 209–219 (1996)
26. Basser, P.J., Mattiello, J., Le Bihan, D.: MR diffusion tensor spectroscopy and imaging. Biophys. J. **66**(1), 259–267 (1994)
27. Basser, P.J., Mattiello, J., Le Bihan, D.: Estimation of the effective self-diffusion tensor from the NMR spin echo. J. Magn. Reson. **103**, 247–254 (1994)
28. Basu, S., Fletcher, T., Whitaker, R.: Rician noise removal in diffusion tensor MRI. In: Larsen, R., Nielsen, M., Sporring, J. (eds.) Proceedings of the 9th International Conference on Medical Image Computing and Computer-Assisted Intervention—MICCAI 2006, Copenhagen, Denmark, October 1–6, 2006. Lecture Notes in Computer Science, vol. 4190–4191, pp. 117–125. Springer, Berlin (2006)
29. Batchelor, P.G., Moakher, M., Atkinson, D., Calamante, F., Connelly, A.: A rigorous framework for diffusion tensor calculus. Magn. Reson. Med. **53**(1), 221–225 (2005)
30. Benjamini, Y., Hochberg, Y.: Controlling the false discovery rate: a practical and powerful approach to multiple testing. J. R. Stat. Soc., Ser. B, Stat. Methodol. **57**(1), 289–300 (1995)
31. Bennett, J., Khotanzad, A.: Multispectral random field models for synthesis and analysis of color images. IEEE Trans. Pattern Anal. Mach. Intell. **20**(3), 327–332 (1998)
32. Bennett, J., Khotanzad, A.: Maximum likelihood estimation methods for multispectral random field image models. IEEE Trans. Pattern Anal. Mach. Intell. **21**(6), 537–543 (1999)
33. Besag, J.: Spatial interaction and the statistical analysis of lattice systems (with discussion). J. R. Stat. Soc. B **36**(2), 192–236 (1974)
34. Beucher, S., Lantuéjoul, C.: Use of watersheds in contour detection. In: Proceedings of the International Workshop on Image Processing, Real-Time Edge and Motion Detection (1979)
35. Bigun, J.: Vision with Direction. Springer, Berlin (2006)
36. Bigün, J., Granlund, G.H.: Optimal orientation detection of linear symmetry. In: Proceedings of the 1st International Conference on Computer Vision, London, UK, June 8–11, 1987, pp. 433–438. IEEE Computer Society, Los Alamitos (1987)
37. Bigun, J., Granlund, G.H., Wiklund, J.: Multidimensional orientation estimation with applications to texture analysis and optical flow. IEEE Trans. Pattern Anal. Mach. Intell. **13**(8), 775–790 (1991)
38. Bijl, P.: Aspects of visual contrast detection. PhD thesis, Utrecht University, Department of Physics, Utrecht, The Netherlands (May 8, 1991)
39. Björck, Å.: Numerical Methods for Least Squares Problems. Society for Industrial and Applied Mathematics, Philadelphia (1996)
40. Bjornemo, M., Brun, A.: White matter fiber tracking diffusion tensor MRI. Master's thesis, Linkøping University, Linkøping (2002)
41. Black, M.J., Sapiro, G., Marimont, D.H., Heeger, D.: Robust anisotropic diffusion. IEEE Trans. Image Process. **7**(3), 421–432 (1998)

42. Blake, A., Zisserman, A.: Visual Reconstruction. The MIT Press Series in Artificial Intelligence. MIT Press, Cambridge (1987)
43. Blomgren, P., Chan, T.F.: Color TV: total variation methods for restoration of vector valued images. IEEE Trans. Image Process. **7**(3), 304–309 (1998)
44. Boissonnat, J.-D., Devillers, O., Pion, S., Teillaud, M., Yvinec, M.: Triangulations in CGAL. Comput. Geom., Theory Appl. **22**(1–3), 5–19 (2002)
45. Borwein, J.M., Lewis, A.S.: Partially-finite programming in L_1 and the existence of maximum entropy estimates. SIAM J. Optim. **3**(2), 248–267 (1993)
46. Borwein, J.M., Lewis, A.S., Noll, D.: Maximum entropy reconstruction using derivative information. I. Fisher information and convex duality. Math. Oper. Res. **21**(2), 442–468 (1996)
47. Bosking, W.H., Zhang, Y., Schofield, B., Fitzpatrick, D.: Orientation selectivity and the arrangement of horizontal connections in tree shrew striate cortex. J. Neurosci. **17**(6), 2112–2127 (1997)
48. Boucher, M., Evans, A.: Dealing with uncertainty in the principal directions of tensors. In: Proceedings of the 2008 Conference on Computer Vision and Pattern Recognition Workshop, CVPRW'08, pp. 1–8. IEEE Computer Society, Los Alamitos (2008)
49. Boucher, M., Evans, A., Siddiqi, K.: Oriented morphometry of folds on surfaces. In: Proceedings of the 21st Conference on Information Processing in Medical Imaging. Lecture Notes in Computer Science, vol. 5636, pp. 614–625. Springer, Berlin (2009)
50. Boucher, M., Whitesides, S., Evans, A.: Depth potential function for folding pattern representation, registration and analysis. Med. Image Anal. **13**(2), 203–214 (2009)
51. Bowman, A.W., Azzalini, A.: Applied Smoothing Techniques for Data Analysis: The Kernel Approach with S-Plus Illustrations. Oxford University Press, Oxford (1997)
52. Boykov, Y., Kolmogorov, V.: An experimental comparison of min-cut/max-flow algorithms for energy minimization in vision. IEEE Trans. Pattern Anal. Mach. Intell. **26**(9), 1124–1137 (2004)
53. Brox, T., Weickert, J., Burgeth, B., Mrázek, P.: Nonlinear structure tensors. Technical report 113, Department of Mathematics, Saarland University, Saarbrücken, Germany (2004)
54. Bryant, R.L., Chern, S.S., Gardner, R.B., Goldschmidt, H.L., Griffiths, P.A.: Exterior Differential Systems. Mathematical Sciences Research Institute Publications, vol. 18. Springer, New York (1991)
55. Buades, A., Coll, B., Morel, J.M.: A review of image denoising algorithms, with a new one. Multiscale Model. Simul. **4**(2), 490–530 (2005)
56. Buades, A., Coll, B., Morel, J.-M.: Nonlocal image and movie denoising. Int. J. Comput. Vis. **76**(2), 123–139 (2008)
57. Burgeth, B., Weickert, J.: An explanation for the logarithmic connection between linear and morphological systems. In: Griffin, L.D., Lillholm, M. (eds.) Scale-Space Methods in Computer Vision: Proceedings of the Fourth International Conference, Scale-Space 2003, Isle of Skye, UK, June 2003. Lecture Notes in Computer Science, vol. 2695, pp. 325–339. Springer, Berlin (2003)
58. Burgeth, B., Didas, S., Weickert, J.: Relativistic scale-spaces. In: Kimmel, R., Sochen, N., Weickert, J. (eds.) Scale Space and PDE Methods in Computer Vision: Proceedings of the 5th International Conference, Scale-Space 2005, Hofgeismar, Germany, April 2005. Lecture Notes in Computer Science, vol. 3459, pp. 1–12. Springer, Berlin (2005)
59. Burgeth, B., Bruhn, A., Didas, S., Weickert, J., Welk, M.: Morphology for matrix-data: ordering versus PDE-based approach. Image Vis. Comput. **25**(4), 496–511 (2007)
60. Burgeth, B., Didas, S., Florack, L., Weickert, J.: A generic approach for singular PDEs for the processing of matrix fields. In: Sgallari, F., Murli, A., Paragios, N. (eds.) Scale Space and Variational Methods in Computer Vision: Proceedings of the 1st International Conference, SSVM 2007, Ischia, Italy, May–June 2007. Lecture Notes in Computer Science, vol. 4485, pp. 556–567. Springer, Berlin (2007)
61. Burgeth, B., Didas, S., Florack, L., Weickert, J.: A generic approach to diffusion filtering of matrix-fields. Computing **81**(2–3), 179–197 (2007)

62. Burgeth, B., Didas, S., Weickert, J.: A general structure tensor concept and coherence-enhancing diffusion filtering for matrix fields. In: Laidlaw, D.H., Weickert, J. (eds.) Visualization and Processing of Tensor Fields. Mathematics and Visualization, pp. 305–323. Springer, Berlin (2009)
63. Burt, P.J., Adelson, E.H.: The Laplacian pyramid as a compact image code. IEEE Trans. Commun. **31**(4), 532–540 (1983)
64. Campbell, J.S.W., Siddiqi, K., Vemuri, B.C., Pike, G.B.: A geometric flow for white matter fiber tract reconstruction. In: Proceedings IEEE International Symposium on Biomedical Imaging: Macro to Nano, pp. 505–508. IEEE Computer Society Press, Los Alamitos (2002)
65. Campilho, A., Kamel, M. (eds.): ICIAR 2004: Proceedings of the 2nd International Conference on Image Analysis and Recognition, Porto, Portugal, September 29–October 1, 2004. Lecture Notes in Computer Science, vols. 3211–3212. Springer, Berlin (2004)
66. Canny, J.: A computational approach to edge detection. IEEE Trans. Pattern Anal. Mach. Intell. **8**, 679–714 (1986)
67. Carr, H.Y., Purcell, E.M.: Effects of diffusion on free precession in nuclear magnetic resonance experiments. Phys. Rev. **94**(3), 630–638 (1954)
68. Catté, F., Lions, P.L., Morel, J.M., Coll, T.: Image selective smoothing and edge detection by nonlinear diffusion. SIAM J. Numer. Anal. **29**(3) (1992)
69. Chassande-Mottin, E., Daubechies, I., Auger, F., Flandrin, P.: Differential reassignment. IEEE Signal Process. Lett. **4**(10), 293–294 (1997)
70. Chaudhuri, P., Marron, J.S.: Scale space view of curve estimation. Ann. Stat. **28**(2), 408–428 (2000)
71. Chefd'Hotel, C., Tschumperlé, D., Deriche, R., Faugeras, O.: Constrained flows on matrix-valued functions: application to diffusion tensor regularization. In: Heyden, A., Sparr, G., Nielsen, M., Johansen, P. (eds.) Proceedings of the 7th European Conference on Computer Vision, Copenhagen, Denmark, May–June 2002. Lecture Notes in Computer Science, vol. 2350–2353, pp. 251–265. Springer, Berlin (2002)
72. Chefd'hotel, C., Tschumperlé, D., Deriche, R., Faugeras, O.: Regularizing flows for constrained matrix-valued images. J. Math. Imaging Vis. **20**(1–2), 147–162 (2004)
73. Chen, K.: Simple learning algorithm for the traveling salesman problem. Phys. Rev. E **55**, 7809–7812 (1997)
74. Cheng, K., Waggoner, R.A., Tanaka, K.: Human ocular dominance columns as revealed by high-field functional magnetic resonance imaging. Neuron **32**(2), 359–374 (2001)
75. Chu, C.K., Glad, I.K., Godtliebsen, F., Marron, J.S.: Edge-preserving smoother for image processing. J. Am. Stat. Assoc. **93**(442), 526–541 (1998)
76. CIBC: Data sets: NCRR Center for Integrative Biomedical Computing (CIBC) data set archive. http://www.sci.utah.edu/cibc/software/login_datasets.html (2008)
77. Cignoni, P., Montani, C., Scopigno, R.: DeWall: a fast divide and conquer Delaunay triangulation algorithm. Comput. Aided Des. **30**(5), 333–341 (1998)
78. Clark, C.A., Le Bihan, D.: Water diffusion compartmentation and anisotropy at high b values in the human brain. Magn. Reson. Med. **44**(6), 852–859 (2000)
79. Clifford, P., Middleton, R.D.: Reconstruction of polygonal images. J. Appl. Stat. **16**, 409–422 (1989)
80. Coifman, R.R., Maggioni, M.: Diffusion wavelets. Appl. Comput. Harmon. Anal. **21**(1), 53–94 (2006)
81. Comaniciu, D., Meer, P.: Mean shift: a robust approach toward feature space analysis. IEEE Trans. Pattern Anal. Mach. Intell. **24**(5), 603–619 (2002)
82. Cottet, G.H., El Ayyadi, M.: A Volterra type model for image processing. IEEE Trans. Image Process. **7**(3), 292–303 (1998)
83. Cottet, G.H., Germain, L.: Image processing through reaction combined with non-linear diffusion. Math. Comput. **61**(204), 659–673 (1993)
84. Coulon, O., Alexander, D.C., Arridge, S.R.: A regularization scheme for diffusion tensor magnetic resonance images. In: Insana, M.F., Leahy, R.M. (eds.) Proceedings of the 17th International Conference on Information Processing in Medical Imaging, IPMI 2001, Davis,

USA, 2001. Lecture Notes in Computer Science, vol. 2082, pp. 92–105. Springer, Berlin (2001)
85. Crandall, M.G., Lions, P.-L.: Viscosity solutions of Hamilton-Jacobi equations. Trans. Am. Math. Soc. **277**(1), 1–42 (1983)
86. Cross, A., Jain, K.: Markov random field texture models. IEEE Trans. Pattern Anal. Mach. Intell. **5**(1), 25–39 (1983)
87. Csiszár, I.: Sanov property, generalized I-projection and a conditional limit theorem. Ann. Probab. **12**(3), 768–793 (1984)
88. Currey, B.N.: Admissibility for a class of quasiregular representations. Can. J. Math. **59**(5), 917–942 (2007)
89. Daley, D.J., Vere-Jones, D.: An Introduction to the Theory of Point Processes: Elementary Theory and Methods. Probability and Its Applications, vol. I. Springer, Berlin (2003)
90. Daley, D.J., Vere-Jones, D.: An Introduction to the Theory of Point Processes: General Theory and Structure. Probability and Its Applications, vol. II. Springer, Berlin (2008)
91. Dana, K.J., Nayar, S.K.: Correlation model for 3D textures. In: Proceedings of the 7th International Conference on Computer Vision, ICCV2007, Corfu, Greece, September 20–27, 1999, pp. 1061–1066. IEEE Computer Society Press, Los Alamitos (1999)
92. Daubechies, I.: Orthonormal bases of compactly supported wavelets. Commun. Pure Appl. Math. **41**(7), 909–996 (1988)
93. Daudet, L., Morvidone, M., Torrésani, B.: Time-frequency and time-scale vector fields for deforming time-frequency and time-scale representations. In: Proceedings of the SPIE Conference on Wavelet Applications in Signal and Image Processing, pp. 2–15. SPIE, Bellingham (1999)
94. De Bonet, J.S.: Multiresolution sampling procedure for analysis and synthesis of textured images. In: ACM SIGGRAPH 97, pp. 361–368. ACM, New York (1997)
95. Decarreau, A., Hilhorst, D., Lemaréchal, C., Navaza, J.: Dual methods in entropy maximization. Application to some problems in crystallography. SIAM J. Optim. **2**(2), 173–197 (1992)
96. Demaret, L., Iske, A.: Adaptive image approximation by linear splines over locally optimal Delaunay triangulations. IEEE Signal Process. Lett. **13**(5), 281–284 (2006)
97. Demaret, L., Dyn, N., Floater, M.S., Iske, A.: Adaptive thinning for terrain modelling and image compression. In: Dodgson, N.A., Floater, M.S., Sabin, M.A. (eds.) Advances in Multiresolution for Geometric Modelling. Mathematics and Visualization, pp. 321–340. Springer, Heidelberg (2005)
98. Demaret, L., Dyn, N., Iske, A.: Image compression by linear splines over adaptive triangulations. Signal Process. **86**(7), 1604–1616 (2006)
99. Dempster, A.P., Laird, N.M., Rubin, D.B.: Maximum likelihood from incomplete data via the EM algorithm. J. R. Stat. Soc. B **39**(1), 1–38 (1977)
100. Descamps, S., Descombes, X., Béchet, A., Zerubia, J.: Automatic flamingo detection using a multiple birth and death process. In: Proceedings of the IEEE International Conference on Acoustics, Speech and Signal Processing, ICASSP, pp. 1113–1116 (2008)
101. Descombes, X., Kruggel, F., Wollny, G., Gertz, H.J.: An object based approach for detecting small brain lesions: application to Virchow-Robin spaces. IEEE Trans. Med. Imaging **23**(2), 246–255 (2004)
102. Descombes, X., Minlos, R., Zhizhina, E.: Object extraction using a stochastic birth-and-death dynamics in continuum. J. Math. Imaging Vis. **33**(3), 347–359 (2009)
103. Descoteaux, M., Angelino, E., Fitzgibbons, S., Deriche, R.: Apparent diffusion coefficients from high angular resolution diffusion imaging: estimation and applications. Magn. Reson. Med. **56**(2), 395–410 (2006)
104. Descoteaux, M., Angelino, E., Fitzgibbons, S., Deriche, R.: Regularized, fast, and robust analytical Q-ball imaging. Magn. Reson. Med. **58**(3), 497–510 (2007)
105. Dieudonné, J.: Treatise on Analysis, vol. 5. Academic Press, New York (1977). Translated by I.G. Macdonald, Pure and Applied Mathematics, vol. 10-V
106. Ding, Z., Gore, J.C., Anderson, A.W.: Reduction of noise in diffusion tensor images using anisotropic smoothing. Magn. Reson. Med. **53**(2), 485–490 (2005)

107. Dong, J., Chantler, M.: Capture and synthesis of 3D surface texture. In: Texture 2002, vol. 1, pp. 41–45. Heriot-Watt University, Edinburgh (2002)
108. Dressel, A.: Die nichtlineare Diffusion in der Bildverarbeitung. Master's thesis, Faculty of Mathematics, University of Heidelberg, Germany (1999)
109. Duits, R.: Perceptual organization in image analysis. PhD thesis, Eindhoven University of Technology, Department of Biomedical Engineering, The Netherlands (2005)
110. Duits, R., Almsick, M.A.: The explicit solutions of linear left-invariant second order stochastic evolution equations on the 2d-Euclidean motion group. Q. Appl. Math. AMS **66**(1), 27–67 (2008)
111. Duits, R., Burgeth, B.: Scale spaces on Lie groups. In: Sgallari, F., Murli, A., Paragios, N. (eds.) Scale Space and Variational Methods in Computer Vision: Proceedings of the 1st International Conference, SSVM 2007, Ischia, Italy, May–June 2007. Lecture Notes in Computer Science, vol. 4485, pp. 300–312. Springer, Berlin (2007)
112. Duits, R., Franken, E.M.: Left invariant parabolic evolution equations on $SE(2)$ and contour enhancement via invertible orientation scores, part II: Nonlinear left-invariant diffusion equations on invertible orientation scores. Q. Appl. Math. **68**, 293–331 (2010)
113. Duits, R., Franken, E.M.: Left invariant parabolic evolution equations on $SE(2)$ and contour enhancement via invertible orientation scores, part I: Linear left-invariant diffusion equations on $SE(2)$. Q. Appl. Math. **68**, 255–292 (2010)
114. Duits, R., Franken, E.M.: Left-invariant diffusions on the space of positions and orientations and their application to crossing-preserving smoothing of HARDI images. Int. J. Comput. Vis. **92**(3), 231–264 (2011)
115. Duits, R., van Almsick, M.: The explicit solutions of linear left-invariant second order stochastic evolution equations on the 2D-Euclidean motion group. Q. Appl. Math. **66**, 27–67 (2008)
116. Duits, R., Florack, L., de Graaf, J., ter Haar Romeny, B.: On the axioms of scale space theory. J. Math. Imaging Vis. **20**(3), 267–298 (2004)
117. Duits, R., Felsberg, M., Granlund, G., ter Haar Romeny, B.M.: Image analysis and reconstruction using a wavelet transform constructed from a reducible representation of the Euclidean motion group. Int. J. Comput. Vis. **72**(1), 79–102 (2007)
118. Duits, R., Führ, H., Janssen, B.J.: Left invariant evolution equations on Gabor transforms. Technical report CASA-report nr. 9, 2009, Department of Mathematics and Computer Science, Eindhoven University of Technology (2009)
119. Duits, R., Führ, H., Janssen, B.J., Bruurmijn, L.C.M.: Left invariant reassignment and diffusion on Gabor transforms. ACHA (2011, submitted)
120. Durot, C., Rozenholc, Y.: An adaptive test for zero mean. Math. Methods Stat. **15**(1), 26–60 (2006)
121. Duval-Destin, M., Muschietti, M., Torrésani, B.: Continuous wavelet decompositions, multiresolution, and contrast analysis. SIAM J. Math. Anal. **24**(3), 739–755 (1993)
122. Dyn, N., Floater, M.S., Iske, A.: Adaptive thinning for bivariate scattered data. J. Comput. Appl. Math. **145**(2), 505–517 (2002)
123. Edelsbrunner, H., Mücke, E.: Simulation of simplicity: a technique to cope with degenerate cases in geometric algorithms. ACM Trans. Graph. **9**(1), 66–104 (1990)
124. Efros, A.A., Freeman, W.T.: Image quilting for texture synthesis and transfer. In: Fiume, E. (ed.) ACM SIGGRAPH 2001, pp. 341–346. ACM, New York (2001)
125. Efros, A.A., Leung, T.K.: Texture synthesis by non-parametric sampling. In: Proceedings of the 7th International Conference on Computer Vision, ICCV2007, Corfu, Greece, September 20–27, 1999, pp. 1033–1038. IEEE Computer Society, Los Alamitos (1999)
126. Elder, J.H., Zucker, S.W.: Local scale control for edge detection and blur estimation. IEEE Trans. Pattern Anal. Mach. Intell. **20**(7), 699–716 (1998)
127. Engl, H.W., Hanke, M., Neubauer, A.: Regularization of Inverse Problems. Mathematics and its Applications, vol. 375. Kluwer Academic, Dordrecht (1996)
128. Evans, L.C.: Partial Differential Equations. Graduate Studies in Mathematics, vol. 19. American Mathematical Society, Providence (2002)
129. Evans, S.N., Stark, P.B.: Inverse problems as statistics. Inverse Probl. **18**(4), 55–97 (2002)

130. Fan, J., Gijbels, I.: Local Polynomial Modelling and Its Applications. Chapman & Hall, London (1996)
131. Feddern, C., Weickert, J., Burgeth, B., Welk, M.: Curvature-driven PDE methods for matrix-valued images. Int. J. Comput. Vis. **69**(1), 91–103 (2006)
132. Felsberg, M.: Wiener channel smoothing: Robust Wiener filtering of images. In: DAGM 2005. Lecture Notes in Computer Science, vol. 3663, pp. 468–475. Springer, Berlin (2005)
133. Felsberg, M.: Extending graph-cut to continuous value domain minimization. In: Proceedings of the 4th Canadian Conference on Computer and Robot Vision, pp. 274–281 (2007)
134. Felsberg, M.: On the relation between anisotropic diffusion and iterated adaptive filtering. In: 30th DAGM Symposium Mustererkennung. Lecture Notes in Computer Science, vol. 5096, pp. 436–445. Springer, Berlin (2008)
135. Felsberg, M.: Spatio-featural scale-space. In: Tai, X.-C., Mørken, K., Lysaker, M., Lie, K.-A. (eds.) Scale Space and Variational Methods in Computer Vision: Proceedings of the 2nd International Conference, SSVM 2009, Voss, Norway, June 1–5, 2009. Lecture Notes in Computer Science, vol. 5567, pp. 808–819. Springer, Berlin (2009)
136. Felsberg, M., Granlund, G.: Anisotropic channel filtering. In: Proceedings of the 13th Scandinavian Conference on Image Analysis. Lecture Notes in Computer Science, vol. 2749, pp. 755–762 (2003)
137. Felsberg, M., Granlund, G.: P-channels: robust multivariate M-estimation of large datasets. In: Proceedings of the 18th International Conference on Pattern Recognition, ICPR'06, Hong Kong, August 20–24, 2006, pp. 262–267 (2006)
138. Felsberg, M., Sommer, G.: The monogenic scale-space: a unifying approach to phase-based image processing in scale-space. J. Math. Imaging Vis. **21**, 5–26 (2004)
139. Felsberg, M., Forssén, P.-E., Scharr, H.: Channel smoothing: efficient robust smoothing of low-level signal features. IEEE Trans. Pattern Anal. Mach. Intell. **28**(2), 209–222 (2006)
140. Felsberg, M., Kalkan, S., Krüger, N.: Continuous dimensionality characterization of image structures. Image Vis. Comput. **27**(6), 628–636 (2009)
141. Fermin, A.K., Loubes, J.M., Ludeña, C.: Bayesian methods in seismic tomography. Int. J. Tomogr. Stat. **4**(6), 1–19 (2006)
142. Fick, A.: Ueber diffusion. Ann. Phys. **170**(1), 59 (1855)
143. Filip, J., Haindl, M.: BTF modelling using BRDF texels. Int. J. Comput. Math. **84**(9), 1267–1283 (2007)
144. Filip, J., Haindl, M., Chetverikov, D.: Fast synthesis of dynamic colour textures. In: Tang, Y., Wang, S., Yeung, D., Yan, H., Lorette, G. (eds.) Proceedings of the 18th International Conference on Pattern Recognition, ICPR 2006, vol. IV, pp. 25–28. IEEE Computer Society, Los Alamitos (2006). http://doi.ieeecomputersociety.org/10.1109/ICPR.2006.550
145. Fillard, P., Arsigny, V., Ayache, N., Pennec, X.: A Riemannian framework for the processing of tensor-valued images. In: Fogh Olsen, O., Florack, L.M.J., Kuijper, A. (eds.) Deep Structure Singularities and Computer Vision. Lecture Notes in Computer Science, vol. 3753, pp. 112–123. Springer, Berlin (2005)
146. Fillard, P., Pennec, X., Arsigny, V., Ayache, N.: Clinical DT-MRI estimation, smoothing, and fiber tracking with log-Euclidean metrics. IEEE Trans. Med. Imaging **26**(11) (2007)
147. Fischl, B., Sereno, M.I., Dale, A.M.: Cortical surface-based analysis. II: Inflation, flattening, and a surface-based coordinate system. Neuroimage **9**(2), 195–207 (1999)
148. Fletcher, P.T.: Statistical variability in nonlinear spaces: application to shape analysis and DT-MRI. PhD thesis, University of North Carolina at Chapel Hill (2004)
149. Fletcher, P.T., Joshi, S.: Riemannian geometry for the statistical analysis of diffusion tensor data. Signal Process. **87**(2), 250–262 (2007)
150. Floquet, G.: Sur les équations différentielles linéaires à coefficients périodiques. Ann. Éc. Norm. Super. **12**, 47–88 (1883)
151. Florack, L.M.J.: Image Structure. Computational Imaging and Vision Series, vol. 10. Kluwer Academic, Dordrecht (1997)
152. Florack, L.M.J.: A geometric model for cortical magnification. In: Lee, S.-W., Bülthoff, H.H., Poggio, T. (eds.) Biologically Motivated Computer Vision: Proceedings of the 1st IEEE In-

ternational Workshop, BMCV 2000, Seoul, Korea, May 2000. Lecture Notes in Computer Science, vol. 1811, pp. 574–583. Springer, Berlin (2000)
153. Florack, L.M.J.: Modeling foveal vision. In: Sgallari, F., Murli, A., Paragios, N. (eds.) Scale Space and Variational Methods in Computer Vision: Proceedings of the 1st International Conference, SSVM 2007, Ischia, Italy, May–June 2007. Lecture Notes in Computer Science, vol. 4485, pp. 919–928. Springer, Berlin (2007)
154. Florack, L.M.J., Salden, A.H., ter Haar Romeny, B.M., Koenderink, J.J., Viergever, M.A.: Nonlinear scale-space. In: ter Haar Romeny, B.M. (ed.) Geometry-Driven Diffusion in Computer Vision. Computational Imaging and Vision Series, vol. 1, pp. 339–370. Kluwer Academic, Dordrecht (1994)
155. Florack, L.M.J., Salden, A.H., ter Haar Romeny, B.M., Koenderink, J.J., Viergever, M.A.: Nonlinear scale-space. Image Vis. Comput. **13**(4), 279–294 (1995). Published in the Promising Research Direction Track
156. Florack, L., Duits, R., Bierkens, J.: Tikhonov regularization versus scale space: a new result. In: Proceedings of the 11th International Conference on Image Processing, Singapore, October 24–27, 2004, pp. 271–274. IEEE Press, New York (2004)
157. Florack, L., Balmashnova, E., Astola, L., Brunenberg, E.: A new tensorial framework for single-shell high angular resolution diffusion imaging. J. Math. Imaging Vis. **3**(38), 171–181 (2010). Published online: doi:10.1007/s10851-010-0217-3
158. Folland, G.B.: Lipschitz classes and Poisson integrals on stratified groups. Stud. Math. **66**(1), 37–55 (1979)
159. Forssén, P.-E.: Low and medium level vision using channel representations. PhD thesis, Linköping University, Sweden (2004)
160. Forssén, P.-E., Granlund, G.: Robust multi-scale extraction of blob features. In: Proceedings of the 13th Scandinavian Conference on Image Analysis. Lecture Notes in Computer Science, vol. 2749, pp. 11–18 (2003)
161. Förstner, W.: Image preprocessing for feature extraction in digital intensity, color and range images. In: Dermanis, A., Grün, A., Sansò, F. (eds.) Proceedings of the International Summer School on Data Analysis and the Statistical Foundation of Geomatics, Chania, Crete, Greece, May 25–30, 1998. Lecture Notes on Earth Sciences, pp. 165–189. Springer, Berlin (1998)
162. Förstner, W., Gülch, E.: A fast operator for detection and precise location of distinct points, corners and centres of circular features. In: Proceedings of the ISPRS Intercomission Conference on Fast Processing of Photogrammetric Data, pp. 281–305 (1987)
163. Forsyth, D., Torr, P., Zisserman, A. (eds.): Proceedings of the 10th European Conference on Computer Vision, Marseille, France, October 12–18, 2008. Lecture Notes in Computer Science, vols. 5302–5305. Springer, Berlin (2008)
164. Frangakis, A.S., Hegerl, R.: Nonlinear anisotropic diffusion in three-dimensional electron microscopy. In: Nielsen, M., Johansen, P., Olsen, O.F., Weickert, J. (eds.) Scale-Space Theories in Computer Vision: Proceedings of the 2nd International Conference, Scale-Space'99, Corfu, Greece, September 1999. Lecture Notes in Computer Science, vol. 1682, pp. 386–397. Springer, Berlin (1999)
165. Frank, L.R.: Anisotropy in high angular resolution diffusion-weighted MRI. Magn. Reson. Med. **45**(6), 935–939 (2001)
166. Franken, E.M.: Enhancement of crossing elongated structures in images. PhD thesis, Eindhoven University of Technology, Department of Biomedical Engineering, Eindhoven, The Netherlands (2008). http://bmia.bmt.tue.nl/people/efranken/PhDThesisErikFranken.pdf
167. Franken, E.M., Duits, R.: Crossing preserving coherence-enhancing diffusion on invertible orientation scores. Int. J. Comput. Vis. **85**(3), 253–278 (2009)
168. Franken, E., Duits, R.: Crossing-preserving coherence-enhancing diffusion on invertible orientation scores. Int. J. Comput. Vis. **85**(3), 253–278 (2009)
169. Franken, E., Duits, R.: Crossing-preserving coherence-enhancing diffusion on invertible orientation scores. Int. J. Comput. Vis. **85**(3), 253–278 (2009)
170. Franken, E., Duits, R., ter Haar Romeny, B.: Nonlinear diffusion on the 2D Euclidean motion group. In: Sgallari, F., Murli, A., Paragios, N. (eds.) Scale Space and Variational Methods in Computer Vision: Proceedings of the 1st International Conference, SSVM 2007, Ischia,

Italy, May–June 2007. Lecture Notes in Computer Science, vol. 4485, pp. 461–472. Springer, Berlin (2007)
171. Freeman, W.T., Adelson, E.H.: The design and use of steerable filters. IEEE Trans. Pattern Anal. Mach. Intell. **13**(9), 891–906 (1991)
172. Friston, K.J., Holmes, A.P., Worsley, K.J., Poline, J.-B., Frith, C.D., Frackowiak, R.S.J.: Statistical parametric maps in functional imaging: a general linear approach. Hum. Brain Mapp. **2**, 189–210 (1995)
173. Fröhlich, J., Weickert, J.: Image processing using a wavelet algorithm for nonlinear diffusion. Technical report 104, Laboratory of Technomathematics, University of Kaiserslautern, P.O. Box 3049, 67653 Kaiserslautern, Germany (1994)
174. Führ, H.: Abstract Harmonic Analysis of Continuous Wavelet Transforms. Lecture Notes in Mathematics, vol. 1863. Springer, Heidelberg (2005)
175. Gabella, M., Guzzi, R., Kisselev, V., Perona, G.: Retrieval of aerosol profile variations in the visible and near infrared: theory and application of the single-scattering approach. Appl. Opt. **36**(6), 1328–1336 (1997)
176. Gabella, M., Kisselev, V., Perona, G.: Retrieval of aerosol profile variations from reflected radiation in the oxygen absorption A band. Appl. Opt. **38**(15), 3190–3195 (1999)
177. Gabor, D.: Theory of communication. J. Inst. Electr. Eng. **93**, 429–457 (1946)
178. Gamboa, F.: New Bayesian methods for ill-posed problems. Stat. Decis. **17**(4), 315–337 (1999)
179. Gamboa, F., Gassiat, E.: Bayesian methods and maximum entropy for ill-posed inverse problems. Ann. Stat. **25**(1), 328–350 (1997)
180. Gayraud, G., Pouet, C.: Adaptive minimax testing in the discrete regression scheme. Probab. Theory Relat. Fields **133**(4), 531–558 (2005)
181. Geller, D., Mayeli, A.: Continuous wavelets and frames on stratified Lie groups. I. J. Fourier Anal. Appl. **12**(5), 543–579 (2006)
182. Geller, D., Mayeli, A.: Nearly tight frames and space-frequency analysis on compact manifolds. Math. Z. **263**(2), 235–264 (2009)
183. Geman, S., Geman, D.: Stochastic relaxation, Gibbs distribution, and the Bayesian restoration of images. IEEE Trans. Pattern Anal. Mach. Intell. **6**(6), 711–741 (1984)
184. Georgiev, T.: Covariant derivatives and vision. In: Leonardis, A., Bischof, H., Prinz, A. (eds.) Proceedings of the 9th European Conference on Computer Vision, Graz, Austria, May 2006. Lecture Notes in Computer Science, vol. 3951–3954, pp. 56–69. Springer, Berlin (2006)
185. Geyer, C.J., Møller, J.: Simulation and likelihood inference for spatial point processes. Scand. J. Stat. **21**(4), 359–373 (1994)
186. Gimel'farb, G.L.: Image Textures and Gibbs Random Fields. Kluwer Academic, Norwell (1999)
187. Godtliebsen, F., Spjøtvoll, E., Marron, J.S.: A nonlinear Gaussian filter applied to images with discontinuities. J. Nonparametr. Stat. **8**(1), 21–43 (1997)
188. Granlund, G.H.: In search of a general picture processing operator. Comput. Graph. Image Process. **8**, 155–173 (1978)
189. Granlund, G.H.: An associative perception-action structure using a localized space variant information representation. In: Proceedings of the International Workshop on Algebraic Frames for the Perception-Action Cycle. Lecture Notes in Computer Science, vol. 1888, pp. 48–68. Springer, Heidelberg (2000)
190. Green, P.: Reversible jump MCMC computation and Bayesian model determination. Biometrika **82**(4), 711–732 (1995)
191. Grim, J.: On numerical evaluation of maximum likelihood estimates for finite mixtures of distributions. Kybernetika **18**, 173–190 (1982)
192. Gröchenig, K.: Foundations of Time-Frequency Analysis. Applied and Numerical Harmonic Analysis. Birkhäuser Boston, Boston (2001)
193. Grossmann, A., Morlet, J., Paul, T.: Transforms associated to square integrable representations. J. Math. Phys. **26**(10), 2473–2479 (1985)
194. Gudbjartsson, H., Patz, S.: The Rician distribution of noisy MRI data. Magn. Reson. Med. **34**(6), 910–914 (1995)

195. Guibas, L., Stolfi, J.: Primitives for the manipulation of general subdivisions and the computation of Voronoi diagrams. ACM Trans. Graph. **4**(2), 74–123 (1985)
196. Gur, Y., Sochen, N.: Coordinate-free diffusion over compact Lie groups. In: Sgallari, F., Murli, A., Paragios, N. (eds.) Scale Space and Variational Methods in Computer Vision: Proceedings of the 1st International Conference, SSVM 2007, Ischia, Italy, May–June 2007. Lecture Notes in Computer Science, vol. 4485, pp. 580–591. Springer, Berlin (2007)
197. Gzyl, H., Zeev, N.: Probabilistic approach to an image reconstruction problem. Methodol. Comput. Appl. Probab. **4**(3), 279–290 (2003)
198. Hahn, K., Prigarin, S., Heim, S., Hasan, K.: Random noise in diffusion tensor imaging, its destructive impact and some corrections. In: Weickert, J., Hagen, H. (eds.) Visualization and Image Processing of Tensor Fields. Mathematics and Visualization, pp. 1–13. Springer, Berlin (2006)
199. Haindl, M.: Identification of the stochastic differential equation of the type ARMA. PhD thesis, ÚTIA Czechoslovak Academy of Sciences, Prague, Czech Republic (1983)
200. Haindl, M.: Texture synthesis. CWI Q. **4**(4), 305–331 (1991)
201. Haindl, M.: Texture segmentation using recursive Markov random field parameter estimation. In: Bjarne, K.E., Peter, J. (eds.) Proceedings of the 11th Scandinavian Conference on Image Analysis, Kangerlussuaq, Greenland, June 7–11, 1999, pp. 771–776 (1999). Lyngby, Denmark
202. Haindl, M.: Recursive square-root filters. In: Sanfeliu, A., Villanueva, J.J., Vanrell, M., Alquezar, R., Jain, A.K., Kittler, J. (eds.) Proceedings of the 15th IAPR International Conference on Pattern Recognition, September 2000, vol. II, pp. 1018–1021. IEEE Press, Los Alamitos (2000)
203. Haindl, M.: Texture modelling. In: Sanchez, B., Pineda, J.M., Wolfmann, J., Bellahse, Z., Ferri, F. (eds.) Proceedings of the World Multiconference on Systemics, Cybernetics and Informatics, July 2000, vol. VII, pp. 634–639. International Institute of Informatics and Systemics, Orlando (2000)
204. Haindl, M., Filip, J.: Fast restoration of colour movie scratches. In: Kasturi, R., Laurendeau, D., Suen, C. (eds.) Proceedings of the 16th International Conference on Pattern Recognition, August 2002, pp. 269–272. IEEE Computer Society, Los Alamitos (2002)
205. Haindl, M., Filip, J.: A fast probabilistic bidirectional texture function model. In: Campilho, A., Kamel, M. (eds.) ICIAR 2004: Proceedings of the 1st International Conference on Image Analysis and Recognition, Porto, Portugal, September 29–October 1, 2004. Lecture Notes in Computer Science, vols. 3211–3212, pp. 298–305. Springer, Berlin (2004)
206. Haindl, M., Filip, J.: Extreme compression and modeling of bidirectional texture function. IEEE Trans. Pattern Anal. Mach. Intell. **29**(10), 1859–1865 (2007)
207. Haindl, M., Havlíček, V.: Multiresolution colour texture synthesis. In: Dobrovodský, K. (ed.) Proceedings of the 7th International Workshop on Robotics in Alpe-Adria-Danube Region, June 1998, pp. 297–302. ASCO Art, Bratislava (1998)
208. Haindl, M., Havlíček, V.: A multiresolution causal colour texture model. In: Ferri, F.J., Inesta, J.M., Amin, A., Pudil, P. (eds.) Advances in Pattern Recognition. Lecture Notes in Computer Science, vol. 1876, pp. 114–122. Springer, Berlin (2000), Chap. 1
209. Haindl, M., Mikeš, S.: Model-based texture segmentation. In: Campilho, A., Kamel, M. (eds.) ICIAR 2004: Proceedings of the 1st International Conference on Image Analysis and Recognition, Porto, Portugal, September 29–October 1, 2004. Lecture Notes in Computer Science, vols. 3211–3212, pp. 306–313. Springer, Berlin (2004)
210. Haindl, M., Mikeš, S.: Colour texture segmentation using modelling approach. In: Singh, S., Singh, M., Apte, C., Perner, P. (eds.) Proceedings of the 3rd International Conference on Advances in Pattern Recognition, ICAPR 2005, Bath, UK, August 22–25, 2005. Lecture Notes in Computer Science, vols. 3686–3687, pp. 484–491. Springer, Berlin (2005)
211. Haindl, M., Mikeš, S.: Unsupervised texture segmentation using multispectral modelling approach. In: Tang, Y.Y., Wang, S.P., Yeung, D.S., Yan, H., Lorette, G. (eds.) Proceedings of the 18th International Conference on Pattern Recognition, Hong Kong, China, August 2006, pp. 203–206. IEEE Computer Society, Los Alamitos (2006). http://doi.ieeecomputersociety.org/10.1109/ICPR.2006.1148

212. Haindl, M., Mikeš, S.: Texture segmentation benchmark. In: Lovell, B., Laurendeau, D., Duin, R. (eds.) Proceedings of the 19th International Conference on Pattern Recognition, Tampa, USA, December 2008, pp. 1–4. IEEE Computer Society, Los Alamitos (2008). doi:10.1109/ICPR.2008.4761118. http://doi.ieeecomputersociety.org/
213. Haindl, M., Šimberová, S.: A multispectral image line reconstruction method. In: Johansen, P., Olsen, S. (eds.) Theory & Applications of Image Analysis. Series in Machine Perception and Artificial Intelligence, vol. 2, pp. 306–315. World Scientific, Singapore (1992). Selected papers from the 7th Scandinavian Conference on Image Analysis
214. Haindl, M., Šimberová, S.: A high-resolution radiospectrograph image reconstruction method. Astron. Astrophys. **115**(1), 189–193 (1996)
215. Haindl, M., Šimberová, S.: A scratch removal method. Kybernetika **34**(4), 423–428 (1998)
216. Haindl, M., Šimberová, S.: Restoration of multitemporal short-exposure astronomical images. In: Kalviainen, H., Parkkinen, J., Kaarna, A. (eds.) Proceedings of the 14th Scandinavian Conference on Image Analysis, Joensuu, Finland, June 19–22, 2005. Lecture Notes in Computer Science, vol. 3540, pp. 1037–1046. Springer, Berlin (2005)
217. Haindl, M., Filip, J., Arnold, M.: BTF image space utmost compression and modelling method. In: Kittler, J., Petrou, M., Nixon, M. (eds.) Proceedings of the 17th International Conference on Pattern Recognition, Cambridge, UK, August 2004, vol. III, pp. 194–197. IEEE Computer Society, Los Alamitos (2004)
218. Hebisch, W.: Estimates on the semigroups generated by left invariant operators on Lie groups. J. Reine Angew. Math. **423**, 1–45 (1992)
219. Heeger, D.J., Bergen, J.R.: Pyramid based texture analysis/synthesis. In: ACM SIGGRAPH 95, pp. 229–238. ACM, New York (1995)
220. Heier, L.A., Bauer, C.J., Schwartz, L., Zimmerman, R.D., Morgelli, S., Deck, M.D.: Large Virchow-Robin spaces: MR-clinical correlation. Am. J. Neuroradiol. **10**(5), 929–936 (1989)
221. Heim, S., Fahrmeir, L., Eilers, P.H.C., Marx, B.D.: 3D space-varying coefficient models with application to diffusion tensor imaging. Comput. Stat. Data Anal. **51**(12), 6212–6228 (2007)
222. Heinlein, P.: Wavelet-Methoden zur Analyse mammographischer Bilddaten. PhD thesis, TU München (2001)
223. Heinlein, P.: Discretizing continuous wavelet transforms using integrated wavelets. Appl. Comput. Harmon. Anal. **14**(3), 238–256 (2003)
224. Helstrom, C.: An expansion of a signal in Gaussian elementary signals. IEEE Trans. Inf. Theory **12**(1), 81–82 (1966)
225. Hermann, U., Noll, D.: Adaptive image reconstruction using information measures. SIAM J. Control Optim. **38**(4), 1223–1240 (2000)
226. Hess, C.P., Mukherjee, P., Tan, E.T., Xu, D., Vigneron, D.B.: Q-ball reconstruction of multimodal fiber orientations using the spherical harmonic basis. Magn. Reson. Med. **56**, 104–117 (2006)
227. Heyden, A., Sparr, G., Nielsen, M., Johansen, P. (eds.): Proceedings of the 7th European Conference on Computer Vision, Copenhagen, Denmark, May–June 2002. Lecture Notes in Computer Science, vol. 2350–2353. Springer, Berlin (2002)
228. Hörmander, L.: Hypoelliptic second order differential equations. Acta Math. **119**, 147–171 (1967)
229. Horn, R.A., Johnson, C.R.: Matrix Analysis. Cambridge University Press, Cambridge (1990). Corrected reprint of the 1985 original
230. Howard, I.P., Rogers, B.J.: Binocular Vision and Stereopsis. Oxford University Press, Oxford (1995)
231. Iijima, T.: Basic theory of pattern observation. Papers of Technical Group on Automata and Automatic Control, IECE, Japan (1959)
232. Iijima, T.: Basic theory of pattern observation. In: Papers of Technical Group on Automata and Automatic Control, IECE, Japan, December, 1959
233. Ishi, H.: Wavelet transforms for semidirect product groups with not necessarily commutative normal subgroups. J. Fourier Anal. Appl. **12**(1), 37–52 (2006)
234. Iverson, L.A., Zucker, S.W.: Logical/linear operators for image curves. IEEE Trans. Pattern Anal. Mach. Intell. **17**(10), 982–996 (1995)

235. Janssen, A.J.E.M.: The Zak transform: a signal transform for sampled time-continuous signals. Philips J. Res. **43**(1), 23–69 (1988)
236. Janssen, B.J.: Representation and manipulation of images based on linear functionals. PhD thesis, Eindhoven University of Technology, Eindhoven, The Netherlands (2009)
237. Jawerth, B., Lin, P., Sinzinger, E.: Lattice Boltzmann models for anisotropic diffusion of images. In: Nielsen, M., Johansen, P., Olsen, O.F., Weickert, J. (eds.) Scale-Space Theories in Computer Vision: Proceedings of the 2nd International Conference, Scale-Space'99, Corfu, Greece, September 1999. Lecture Notes in Computer Science, vol. 1682, pp. 283–293. Springer, Berlin (1999)
238. Jian, B., Vemuri, B.C., Özarslan, E., Carney, P.R., Mareci, T.H.: A novel tensor distribution model for the diffusion-weighted MR signal. NeuroImage **37**, 164–176 (2007)
239. Johnson, A.R., Cézilly, F.: The Greater Flamingo. T & AD Poyse (2007)
240. Jones, D.K., Basser, P.J.: "Squashing peanuts and smashing pumpkins": How noise distorts diffusion-weighted MR data. Magn. Reson. Med. **52**(5), 979–993 (2004)
241. Jonsson, E.: Channel-coded feature maps for computer vision and machine learning. PhD thesis, Linköping University, Sweden, SE-581 83 Linköping, Sweden (February 2008). Dissertation No. 1160, ISBN 978-91-7393-988-1
242. Jonsson, E., Felsberg, M.: Reconstruction of probability density functions from channel representations. In: Proceedings of the 14th Scandinavian Conference on Image Analysis (2005)
243. Jonsson, E., Felsberg, M.: Accurate interpolation in appearance-based pose estimation. In: Ersbøll, B.K., Steenstrup Pedersen, K. (eds.) Proceedings of the 15th Scandinavian Conference on Image Analysis, Aalborg, Denmark, June 10–14, 2007. Lecture Notes in Computer Science, vol. 4522, pp. 1–10. Springer, Berlin (2007)
244. Jonsson, E., Felsberg, M.: Efficient computation of channel-coded feature maps through piecewise polynomials. Image Vis. Comput. **27**(11), 1688–1694 (2009)
245. Jorgenson, J., Lang, S.: $Pos_n(R)$ and Eisenstein Series. Lecture Notes in Mathematics, vol. 1868. Springer, Berlin (2005)
246. Jost, J.: Riemannian Geometry and Geometric Analysis. Springer, New York (2001)
247. Kashyap, R.L.: Analysis and synthesis of image patterns by spatial interaction models. In: Kanal, L.N., Rosenfeld, A. (eds.) Progress in Pattern Recognition, vol. 1, pp. 149–186. North-Holland, Amsterdam (1981)
248. Kashyap, R.L.: Image models. In: Young, T.Y., Fu, K.S. (eds.) Handbook of Pattern Recognition and Image Processing, pp. 281–310. Academic Press, New York (1986)
249. Kim, J.S., Singh, V., Lee, J.K., Lerch, J., Ad-Dab'bagh, Y., MacDonald, D., Lee, J.M., Kim, S.I., Evans, A.C.: Automated 3-D extraction and evaluation of the inner and outer cortical surfaces using a Laplacian map and partial volume effect classification. Neuroimage **27**(1), 210–221 (2005)
250. Kimmel, R., Malladi, R., Sochen, N.: Image processing via the Beltrami operator. In: Chin, R.T., Pong, T.-C. (eds.) Proceedings of the 3rd Asian Conference on Computer Vision, Hong Kong, China, January 8–10, 1998. Lecture Notes in Computer Science, vol. 1351–1352, pp. 574–581. Springer, Berlin (1997)
251. Kimmel, R., Sochen, N., Malladi, R.: From high energy physics to low level vision. In: Scale-Space Theory in Computer Vision. Lecture Notes in Computer Science, vol. 1252, pp. 236–247. Springer, Berlin (1997)
252. Kingsley, P.B.: Introduction to diffusion tensor imaging mathematics: Part II. Anisotropy, diffusion-weighting factors, and gradient encoding schemes. Concepts Magn. Reson., Part A **28A**(2), 123–154 (2006)
253. Kingsley, P.B.: Introduction to diffusion tensor imaging mathematics: Part III. Tensor calculation, noise, simulations, and optimization. Concepts Magn. Reson., Part A **28A**(2), 155–179 (2006)
254. Kingsley, P.B.: Introduction to diffusion tensor imaging mathematics: Part I. Tensors, rotations, and eigenvectors. Concepts Magn. Reson., Part A **28A**(2), 101–122 (2006)
255. Kleinschmidt, A.: Different analysis solutions for different spatial resolutions? Moving towards a mesoscopic mapping of functional architecture in the human brain. NeuroImage **38**(4), 663–665 (2007)

256. Kluszczyński, R., van Lieshout, M.N.M., Schreiber, T.: An algorithm for binary image segmentation using polygonal Markov fields. In: Roli, F., Vitulano, S. (eds.) Proceedings of the 13th International Conference on Image Analysis and Processing. Lecture Notes in Computer Science, vol. 3617, pp. 383–390 (2005)
257. Kluszczyński, R., van Lieshout, M.N.M., Schreiber, T.: Image segmentation by polygonal Markov fields. Ann. Inst. Stat. Math. **59**(3), 465–486 (2007)
258. Knutsson, H., Westin, C.-F.: Normalized convolution: technique for filtering incomplete and uncertain data. In: Høgda, K.A., Braathen, B., Heia, K. (eds.) Proceedings of the 8th Scandinavian Conference on Image Analysis, Tromsø, Norway, May 25–28, 1993, pp. 997–1006. NOBIM, Tromsø (1993)
259. Knutsson, H., Wilson, R., Granlund, G.H.: Anisotropic non-stationary image estimation and its applications: Part I—restoration of noisy images. IEEE Trans. Commun. **COM–31**(3), 388–397 (1983)
260. Kodera, K., de Villedary, C., Gendrin, R.: A new method for the numerical analysis of nonstationary signals. Phys. Earth Planet. Inter. **12**(2–3), 142–150 (1976)
261. Koenderink, J.J.: The structure of images. Biol. Cybern. **50**, 363–370 (1984)
262. Koenderink, J.J.: The brain a geometry engine. Psychol. Res. **52**, 122–127 (1990)
263. Koenderink, J.J.: Image space. In: Ablamwicz, R. (ed.) Clifford Algebras; Applications to Mathematics, Physics, and Engineering, pp. 577–596. Birkhaüser, Boston (2004)
264. Koenderink, J.J., van Doorn, A.D.: Image processing done right. In: Heyden, A., Sparr, G., Nielsen, M., Johansen, P. (eds.) Proceedings of the 7th European Conference on Computer Vision, Copenhagen, Denmark, May–June 2002. Lecture Notes in Computer Science, vols. 2350–2353, pp. 158–172. Springer, Berlin (2002)
265. Koffka, K.: Principles of Gestalt Psychology. Harcourt, New York (1935)
266. Krajsek, K., Menzel, M.I., Zwanger, M., Scharr, H.: Riemannian anisotropic diffusion for tensor valued images. In: Forsyth, D., Torr, P., Zisserman, A. (eds.) Proceedings of the 10th European Conference on Computer Vision, Marseille, France, October 12–18, 2008. Lecture Notes in Computer Science, vols. 5302–5305, pp. 326–339. Springer, Berlin (2008)
267. Krajsek, K., Mester, R., Scharr, H.: Statistically optimal averaging for image restoration and optical flow estimation. In: Rigoll, G. (ed.) Pattern Recognition. Proceedings of the 30th DAGM-Symposium, Munich, Germany, June 10–13, 2008. Lecture Notes in Computer Science, vol. 5096, pp. 466–475. Springer, Berlin (2008)
268. Kriegeskorte, N., Bandettini, P.: Analyzing for information, not activation, to exploit high-resolution fMRI. NeuroImage **38**(4), 649–662 (2007)
269. Lacoste, C., Descombes, X., Zerubia, J.: Point processes for unsupervised line network extraction in remote sensing. IEEE Trans. Pattern Anal. Mach. Intell. **27**(1), 1568–1579 (2005)
270. Lafarge, F., Descombes, X., Zerubia, J., Pierrot-Deseilligny, M.: Automatic building extraction from DEMs using an object approach and application to the 3D-city modeling. J. Photogramm. Remote Sens. **63**(3), 365–381 (2008)
271. Lang, S.: Fundamentals of Differential Geometry. Springer, New York (1999)
272. Lange, N.: Statistical approaches to human brain mapping by functional magnetic resonance imaging. Stat. Med. **15**(4), 389–428 (1996)
273. Lange, N., Zeger, S.L.: Non-linear fourier time series analysis for human brain mapping by functional magnetic resonance imaging. J. R. Stat. Soc., Ser. C, Appl. Stat. **46**(1), 1–29 (1997)
274. Larsen, R., Nielsen, M., Sporring, J. (eds.): Proceedings of the 9th International Conference on Medical Image Computing and Computer-Assisted Intervention, MICCAI 2006, Copenhagen, Denmark, October 1–6, 2006. Lecture Notes in Computer Science, vols. 4190–4191. Springer, Berlin (2006)
275. Lazar, N.A.: The Statistical Analysis of Functional MRI Data. Statistics for Biology and Health. Springer, Berlin (2008)
276. Le Bihan, D., Breton, E.: Imagerie de diffusion in vivo par résonance magnétique nucléaire. C. R. Acad. Sci. **301**, 1109–1112 (1985)

277. Le Bihan, D., Mangin, J.-F., Poupon, C., Clark, C.A., Pappata, S., Molko, N., Chabriat, H.: Diffusion tensor imaging: concepts and applications. J. Magn. Reson. Imaging **13**, 534–546 (2001)
278. Ledoux, H., Gold, C.M., Baciu, G.: Flipping to robustly delete a vertex in a Delaunay tetrahedralization. In: Gervasi, O., Gavrilova, M.L., Kumar, V., Laganà, A., Lee, H.P., Mun, Y., Taniar, D., Tan, C.J.K. (eds.) Proceedings of the International Conference on Computational Science and Its Applications, ICCSA 2005, Singapore, May 9–12, 2005. Lecture Notes in Computer Science, vols. 3480–3483, pp. 737–747. Springer, Berlin (2005)
279. Lee, J.S.: Digital image enhancement and noise filtering by use of local statistics. IEEE Trans. Pattern Anal. Mach. Intell. **2**(2), 165–168 (1980)
280. Lee, J.H., Chung, M.K., Oakes, T.E., Alexander, A.L.: Anisotropic Gaussian kernel smoothing of DTI data. In: Proceedings of the 13th International Society for Magnetic Resonance in Medicine, ISMRM, Miami, USA, May 7–13, 2005, p. 2253. International Society for Magnetic Resonance in Medicine, Berkeley (2005)
281. Lehner, B., Umlauf, G., Hamann, B.: Video compression using data-dependent triangulations. In: Proceedings of the IADIS Multi Conference on Computer Science and Information Systems, MMCIS 2008, Amsterdam, The Netherlands, July 22–27, 2008, pp. 244–248. IADIS Press, Salamanca (2008)
282. Leonardis, A., Bischof, H., Prinz, A. (eds.): Proceedings of the 9th European Conference on Computer Vision, Graz, Austria, May 2006. Lecture Notes in Computer Science, vols. 3951–3954. Springer, Berlin (2006)
283. Leow, A.D., Zhu, S., McMahon, K., de Zubicaray, G.I., Meredith, M.J., Wright, M.J., Toga, A.W., Thompson, P.M.: The tensor distribution function. Magn. Reson. Med. **61**(1), 205–214 (2009)
284. Lepore, N., Brun, C., Chou, Y.Y., Chiang, M.C., Dutton, R.A., Hayashi, K.M., Luders, E., Lopez, O.L., Aizenstein, H.J., Toga, A.W., Becker, J.T., Thompson, P.M.: Generalized tensor-based morphometry of HIV/AIDS using multivariate statistics on deformation tensors. IEEE Trans. Med. Imaging **27**(1), 129–141 (2008)
285. Lepskiĭ, O.V.: A problem of adaptive estimation in Gaussian white noise. Theory Probab. Appl. **35**(3), 454–466 (1990)
286. Lepskiĭ, O.V.: Asymptotically minimax adaptive estimation. I. Upper bounds. Optimally adaptive estimates. Theory Probab. Appl. **36**(4), 682–697 (1991/1992)
287. Lepskiĭ, O.V., Spokoiny, V.G.: Optimal pointwise adaptive methods in nonparametric estimation. Ann. Stat. **25**(6), 2512–2546 (1997)
288. Liang, L., Liu, C., Xu, Y.-Q., Guo, B., Shum, H.-Y.: Real-time texture synthesis by patch-based sampling. ACM Trans. Graph. **20**(3), 127–150 (2001)
289. Liggett, T.: Interacting Particle Systems. Springer, New York (1985)
290. Lindeberg, T.: Scale-Space Theory in Computer Vision. The Kluwer International Series in Engineering and Computer Science. Kluwer Academic, Dordrecht (1994)
291. Lions, J.L., Magenes, E.: Non-homogeneous Boundary Value Problems and Applications, vols. 1–3. Springer, Berlin (1972–1973)
292. Liu, F., Picard, R.W.: Periodicity, directionality, and randomness: Wold features for image modeling and retrieval. IEEE Trans. Pattern Anal. Mach. Intell. **18**(7), 722–733 (1996)
293. Lloyd, S.P.: Least square quantization in PCM. IEEE Trans. Inf. Theory **28**(2), 129–137 (1982)
294. Loubes, J.-M., Pelletier, B.: Maximum entropy solution to ill-posed inverse problems with approximately known operator. J. Math. Model. Algorithms **344**(1), 260–273 (2008)
295. Lüdtke, N.L., Wilson, R.C., Hancock, E.R.: Probabilistic population coding of multiple edge orientation. In: Proceedings of the 9th International Conference on Image Processing, Rochester, NY, USA, September 22–25, 2002, pp. 865–868. IEEE Press, New York (2002)
296. Lyttelton, O., Boucher, M., Robbins, S., Evans, A.C.: An unbiased iterative group registration template for cortical surface analysis. Neuroimage **34**(4), 1535–1544 (2007)
297. MacDonald, D., Kabani, N., Avis, D., Evans, A.C.: Automated 3-D extraction of inner and outer surfaces of cerebral cortex from MRI. NeuroImage **12**(3), 340–356 (2000)

298. Mallat, S.G.: A theory for multiresolution signal decomposition: the wavelet representation. IEEE Trans. Pattern Anal. Mach. Intell. **11**, 674–693 (1989)
299. Manjunath, B.S., Chellapa, R.: Unsupervised texture segmentation using Markov random field models. IEEE Trans. Pattern Anal. Mach. Intell. **13**(5), 478–482 (1991)
300. Mantini, D., Perrucci, M.G., Del Gratta, C., Romani, G.L., Corbetta, M.: Electrophysiological signatures of resting state networks in the human brain. Proc. Natl. Acad. Sci. **104**(32), 13170–13175 (2007)
301. Marpe, D., Schwarz, H., Wiegand, T.: Context-based adaptive binary arithmetic coding in the H.264/AVC video compression standard. IEEE Trans. Circuits Syst. Video Technol. **13**(7), 620–636 (2003)
302. Mayeli, A.: Discrete and continuous wavelet transformations on the Heisenberg group. PhD thesis, TU München (2005)
303. Mayeli, A.: Mexican hat wavelet on the Heisenberg group. Technical report, TU München, (2007). arXiv:0705.3364v1
304. Mazziotta, J.C., Toga, A.W., Evans, A., Fox, P., Lancaster, J.: A probabilistic atlas of the human brain: theory and rationale for its development, the International Consortium for Brain Mapping (ICBM). NeuroImage **2**(2), 89–101 (1995)
305. McCullagh, P., Nelder, J., McCullagh, M.C.: Generalized Linear Models. Chapman & Hall/CRC Press, London/Boca Raton (1989)
306. McGraw, T., Vemuri, B.C., Chen, Y., Rao, M., Mareci, T.: DT-MRI denoising and neuronal fiber tracking. Med. Image Anal. **8**(2), 95–111 (2004)
307. Meek, C., Patel, J.M., Kasetty, S.: Oasis: an online and accurate technique for local-alignment searches on biological sequences. In: Proceedings of the 29th International Conference on Very Large Data Bases, vol. 29, pp. 910–921. VLDB Endowment, Lyon (2003)
308. Mehrseresht, M., Taubman, D.: An efficient content-adaptive motion-compensated 3-D DWT with enhanced spatial and temporal scalability. IEEE Trans. Image Process. **15**(6), 1397–1412 (2006)
309. Merboldt, K.-D., Hanicke, W., Frahm, J.: Self-diffusion NMR imaging using stimulated echoes. J. Magn. Reson. **64**(3), 479–486 (1985)
310. Meseth, J., Müller, G., Klein, R.: Preserving realism in real-time rendering. In: Reiners, D. (ed.) OpenGL Symposium, pp. 89–96. Eurographics Association, Geneva (2003)
311. Moakher, M.: A differential geometric approach to the geometric mean of symmetric positive-definite matrices. SIAM J. Matrix Anal. Appl. **26**(3), 735–747 (2005)
312. Moakher, M., Zerai, M.: Riemannian curvature-driven flows for tensor-valued data. In: Sgallari, F., Murli, A., Paragios, N. (eds.) Scale Space and Variational Methods in Computer Vision: Proceedings of the 1st International Conference, SSVM 2007, Ischia, Italy, May–June 2007. Lecture Notes in Computer Science, vol. 4485, pp. 592–602. Springer, Berlin (2007)
313. Møller, J., Skare, Ø.: Bayesian image analysis with coloured Voronoi tesselations and a view to applications in reservoir modelling. Stat. Model. **1**, 213–232 (2001)
314. Mori, S., Wakana, S., van Zijl, P.C.M., Nagae-Poetscher, L.M.: MRI Atlas of Human White Matter. Elsevier, Amsterdam (2005)
315. Mücke, E.P.: A robust implementation for three-dimensional Delaunay triangulations. Int. J. Comput. Geom. Appl. **8**(2), 255–276 (1998)
316. Mumford, D., Shah, J.: Optimal approximations by piecewise smooth functions and associated variational problems. Commun. Pure Appl. Math. **42**(5), 577–685 (1989)
317. Muschietti, M., Torrésani, B.: Pyramidal algorithms for Littlewood-Paley decompositions. SIAM J. Math. Anal. **26**(4), 925–943 (1995)
318. Nicholls, G.K.: Bayesian image analysis with Markov chain Monte Carlo and coloured continuum triangulation models. J. R. Stat. Soc., Ser. B, Stat. Methodol. **60**, 643–659 (1998)
319. Nicholls, G.K.: Spontaneous magnetization in the plane. J. Stat. Phys. **102**(5–6), 1229–1251 (2001)
320. Nielsen, M., Johansen, P., Olsen, O.F., Weickert, J. (eds.): Scale-Space Theories in Computer Vision: Proceedings of the 2nd International Conference, Scale-Space'99, Corfu, Greece, September 1999. Lecture Notes in Computer Science, vol. 1682. Springer, Berlin (1999)

321. Ogawa, S., Lee, T.M., Kay, A.R., Tank, D.W.: Brain magnetic resonance imaging with contrast dependent on blood oxygenation. Proc. Natl. Acad. Sci. **87**(24), 9868–9872 (1990)
322. Ogawa, S., Tank, D.W., Menon, R., Ellermann, J.M., Kim, S., Merkle, H., Ugurbil, K.: Intrinsic signal changes accompanying sensory stimulation: functional brain mapping with magnetic resonance imaging. Proc. Natl. Acad. Sci. **89**(13), 5951–5955 (1992)
323. Ortner, M., Descombes, X., Zerubia, J.: Building outline extraction from digital elevation models using marked point processes. Int. J. Comput. Vis. **72**(2), 107–132 (2007)
324. Özarslan, E., Mareci, T.H.: Generalized diffusion tensor imaging and analytical relationships between diffusion tensor imaging and high angular resolution imaging. Magn. Reson. Med. **50**(5), 955–965 (2003)
325. Özarslan, E., Shepherd, T.M., Vemuri, B.C., Blackband, S.J., Mareci, T.H.: Resolution of complex tissue microarchitecture using the diffusion orientation transform (DOT). NeuroImage **31**, 1086–1103 (2006)
326. Paget, R., Longstaff, I.D.: Texture synthesis via a noncausal nonparametric multiscale Markov random field. IEEE Trans. Image Process. **7**(8), 925–932 (1998)
327. Pajevic, S., Pierpaoli, C.: Color schemes to represent the orientation of anisotropic tissues from diffusion tensor data: application to white matter fiber tract mapping in the human brain. Magn. Reson. Med. **42**, 526–540 (1999)
328. Panjwani, D.K., Healey, G.: Markov random field models for unsupervised segmentation of textured color images. IEEE Trans. Pattern Anal. Mach. Intell. **17**(10), 939–954 (1995)
329. Paris, S., Durand, F.: A fast approximation of the bilateral filter using a signal processing approach. Int. J. Comput. Vis. **81**(1), 24–52 (2009)
330. Parker, G.J., Schnabel, J.A., Symms, M.R., Werring, D.J., Barker, G.J.: Nonlinear smoothing for reduction of systematic and random errors in diffusion tensor imaging. J. Magn. Reson. Imaging **11**(6), 702–710 (2000)
331. Pasternak, O., Sochen, N., Intrator, N., Assaf, Y.: Variational multiple-tensors fitting of fiber-ambiguous DW-MRI voxels. J. Magn. Reson. Imaging **26**(8), 1133–1144 (2008)
332. Pennec, X., Fillard, P., Ayache, N.: A Riemannian framework for tensor computing. Int. J. Comput. Vis. **66**(1), 41–66 (2006)
333. Peretto, P.: An introduction to the modeling of neural networks. Collect. Alia-Saclay, Monogr. Texts Stat.. Phys. **2** (1992)
334. Perona, P., Malik, J.: Scale-space and edge detection using anisotropic diffusion. IEEE Trans. Pattern Anal. Mach. Intell. **12**(7), 629–639 (1990)
335. Perrin, G., Descombes, X., Zerubia, J.: 2D and 3D vegetation resource parameters assessment using marked point processes. In: Tang, Y.Y., Wang, S.P., Yeung, D.S., Yan, H., Lorette, G. (eds.) Proceedings of the 18th International Conference on Pattern Recognition, Hong Kong, China, August 2006, vol. 1, pp. 1–4. IEEE Computer Society, Los Alamitos (2006)
336. Pierpaoli, C., Jezzard, P., Basser, P.J., Barnett, A., Chiro, G.D.: Diffusion tensor MRI of the human brain. Radiology **201**, 637–648 (1996)
337. Pizarro, L., Burgeth, B., Didas, S., Weickert, J.: A generic neighbourhood filtering framework for matrix fields. In: Forsyth, D., Torr, P., Zisserman, A. (eds.) Proceedings of the 10th European Conference on Computer Vision, Marseille, France, October 12–18, 2008. Lecture Notes in Computer Science, vols. 5302–5305, pp. 521–532. Springer, Berlin (2008)
338. Polyakov, A.M.: Quantum geometry of bosonic strings. Phys. Lett. B **103**(3), 207–210 (1981)
339. Polzehl, J., Spokoiny, V.: Adaptive weights smoothing with applications to image restoration. J. R. Stat. Soc., Ser. B, Stat. Methodol. **62**(2), 335–354 (2000)
340. Polzehl, J., Spokoiny, V.: Propagation-separation approach for local likelihood estimation. Probab. Theory Relat. Fields **135**(3), 335–362 (2006)
341. Polzehl, J., Spokoiny, V.: Structural adaptive smoothing by propagation-separation methods. In: Chen, C., Härdle, W., Unwin, A. (eds.) Handbook of Data Visualization. Springer Handbooks of Computational Statistics, pp. 471–492. Springer, Berlin (2008)
342. Polzehl, J., Tabelow, K.: Adaptive smoothing of digital images: the R package adimpro. J. Stat. Softw. **19**(1), 1–17 (2007)

343. Polzehl, J., Tabelow, K.: fMRI: a package for analyzing fMRI data. R News **7**(2), 13–17 (2007)
344. Polzehl, J., Tabelow, K.: Structure adaptive smoothing diffusion tensor imaging data: the R package DTI. Preprint 1382, WIAS (2008)
345. Portilla, J., Simoncelli, E.P.: A parametric texture model based on joint statistics of complex wavelet coefficients. Int. J. Comput. Vis. **40**(1), 49–71 (2000)
346. Portilla, J., Strela, V., Wainwright, J., Simoncelli, E.P.: Image denoising using scale mixtures of Gaussians in the wavelet domain. IEEE Trans. Image Process. **12**(11), 1338–1351 (2003)
347. Pouget, A., Dayan, P., Zemel, R.: Information processing with population codes. Nat. Rev., Neurosci. **1**, 125–132 (2000)
348. Preparata, F.P., Shamos, M.I.: Computational Geometry. Springer, NewYork (1988)
349. Preston, C.: Spatial birth-and-death processes. Bull. Int. Stat. Inst. **46**(2), 371–391 (1977)
350. Preußer, T., Rumpf, M.: An adaptive finite element method for large scale image processing. In: Nielsen, M., Johansen, P., Olsen, O.F., Weickert, J. (eds.) Scale-Space Theories in Computer Vision: Proceedings of the 2nd International Conference, Scale-Space'99, Corfu, Greece, September 1999. Lecture Notes in Computer Science, vol. 1682, pp. 223–234. Springer, Berlin (1999)
351. Proceedings of the 7th International Conference on Computer Vision, Corfu, Greece, September 20–27, 1999. IEEE Computer Society (1999)
352. R Development Core Team: R: A Language and Environment for Statistical Computing. Foundation for Statistical Computing, Vienna (2005). ISBN3-900051-07-0
353. Ranjan, U.S., Ramakrishnan, K.R.: A stochastic scale space for multiscale image representation. In: Nielsen, M., Johansen, P., Olsen, O.F., Weickert, J. (eds.) Scale-Space Theories in Computer Vision: Proceedings of the 2nd International Conference, Scale-Space'99, Corfu, Greece, September 1999. Lecture Notes in Computer Science, vol. 1682, pp. 441–446. Springer, Berlin (1999)
354. Rao, A.R., Schunck, B.G.: Computing oriented texture fields. CVGIP, Graph. Models Image Process. **53**(2), 157–185 (1991)
355. Reed, T.R., du Buf, J.M.H.: A review of recent texture segmentation and feature extraction techniques. CVGIP, Image Underst. **57**(3), 359–372 (1993)
356. Reiß, M., Rozenholc, Y., Cuenod, C.-A.: Pointwise adaptive estimation for robust and quantile regression. Technical report, ArXiv:0904.0543v1 (2009)
357. Rettmann, M.E., Han, X., Xu, C., Prince, J.L.: Automated sulcal segmentation using watersheds on the cortical surface. NeuroImage **15**(2), 329–344 (2002)
358. Robbins, S.: Anatomical standardization of the human brain in Euclidean 3-space and on the cortical 2-manifold. PhD thesis, School of Computer Science, McGill University (2004)
359. Rodieck, R.W.: The First Steps in Seeing. Sinauer Associates, Sunderland (1998)
360. Rosenfeld, B.: The Geometry of Lie Groups. Kluwer Academic, Dordrecht (1997)
361. Roth, S., Black, M.J.: Fields of experts: a framework for learning image priors. In: Proceedings of the IEEE Computer Society Conference on Computer Vision and Pattern Recognition, San Diego, California, USA, June 20–25, 2005, pp. 860–867. IEEE Computer Society, Los Alamitos (2005)
362. Roth, S., Black, M.J.: Steerable random fields. In: Proceedings of the 11th International Conference on Computer Vision, Rio de Janeiro, Brazil, October 14–20, 2007, pp. 1–8. IEEE Computer Society, Los Alamitos (2007). Digital proceedings by Omnipress
363. Rozenholc, Y., Reiß, M., Balvay, D., Cuenod, C.-A.: Growing time homogeneous neighborhoods for denoising and clustering dynamic contrast enhanced-CT sequences. Technical report, University Paris Descartes (2009)
364. Rudin, L.I., Osher, S., Fatemi, E.: Nonlinear total variation based noise removal algorithms. In: Proceedings of the 11th Annual International Conference of the Center for Nonlinear Studies on Experimental Mathematics: Computational Issues in Nonlinear Science, pp. 259–268. Elsevier/North-Holland, Amsterdam (1992)
365. Rue, H., Hurn, M.: Bayesian object identification. Biometrika **86**(3), 649–660 (1999)

366. Scharr, H.: Diffusion-like reconstruction schemes from linear data models. In: Franke, K., Müller, K.-R., Nickolay, B., Schäfer, R. (eds.) Mustererkennung 2006. Proceedings of the 28th DAGM-Symposium, Berlin, Germany, September 12–14, 2000. Lecture Notes in Computer Science, vol. 4174, pp. 51–60. Springer, Berlin (2006)
367. Scharr, H.: Optimal second order derivative filter families for transparent motion estimation. In: Proceedings of the 15th European Signal Processing Conference, Poznan, Poland, September 3–7, 2007, pp. 302–306. EURASIP (2007)
368. Scharr, H.: Optimal filters for extended optical flow. In: Jähne, B., Barth, E., Mester, R., Scharr, H. (eds.) Proceedings of the 1st International Workshop on Complex Motion, IWCM 2004, Gunzburg, Germany, October 12–14, 2004. Lecture Notes in Computer Science, vol. 3417, pp. 14–29. Springer, Berlin (2007)
369. Scharr, H., Spies, H.: Accurate optical flow in noisy image sequences using flow adapted anisotropic diffusion. Signal Process. Image Commun. **20**(6), 537–553 (2005)
370. Scharr, H., Weickert, H.: An anisotropic diffusion algorithm with optimized rotation invariance. In: Sommer, G., Krüger, N., Perwass, C. (eds.) Mustererkennung 2000. Proceedings of the 22nd DAGM-Symposium, Kiel, Germany, September 13–15, 2000. Informatik Aktuell, pp. 460–467. Springer, Berlin (2000)
371. Scharr, H., Black, M.J., Haussecker, H.W.: Image statistics and anisotropic diffusion. In: Proceedings of the 9th International Conference on Computer Vision, Nice, France, October 13–16, 2003, pp. 840–847. IEEE Computer Society, Los Alamitos (2003)
372. Scharr, H., Felsberg, M., Forssén, P.-E.: Noise adaptive channel smoothing of low-dose images. In: Proceedings of the 2003 Conference on Computer Vision and Pattern Recognition Workshop, CVPRW'03, pp. 1–8 (2003)
373. Scheltens, P., Erkinjunti, T., Leys, D., Wahlund, L.O., del Ser, T., Pasquier, F., Barkhof, F., Mantyla, R., Bowler, J., Wallin, A., Ghika, J., Fazekas, F., Pantoni, L.: White matter changes on CT and MRI: an overview of visual rating scales. Eur. J. Neurol. **39**(2), 80–89 (1998)
374. Schlesinger, M.I.: Relation between learning and self-learning in pattern recognition. Kibernetika **2**, 81–88 (1968) (in Russian)
375. Schnörr, C.: A study of a convex variational diffusion approach for image segmentation and feature extraction. J. Math. Imaging Vis. **8**(3), 271–292 (1998)
376. Schnörr, C., Weickert, J.: Variational image motion computation: theoretical framework, problems and perspectives. In: Sommer, G., Krüger, N., Perwass, C. (eds.) Mustererkennung 2000. Proceedings of the 22nd DAGM-Symposium, Kiel, Germany, September 13–15, 2000. Informatik Aktuell, pp. 476–488. Springer, Berlin (2000)
377. Schreiber, T.: Random dynamics and thermodynamic limits for polygonal Markov fields in the plane. Adv. Appl. Probab. **37**, 884–907 (2005)
378. Schreiber, T.: Dobrushin-Kotecký-Shlosman theorem for polygonal Markov fields in the plane. J. Stat. Phys. **123**, 631–684 (2006)
379. Schreiber, T.: Non-homogeneous polygonal Markov fields in the plane: graphical representations and geometry of higher order correlations. J. Stat. Phys. **132**(4), 669–705 (2008)
380. Schreiber, T., van Lieshout, M.N.M.: Disagreement loop and path creation/annihilation algorithms for binary planar Markov fields with applications to image segmentation. Scand. J. Stat. **37**(2), 264–285 (2010)
381. Schultz, T., Burgeth, B., Weickert, J.: Flexible segmentation and smoothing of DT-MRI fields through a customizable structure tensor. In: Bebis, G., Boyle, R., Parvin, B., Koracin, D., Remagnino, P., Nefian, A., Meenakshisundaram, G., Pascucci, V., Zara, J., Molineros, J., Theisel, H., Malzbender, T. (eds.) Advances in Visual Computing. Proceedings of the 2nd International Symposium, ISVC 2006, Lake Tahoe, NV, USA, November 6–8, 2006. Lecture Notes in Computer Science, vols. 4291–4292, pp. 454–464. Springer, Berlin (2006)
382. Sgallari, F., Murli, A., Paragios, N. (eds.): Scale Space and Variational Methods in Computer Vision: Proceedings of the 1st International Conference, SSVM 2007, Ischia, Italy, May–June 2007. Lecture Notes in Computer Science, vol. 4485. Springer, Berlin (2007)
383. Shafrir, D., Sochen, N., Deriche, R.: Regularization of mappings between implicit manifolds of arbitrary dimension and codimension. In: Paragios, N., Faugeras, O., Chan, T., Schnör, C.

(eds.) Proceedings of the 3rd IEEE Workshop on Variational, Geometric and Level-Set Methods in Computer Vision, VLSM 2005, Beijing, China, October 16, 2005. Lecture Notes in Computer Science, vol. 3752, pp. 344–355. Springer, Berlin (2005)
384. Sharon, E., Brandt, A., Basri, R.: Completion energies and scale. IEEE Trans. Pattern Anal. Mach. Intell. **22**(10), 1117–1131 (2000)
385. Shewchuk, J.R.: Constrained Delaunay tetrahedralizations and provably good boundary recovery. In: Proceedings of the 11th International Meshing Roundtable, IMR 11, Ithaka, NY, USA, September 15–18, 2002, pp. 193–204 (2002)
386. Simonoff, J.: Smoothing Methods in Statistics. Springer, New York (1996)
387. Skilling, J., Gull, S.F.: Bayesian maximum entropy image reconstruction. In: Spatial Statistics and Imaging. Lecture Notes—Monograph Series, vol. 20, pp. 341–367. Institute of Mathematical Statistics, Hayward (1991)
388. Sochen, N., Kimmel, R., Malladi, R.: A general framework for low level vision. IEEE Trans. Image Process. **7**(3), 310–318 (1998)
389. Sommer, G., Krüger, N., Perwass, C. (eds.): Mustererkennung 2000. Proceedings of the 22nd Dagm-Symposium, Kiel, Germany, September 13–15, 2000. Informatik Aktuell. Springer, Berlin (2000)
390. Spivak, M.: Differential Geometry, vols. 1–5. Perish, Berkeley (1975)
391. Sporring, J., Nielsen, M., Florack, L.M.J., Johansen, P. (eds.): Gaussian Scale-Space Theory. Computational Imaging and Vision Series, vol. 8. Kluwer Academic, Dordrecht (1997)
392. Stejskal, E.O., Tanner, J.E.: Spin diffusion measurements: Spin echoes in the presence of a time-dependent field gradient. J. Comput. Phys. **42**, 288–292 (1965)
393. Stoica, R., Descombes, X., van Lieshout, M.N.M., Zerubia, J.: An application of marked point processes to the extraction of linear networks for images. In: Mateu, J., Montes, F. (eds.) Spatial Statitics Through Applications, pp. 289–314. WIT Press, Southampton (2002)
394. Stuke, I., Aach, T., Barth, E., Mota, C.: Analysing superimposed oriented patterns. In: Proceedings of the 6th IEEE Southwest Symposium on Image Analysis and Interpretation, SSIAI 2004, Lake Tahoe, Nevada, USA, March 28–30, 2004, pp. 133–137. IEEE Computer Society, Los Alamitos (2004)
395. Surgailis, D.: Thermodynamic limit of polygonal models. Acta Appl. Math. **22**(1), 77–102 (1991)
396. Tabelow, K., Polzehl, J., Voss, H.U., Spokoiny, V.: Analyzing fMRI experiments with structural adaptive smoothing procedures. NeuroImage **33**(1), 55–62 (2006)
397. Tabelow, K., Polzehl, J., Spokoiny, V., Voss, H.U.: Diffusion tensor imaging: structural adaptive smoothing. NeuroImage **39**(4), 1763–1773 (2008)
398. Tabelow, K., Polzehl, J., Ulug, A.M., Dyke, J.P., Heier, L.A., Voss, H.U.: Accurate localization of functional brain activity using structure adaptive smoothing. IEEE Trans. Med. Imaging **27**(4), 531–537 (2008)
399. Tabelow, K., Piëch, V., Polzehl, J., Voss, H.U.: High-resolution fMRI: overcoming the signal-to-noise problem. J. Neurosci. Methods **178**(2), 357–365 (2009)
400. Tang, Y.Y., Wang, S.P., Yeung, D.S., Yan, H., Lorette, G. (eds.): Proceedings of the 18th International Conference on Pattern Recognition, Hong Kong, China, August 2006. IEEE Computer Society, Los Alamitos (2006)
401. Taylor, T.: A parametrix for step-two hypoelliptic diffusion equations. Trans. Am. Math. Soc. **296**, 191–215 (1986)
402. Taylor, D.G., Bushell, M.C.: The spatial mapping of translational diffusion coefficients by the NMR imaging technique. Phys. Med. Biol. **30**(4), 345–349 (1985)
403. ter Haar Romeny, B.M. (ed.): Geometry-Driven Diffusion in Computer Vision. Computational Imaging and Vision Series, vol. 1. Kluwer Academic, Dordrecht (1994)
404. ter Haar Romeny, B.M.: Front-End Vision and Multi-Scale Image Analysis: Multi-Scale Computer Vision Theory and Applications, Written in Mathematica. Computational Imaging and Vision Series, vol. 27. Kluwer Academic, Dordrecht (2003)
405. Terras, A.: Harmonic Analysis on Symmetric Spaces and Applications, vol. 2. Springer, Berlin (1988)

406. Therrien, C.W.: Decision, Estimation, and Classification: An Introduction Into Pattern Recognition and Related Topics. Wiley, New York (1989)
407. Thompson, P.M., Toga, A.W.: A framework for computational anatomy. Comput. Vis. Sci. **5**(1), 13–34 (2002)
408. Thompson, P.M., Hayashi, K.M., De Zubicaray, G., Janke, A.L., Rose, S.E., Semple, J., Herman, D., Hong, M.S., Dittmer, S.S., Doddrell, D.M., et al.: Dynamics of gray matter loss in Alzheimer's disease. J. Neurosci. **23**(3), 994 (2003)
409. Tomasi, C., Manduchi, R.: Bilateral filtering for gray and color images. In: Proceedings of the 6th International Conference on Computer Vision, Bombay, India, January 4–7, 1998, pp. 839–846. IEEE Computer Society, Los Alamitos (1998)
410. Tong, X., Zhang, J., Liu, L., Wang, X., Guo, B., Shum, H.-Y.: Synthesis of bidirectional texture functions on arbitrary surfaces. ACM Trans. Graph. **21**(3), 665–672 (2002)
411. Tosun, D., Rettmann, M.E., Han, X., Tao, X., Xu, C., Resnick, S.M., Pham, D.L., Prince, J.L.: Cortical surface segmentation and mapping. Neuroimage **23**(Supplement 1), S108–S118 (2004)
412. Tschumperlé, D., Deriche, R.: Regularization of orthonormal vector sets using coupled PDE's. In: Proceedings 1st IEEE Workshop on Variational and Level Set Methods, pp. 3–10. IEEE Computer Society, Los Alamitos (2001)
413. Tschumperlé, D., Deriche, R.: Orthonormal vector sets regularization with PDE's and applications. Int. J. Comput. Vis. **50**(3), 237–252 (2002)
414. Tschumperle, D., Deriche, R.: Vector-valued image regularization with PDEs: a common framework for different applications. In: Proceedings of the IEEE Computer Society Conference on Computer Vision and Pattern Recognition, Madison, Wisconsin, June 16–22, 2003, pp. 651–656. IEEE Computer Society, Los Alamitos (2003)
415. Tuch, D.S.: Q-ball imaging. Magn. Reson. Med. **52**, 1358–1372 (2004)
416. Tuch, D.S., Weisskoff, R.M., Belliveau, J.W., Wedeen, V.J.: High angular resolution diffusion imaging of the human brain. In: Proceedings of the 7th International Society for Magnetic Resonance in Medicine, ISMRM, Pennsylvania, USA, May 24–28, 1999, p. 321. International Society for Magnetic Resonance in Medicine, Berkeley (1999)
417. Tuch, D.S., Reese, T.G., Wiegell, M.R., Wedeen, V.J.: Diffusion MRI of complex neural architecture. Neuron **40**(5), 885–895 (2003)
418. Tupin, F., Maitre, H., Mangin, J.-M., Nicolas, J.-M., Pechersky, E.: Detection of linear features in SAR images: application to the road network extraction. IEEE Trans. Geosci. Remote Sens. **36**(2), 434–453 (1998)
419. van Almsick, M.A.: Context models of line and contours. PhD thesis, Eindhoven University of Technology (September 2007)
420. van den Boomgaard, R.: Nonlinear diffusion in computer vision. http://staff.science.uva.nl/~rein/nldiffusionweb/material.html
421. van der Vaart, A.W.: Asymptotic Statistics. Cambridge Series in Statistical and Probabilistic Mathematics, vol. 3. Cambridge University Press, Cambridge (1998)
422. van Essen, D.C.: A population-average, landmark- and surface-based (PALS) atlas of human cerebral cortex. NeuroImage **28**(3), 635–662 (2005)
423. van Lieshout, M.N.M.: Markov Point Processes and Their Applications. Imperial College Press, London (2000)
424. van Lieshout, M.N.M., Schreiber, T.: Perfect simulation for length-interacting polygonal Markov fields in the plane. Scand. J. Stat. **34**(3), 615–625 (2007)
425. Varma, M., Zisserman, A.: A statistical approach to texture classification from single images. Int. J. Comput. Vis. **62**(1–2), 61–81 (2005)
426. Voss, H.U., Tabelow, K., Polzehl, J., Tchernichovski, O., Maul, K.K., Salgado-Commissariat, D., Ballon, D., Helekar, S.A.: Functional MRI of the zebra finch brain during song stimulation suggests a lateralized response topography. Proc. Natl. Acad. Sci. **104**(25), 10667–10672 (2007)
427. Wand, M.P., Jones, M.C.: Kernel Smoothing. Chapman & Hall, London (1995)

428. Wang, Z., Vemuri, B.C., Chen, Y., Mareci, T.H.: A constrained variational principle for direct estimation and smoothing of the diffusion tensor field from complex DWI. IEEE Trans. Med. Imaging **23**(8), 930–939 (2004)
429. Watanabe, M., Rodieck, R.W.: Parasol and midget ganglion cells of the primate retina. J. Comp. Neurol. **289**, 434–454 (1989)
430. Wei, L., Levoy, M.: Texture synthesis using tree-structure vector quantization. In: ACM SIGGRAPH 2000, pp. 479–488. ACM/Addison-Wesley/Longman, New York/Reading/Harlow (2000)
431. Weickert, J.: Scale-space properties of nonlinear diffusion filtering with a diffusion tensor. Technical report, Laboratory of Technomathematics, University of Kaiserslautern, Kaiserslautern, Germany (1994)
432. Weickert, J.: Theoretical foundations of anisotropic diffusion in image processing. Comput. Suppl. **11**, 221–236 (1996)
433. Weickert, J.A.: Anisotropic Diffusion in Image Processing. ECMI Series. Teubner, Stuttgart (1998)
434. Weickert, J.: Nonlinear diffusion filtering. In: Jähne, B., Haußecker, H., Geißler, P. (eds.) Handbook on Computer Vision and Applications, Signal Processing and Pattern Recognition, vol. 2, pp. 423–450. Academic Press, San Diego (1999)
435. Weickert, J.A.: Coherence-enhancing diffusion filtering. Int. J. Comput. Vis. **31**(2–3), 111–127 (1999)
436. Weickert, J.: Design of Nonlinear Diffusion Filters. In: Jähne, B.J., Haußecker, H. (eds.) Computer Vision and Applications, pp. 439–458. Academic Press, San Diego (2000)
437. Weickert, J., Brox, T.: Diffusion and regularization of vector- and matrix-valued images. In: Nashed, M.Z., Scherzer, O. (eds.) Inverse Problems, Image Analysis, and Medical Imaging. Contemporary Mathematics, vol. 313, pp. 251–268. AMS, Providence (2002)
438. Weickert, J., Hagen, H. (eds.): Visualization and Processing of Tensor Fields. Mathematics and Visualization. Springer, Berlin (2006)
439. Weickert, J., Scharr, H.: A scheme for coherence-enhancing diffusion filtering with optimized rotation invariance. J. Vis. Commun. Image Represent. **13**(1–2), 103–118 (2002). Special Issue on Partial Differ. Equ. Image Process., Comput. Vis. Comput. Graph.
440. Weickert, J., Schnörr, C.: A theoretical framework for convex regularizers in PDE—based computation of image motion. Int. J. Comput. Vis. **45**(3), 245–264 (2001)
441. Weickert, J., ter Haar Romeny, B.M., Viergever, M.A.: Efficient and reliable schemes for nonlinear diffusion filtering. IEEE Trans. Image Process. **7**(3), 398–410 (1998)
442. Weickert, J., Ishikawa, S., Imiya, A.: Linear scale-space has first been proposed in Japan. J. Math. Imaging Vis. **10**(3), 237–252 (1999)
443. Welk, M., Weickert, J., Steidl, G.: From tensor-driven diffusion to anisotropic wavelet shrinkage. In: Leonardis, A., Bischof, H., Prinz, A. (eds.) Proceedings of the 9th European Conference on Computer Vision, Graz, Austria, May 2006. Lecture Notes in Computer Science, vols. 3951–3954, pp. 391–403. Springer, Berlin (2006)
444. Westin, C.-F., Maier, S.E., Khidhir, B., Everett, P., Jolesz, F.A., Kikinis, R.: Image processing for diffusion tensor magnetic resonance imaging. In: Taylor, C.J., Colchester, A.C.F. (eds.) Proceedings of the 3rd International Conference on Medical Image Computing and Computer-Assisted Intervention, MICCAI 1999, Cambridge, UK, September 19–22, 1999. Lecture Notes in Computer Science, vol. 1679, pp. 441–452. Springer, Berlin (1999)
445. Weule, J.: Iteration Nichtlinearer Gauß-filter in der Bildverarbeitung. PhD thesis, Heinrich Heine Universität Düsseldorf (1994)
446. Williams, L.R., Jacobs, D.: Stochastic completion fields: a neural model of illusory contour shape and salience. Neural Comput. **9**(4), 837–858 (1997)
447. Williams, L., Zweck, J., Wang, T., Thornber, K.: Computing stochastic completion fields in linear-time using a resolution pyramid. Comput. Vis. Image Underst. **76**(3), 289–297 (1999)
448. Winkler, G.: Image Analysis, Random Fields and Markov Chain Monte Carlo Methods: A Mathematical Introduction, 2nd edn. Applications of Mathematics, Stochastic Modelling and Applied Probability, vol. 27. Springer, Berlin (2003)

449. Winkler, G., Aurich, V., Hahn, K., Martin, A., Rodenacker, K.: Noise reduction in images: some recent edge-preserving methods. Technical report, Sonderforschungsbereich 386, Paper 138 (1998)
450. Witkin, A.P.: Scale-space filtering. In: Proceedings of the International Joint Conference on Artificial Intelligence, Karlsruhe, Germany, pp. 1019–1022 (1983)
451. Worsley, K.J.: Local maxima and the expected Euler characteristic of excursion sets of χ^2, f and t fields. Adv. Appl. Probab. **26**(1), 13–42 (1994)
452. Worsley, K.J.: Detecting activation in fMRI data. Stat. Methods Med. Res. **12**(5), 401–418 (2003)
453. Worsley, K.J., Friston, K.J.: Analysis of fMRI time series revisited—again. NeuroImage **2**(3), 173–181 (1995)
454. Worsley, K.J., Andermann, M., Koulis, T., MacDonald, D., Evans, A.C.: Detecting changes in nonisotropic images. Hum. Brain Mapp. **8**(2–3), 98–101 (1999)
455. Worsley, K.J., Liao, C., Aston, J.A.D., Petre, V., Duncan, G.H., Morales, F., Evans, A.C.: A general statistical analysis for fMRI data. NeuroImage **15**(1), 1–15 (2002)
456. Xu, Y., Guo, B., Shum, H.: Chaos mosaic: fast and memory efficient texture synthesis. Technical report MSR-TR-2000-32, Microsoft Research, Redmont (2000)
457. Yang, F., Jiang, T.: Pixon based image segmentation with Markov random fields. IEEE Trans. Image Process. **12**(12), 1552–1559 (2003)
458. Yoshiura, T., Mihara, F., Tanaka, A., Ogomori, K., Ohyagi, Y., Taniwaki, T., Yamada, T., Yamasaki, T., Ichimiya, A., Kinukawa, N., Kuwabara, Y., Honda, H.: High b value diffusion-weighted imaging is more sensitive to white matter degeneration in Alzheimer's disease. NeuroImage **20**(1), 413–419 (2003)
459. Zelinka, S., Garland, M.: Towards real-time texture synthesis with the jump map. In: Proceedings of the 13th Eurographics Workshop on Rendering, Pisa, Italy, June 26–28, 2002, pp. 99–104 (2002)
460. Zemel, R.S., Dayan, P., Pouget, A.: Probabilistic interpretation of population codes. Neural Comput. **10**(2), 403–430 (1998)
461. Zhang, Y.J.: Evaluation and comparison of different segmentation algorithms. Pattern Recognit. Lett. **18**(10), 963–974 (1997)
462. Zhang, F., Hancock, E.R.: Riemannian graph diffusion for DT-MRI regularization. In: Larsen, R., Nielsen, M., Sporring, J. (eds.) Proceedings of the 9th International Conference on Medical Image Computing and Computer-Assisted Intervention, MICCAI 2006, Copenhagen, Denmark, October 1–6, 2006. Lecture Notes in Computer Science, vols. 4190–4191, pp. 234–242. Springer, Berlin (2006)
463. Zhang, F., Hancock, E.R.: Smoothing tensor-valued images using anisotropic geodesic diffusion. In: Yeung, D.-Y., Kwok, J.T., Roli, F., Fred, A., de Ridder, D. (eds.) Proceedings of the Joint IAPR International Workshops on Structural, Syntactic, and Statistical Pattern Recognition, SSPR 2006 and SPR 2006, Hong Kong, China, August 17–19, 2006. Lecture Notes in Computer Science, vol. 4109. Springer, Berlin (2006)
464. Zhang, S., Demiralp, C., Laidlaw, D.H.: Visualizing diffusion tensor MR images using streamtubes and streamsurfaces. IEEE Trans. Vis. Comput. Graph. **9**(4), 454–462 (2003)
465. Zhu, S.C., Mumford, D.: Prior learning and Gibbs reaction-diffusion. IEEE Trans. Pattern Anal. Mach. Intell. **19**(11), 1236–1250 (1997)
466. Zhu, S.C., Liu, X.W., Wu, Y.N.: Exploring texture ensembles by efficient Markov chain Monte Carlo—toward a "trichromacy" theory of texture. IEEE Trans. Pattern Anal. Mach. Intell. **22**(6), 554–569 (2000)
467. Zhukov, L., Barr, A.H.: Oriented tensor reconstruction: tracing neural pathways from diffusion tensor MRI. In: Proceedings IEEE Visualization 2002, pp. 387–394. IEEE Computer Society, Los Alamitos (2002)
468. Zweck, J.W., Williams, L.R.: Euclidean group invariant computation of stochastic completion fields using shiftable-twistable functions. J. Math. Imaging Vision, 1–31 (2003)

Index

L. Florack et al. (eds.), *Mathematical Methods for Signal and Image Analysis and Representation*, Computational Imaging and Vision 41,
DOI 10.1007/978-1-4471-2353-8,

MIX
Papier aus verantwortungsvollen Quellen
Paper from responsible sources
FSC® C105338

If you have any concerns about our products,
you can contact us on
ProductSafety@springernature.com

In case Publisher is established outside the EU,
the EU authorized representative is:
Springer Nature Customer Service Center GmbH
Europaplatz 3, 69115 Heidelberg, Germany

Printed by Libri Plureos GmbH
in Hamburg, Germany